AF388946

Current Advances and Challenges in Fisheries and Aquaculture Science: Feature Papers for the New Journey of *Fishes*

Current Advances and Challenges in Fisheries and Aquaculture Science: Feature Papers for the New Journey of *Fishes*

Editors

Maria Angeles Esteban
Bernardo Baldisserotto
Eric Hallerman

MDPI • Basel • Beijing • Wuhan • Barcelona • Belgrade • Manchester • Tokyo • Cluj • Tianjin

Editors
Maria Angeles Esteban
University of Murcia
Spain

Bernardo Baldisserotto
Universidade Federal de
Santa Maria
Brazil

Eric Hallerman
Virginia Polytechnic Institute
and State University
USA

Editorial Office
MDPI
St. Alban-Anlage 66
4052 Basel, Switzerland

This is a reprint of articles from the Special Issue published online in the open access journal *Fishes* (ISSN 2410-3888) (available at: https://www.mdpi.com/journal/fishes/special_issues/Fisheries_Aquaculture_Science).

For citation purposes, cite each article independently as indicated on the article page online and as indicated below:

LastName, A.A.; LastName, B.B.; LastName, C.C. Article Title. *Journal Name* **Year**, *Volume Number*, Page Range.

ISBN 978-3-0365-4075-7 (Hbk)
ISBN 978-3-0365-4076-4 (PDF)

Cover image courtesy of Kathryn McBaine

Contents

About the Editors vii

Eric Hallerman, Maria Angeles Esteban and Bernardo Baldisserotto
Current Advances and Challenges in Fisheries and Aquaculture Science
Reprinted from: *Fishes* 2022, 7, 87, doi:10.3390/fishes7020087 1

Simrith E. Cordova-de la Cruz, Marta F. Riesco, Gil Martínez-Bautista, Daniel Calzada-Ruiz, Talhia Martínez-Burguete, Emyr S. Peña-Marín, Carlos Alfonso Álvarez-Gonzalez and Ignacio Fernández
Larval Development in Tropical Gar (*Atractosteus tropicus*) Is Dependent on the Embryonic Thermal Regime: Ecological Implications under a Climate Change Context
Reprinted from: *Fishes* 2022, 7, 16, doi:10.3390/fishes7010016 5

Kathryn E. McBaine, Eric M. Hallerman and Paul L. Angermeier
Direct and Molecular Observation of Movement and Reproduction by Candy Darter, *Etheostoma osburni*, an Endangered Benthic Stream Fish in Virginia, USA
Reprinted from: *Fishes* 2022, 7, 30, doi:10.3390/fishes7010030 21

Logan W. Sikora, Joseph T. Mrnak, Rebecca Henningsen, Justin A. VanDeHey and Greg G. Sass
Demographic and Life History Characteristics of Black Bullheads *Ameiurus melas* in a North Temperate USA Lake
Reprinted from: *Fishes* 2022, 7, 21, doi:10.3390/fishes7010021 41

Delphine Mallet, Marion Olivry, Sophia Ighiouer, Michel Kulbicki and Laurent Wantiez
Nondestructive Monitoring of Soft Bottom Fish and Habitats Using a Standardized, Remote and Unbaited 360° Video Sampling Method
Reprinted from: *Fishes* 2021, 6, 50, doi:10.3390/fishes6040050 61

Imanol Miqueleiz, Rafael Miranda, Arturo Hugo Ariño and Elena Ojea
Conservation-Status Gaps for Marine Top-Fished Commercial Species
Reprinted from: *Fishes* 2022, 7, 2, doi:10.3390/fishes7010002 79

Leandro Rodríguez-Viera, Ignacio Martí, Rebeca Martínez, Erick Perera, Mario Pablo Estrada, Juan Miguel Mancera and Juan Antonio Martos-Sitcha
Feed Supplementation with the GHRP-6 Peptide, a Ghrelin Analog, Improves Feed Intake, Growth Performance and Aerobic Metabolism in the Gilthead Sea Bream *Sparus aurata*
Reprinted from: *Fishes* 2022, 7, 31, doi:10.3390/fishes7010031 91

Catarina Basto-Silva, Irene García-Meilán, Ana Couto, Cláudia R. Serra, Paula Enes, Aires Oliva-Teles, Encarnación Capilla and Inês Guerreiro
Effect of Dietary Plant Feedstuffs and Protein/Carbohydrate Ratio on Gilthead Seabream (*Sparus aurata*) Gut Health and Functionality
Reprinted from: *Fishes* 2022, 7, 59, doi:10.3390/fishes7020059 105

Francisca P. Martínez-Antequera, Isabel Barranco-Ávila, Juan A. Martos-Sitcha and Francisco J. Moyano
Solid-State Hydrolysis (SSH) Improves the Nutritional Value of Plant Ingredients in the Diet of *Mugil cephalus*
Reprinted from: *Fishes* 2022, 7, 4, doi:10.3390/fishes7010004 123

Rosana Oliveira Batista, Renata Oselame Nobrega, Delano Dias Schleder, James Eugene Pettigrew and Débora Machado Fracalossi
Aurantiochytrium sp. Meal Improved Body Fatty Acid Profile and Morphophysiology in Nile Tilapia Reared at Low Temperature
Reprinted from: *Fishes* **2021**, *6*, 45, doi:10.3390/fishes6040045 . . . . **135**

Omid Safari, Mehrdad Sarkheil, Davar Shahsavani and Marina Paolucci
Effects of Single or Combined Administration of Dietary Synbiotic and Sodium Propionate on Humoral Immunity and Oxidative Defense, Digestive Enzymes and Growth Performances of African Cichlid (*Labidochromis lividus*) Challenged with *Aeromonas hydrophila*
Reprinted from: *Fishes* **2021**, *6*, 63, doi:10.3390/fishes6040063 . . . . **155**

About the Editors

Maria Angeles Esteban currently works at the Department of Cell Biology and Histology, University of Murcia. Maria conducts research in cell biology, immunology, and biotechnology of fish and is the head of the team Immunobiology for Aquaculture. Their current projects are related to farmed fish immunology. She is the Editor-in-Chief of the journal *Fishes*.

Bernardo Baldisserotto currently works at the Department of Physiology and Pharmacology, Universidade Federal de Santa Maria. Bernardo conducts research in the physiology and pharmacology of aquatic animals, mainly fish. His current main project is 'Essential oils as anesthetics, antibacterials, antiparasitic agents, and food additives for aquatic animals'. He is an Associate Editor of the journal *Fishes*.

Eric Hallerman is a professor in the Department of Fish and Wildlife Conservation at Virginia Polytechnic Institute and State University. He conducts research in applied population genetics, aquaculture, and biotechnology; teaches courses in genetics and ichthyology; and is actively engaged with governmental and non-governmental organizations in his areas of expertise. He is an Associate Editor of the journal *Fishes*.

Editorial

Current Advances and Challenges in Fisheries and Aquaculture Science

Eric Hallerman [1], Maria Angeles Esteban [2] and Bernardo Baldisserotto [3,*]

1   Department of Fish and Wildlife Conservation, Virginia Polytechnic Institute and State University, 100 Cheatham Hall, Blacksburg, VA 24061, USA; ehallerm@vt.edu
2   Department of Cell Biology and Histology, Faculty of Biology, University of Murcia, 30100 Murcia, Spain; aesteban@um.es
3   Department of Physiology and Pharmacology, Universidade Federal de Santa Maria, Santa Maria 97105-900, Brazil
*   Correspondence: bernardo.baldisserotto@ufsm.br; Tel.: +55-3220-9382

Citation: Hallerman, E.; Esteban, M.A.; Baldisserotto, B. Current Advances and Challenges in Fisheries and Aquaculture Science. *Fishes* **2022**, *7*, 87. https://doi.org/10.3390/fishes7020087

Received: 1 April 2022
Accepted: 7 April 2022
Published: 8 April 2022

Publisher's Note: MDPI stays neutral with regard to jurisdictional claims in published maps and institutional affiliations.

Advances in fisheries and aquaculture science often follow the introduction of new tools or analytic methods. For example, the introduction of geographic information systems led to advances in spatially explicit conservation planning and the siting of aquaculture operations. Advances in genetic marker technologies led to whole-genome sequencing and improved the detection of performance- or fitness-related loci, in turn leading to advances in marker-assisted breeding and conservation planning. Among the keys to successful modern aquaculture are advancements in understanding the biology of cultivated species, leading to improved diet and health management. Hence, we designed this Special Issue to address current advances and challenges in fisheries and aquaculture science.

Achieving a greater understanding of biology and ecology is critical to the management and conservation of fishes. Precise determination of how temperature affects fish populations is important for assessing the impacts of climate change. Cordova de la Cruz et al. [1] subjected tropical gar *Atractosteus tropicus* to elevated temperatures during embryological development. They found that elevated temperatures may induce craniofacial and morphological alterations, suggesting that global warming may affect the expression of morphological traits, thereby impacting the species. Traditional mark-recapture or telemetry methods for tracking the movement of fish are labor-intensive, limited to sufficiently large individuals, and yielding results only for those individuals handled. McBaine et al. [2] compared the efficacy of direct and molecular marker-based observation of movement and reproduction by candy darter, *Etheostoma osburni*, an endangered fish in the southeastern United States. Molecular markers allowed the tracking of more individuals and provided new insights into the spawning ecology and early life history of the species. Black bullhead *Ameiurus melas*, a catfish native to eastern North America, was introduced outside of its range, where it frequently proves invasive and a nuisance. Reasons for its invasiveness are poorly understood because the species is understudied. By demonstrating relatively fast growth rates, early age at maturity, moderate fecundity, and a diverse omnivorous diet, Sikora et al. [3] explained the potential for black bullheads to dominate fish community biomass in both their native and introduced range.

Cost-effective monitoring of marine systems and conservation of highly exploited species remain technical challenges. While soft-bottom habitats constitute a major part of the coral reef seascape, their fish assemblages are difficult to sample, as individuals are scattered over very large areas and often at significant depth. Existing soft-bottom sampling methods—including trawling, long-lining, and hook-and-line—are destructive. Mallet et al. [4] developed a remote, unbaited 360° video sampling method to monitor fish species assemblages on soft bottoms. They demonstrated that the method was effective for sampling bare soft-bottoms, seagrass beds, macroalgae meadows and mixed soft-bottoms and provided future users with general recommendations for estimating total species richness. Reliable and current information on the conservation status of key commercial marine fishes is crucial for their conservation. Miqueleiz et al. [5] assembled fisheries statistics from the FAO, IUCN Red List, FishBase, and RAM Legacy databases to determine the extent to which the conservation status of the top commercial species has been assessed. While levels of assessment for top-fished species were higher than those for fished species in general, almost half of the species had outdated assessments. Future evaluations for commercial fish species should integrate new parameters from fisheries sources and improve collaboration among fisheries stakeholders and managers.

The continuing growth of aquaculture will depend upon developing feeds that improve the growth, oxidative status, and immune response of fed cultured organisms. Rodriguez-Viera et al. [6] studied the effect of adding the GHRP-6 peptide, a ghrelin analog, at two levels to a commercial diet for gilthead sea bream *Sparus aurata*. Both experimental diets led to increased growth and feed conversion efficiency over the course of 97 days. The lower level of inclusion of GHRP-6 resulted in better aerobic metabolism, while the higher level increased plasma growth hormone levels, indicating that a better understanding of its dose-specific effects is still required. While fish meal has traditionally been used as the main protein source for carnivorous fish diets, its scarcity has led to an increased evaluation of plant-derived feedstuffs. Basto-Silva et al. [7] evaluated the effects of dietary protein sources and the protein/carbohydrate ratio on gilthead seabream gut function and health, assessing gut histomorphology, gut microbiota composition, digestive enzyme activity, and gut immunological and oxidative stress gene expression. Plant-based diets compromised gut absorptive and digestive metabolism, but decreasing the dietary protein/carbohydrate ratio had little effect on the measured parameters. Martínez-Antequera et al. [8] assessed the possibility of improving the nutritional quality of plant byproducts, such as brewers' spent grain and rice bran, through solid-state hydrolysis using carbohydrases and phytase for use in a feed for grey mullet *Mugil cephalus*. Growth and feed conversion efficiency over the course of a 148-day trial were similar among groups fed the experimental and commercial diets, demonstrating that enzyme pretreatment of plant ingredients may improve the nutritive value of high-fiber plant byproducts in practical diets. The culture of Nile tilapia *Oreochromis niloticus* at subtropical temperatures decreases growth, and several studies reported that the fish efficiently stores dietary n-3 long-chain polyunsaturated fatty acids at cold temperatures, increasing fatty acid unsaturation in cell membranes to maintain their fluidity and permeability. Batista et al. [9] investigated the effect of incorporating meal from the microbe *Aurantiochytrium* sp. into diets at different rates on the tilapia body and hepatopancreas fatty-acid profile, body fatty-acid retention, somatic indices, and morphophysiological changes in the intestine and hepatopancreas through 87 days at 22 °C. The use of *Aurantiochytrium* meal improved the body fatty-acid profile and morphophysiology in Nile tilapia reared at low temperatures. The administration of immunostimulants was found to promote growth and reduce microbial infections in aquatic animals. Safari et al. [10] investigated the effects of the single or combined administration of dietary symbiotics and sodium propionate on humoral immunity and oxidative defense, digestive enzymes, and growth performance of the African cichlid *Labidochromis lividus* challenged with *Aeromonas hydrophila*. The single administration of the synbiotic *Pediococcus acidilactici* and galacto-oligosaccharides combined with sodium propionate enhanced survival, growth, humoral immune response, antioxidant and digestive enzymes.

We trust that you will come to agree that the application of new tools in each of these case studies led to compelling new advances. Each advance raises new questions, which go on to become the subject of innovative future work!

Conflicts of Interest: The authors declare no conflict of interest.

References

1. Cordova-de la Cruz, S.; Riesco, M.; Martínez-Bautista, G.; Calzada-Ruiz, D.; Martínez-Burguete, T.; Peña-Marín, E.; Álvarez-Gonzalez, C.; Fernández, I. Larval Development in Tropical Gar (*Atractosteus tropicus*) Is Dependent on the Embryonic Thermal Regime: Ecological Implications under a Climate Change Context. *Fishes* **2022**, *7*, 16. [CrossRef]
2. McBaine, K.; Hallerman, E.; Angermeier, P. Direct and Molecular Observation of Movement and Reproduction by Candy Darter, *Etheostoma osburni*, an Endangered Benthic Stream Fish in Virginia, USA. *Fishes* **2022**, *7*, 30. [CrossRef]
3. Sikora, L.; Mrnak, J.; Henningsen, R.; Van De Hey, J.; Sass, G. Demographic and Life History Characteristics of Black Bullheads *Ameiurus melas* in a North Temperate USA Lake. *Fishes* **2022**, *7*, 21. [CrossRef]
4. Mallet, D.; Olivry, M.; Ighiouer, S.; Kulbicki, M.; Wantiez, L. Nondestructive Monitoring of Soft Bottom Fish and Habitats Using a Standardized, Remote and Unbaited 360° Video Sampling Method. *Fishes* **2021**, *6*, 50. [CrossRef]
5. Miqueleiz, I.; Miranda, R.; Ariño, A.; Ojea, E. Conservation-Status Gaps for Marine Top-Fished Commercial Species. *Fishes* **2022**, *7*, 2. [CrossRef]
6. Rodríguez-Viera, L.; Martí, I.; Martínez, R.; Perera, E.; Estrada, M.; Mancera, J.; Martos-Sitcha, J. Feed Supplementation with the GHRP-6 Peptide, a Ghrelin Analog, Improves Feed Intake, Growth Performance and Aerobic Metabolism in the Gilthead Sea Bream *Sparus aurata*. *Fishes* **2022**, *7*, 31. [CrossRef]
7. Basto-Silva, C.; García-Meilán, I.; Couto, A.; Serra, C.; Enes, P.; Oliva-Teles, A.; Capilla, E.; Guerreiro, I. Effect of Dietary Plant Feedstuffs and Protein/Carbohydrate Ratio on Gilthead Seabream (*Sparus aurata*) Gut Health and Functionality. *Fishes* **2022**, *7*, 59. [CrossRef]
8. Martínez-Antequera, F.; Barranco-Ávila, I.; Martos-Sitcha, J.; Moyano, F. Solid-State Hydrolysis (SSH) Improves the Nutritional Value of Plant Ingredients in the Diet of *Mugil cephalus*. *Fishes* **2022**, *7*, 4. [CrossRef]
9. Batista, R.; Nobrega, R.; Schleder, D.; Pettigrew, J.; Fracalossi, D. *Aurantiochytrium* sp. Meal Improved Body Fatty Acid Profile and Morphophysiology in Nile Tilapia Reared at Low Temperature. *Fishes* **2021**, *6*, 45. [CrossRef]
10. Safari, O.; Sarkheil, M.; Shahsavani, D.; Paolucci, M. Effects of Single or Combined Administration of Dietary Synbiotic and Sodium Propionate on Humoral Immunity and Oxidative Defense, Digestive Enzymes and Growth Performances of African Cichlid (*Labidochromis lividus*) Challenged with *Aeromonas hydrophila*. *Fishes* **2021**, *6*, 63. [CrossRef]

Article

Larval Development in Tropical Gar (*Atractosteus tropicus*) Is Dependent on the Embryonic Thermal Regime: Ecological Implications under a Climate Change Context

Simrith E. Cordova-de la Cruz [1], Marta F. Riesco [2], Gil Martínez-Bautista [3], Daniel Calzada-Ruiz [1], Talhia Martínez-Burguete [1], Emyr S. Peña-Marín [1,†], Carlos Alfonso Álvarez-Gonzalez [1] and Ignacio Fernández [4,*]

[1] División Académica de Ciencias Biológicas, Universidad Juárez Autónoma de Tabasco, Villahermosa 86040, Mexico; simcorcz@outlook.com (S.E.C.-d.l.C.); calzada.rd@gmail.com (D.C.-R.); talhia.mb@gmail.com (T.M.-B.); ocemyr@yahoo.com.mx (E.S.P.-M.); alvarez_alfonso@hotmail.com (C.A.Á.-G.)
[2] Cell Biology Area, Molecular Biology Department, University of Leon, Campus de Vegazana s/n, 24071 Leon, Spain; mferrs@unileon.es
[3] Developmental Physiology Laboratory, Developmental Integrative Biology Research Group, Department of Biological Sciences, University of North Texas, Denton, TX 76203-5017, USA; gil.martinezbautista@unt.edu
[4] Centro Oceanográfico de Vigo, Instituto Español de Oceanografía (IEO-CSIC), 36390 Vigo, Spain
* Correspondence: nacfm@hotmail.com or ignacio.fernandez@ieo.es; Tel.: +34-986492111
† Current Address: Cátedra CONACYT-UJAT, Villahermosa 86040, Mexico.

Citation: Cordova-de la Cruz, S.E.; Riesco, M.F.; Martínez-Bautista, G.; Calzada-Ruiz, D.; Martínez-Burguete, T.; Peña-Marín, E.S.; Álvarez-Gonzalez, C.A.; Fernández, I. Larval Development in Tropical Gar (*Atractosteus tropicus*) Is Dependent on the Embryonic Thermal Regime: Ecological Implications under a Climate Change Context. *Fishes* **2022**, 7, 16. https://doi.org/10.3390/fishes7010016

Academic Editor: Eric Hallerman

Received: 14 December 2021
Accepted: 8 January 2022
Published: 11 January 2022

Publisher's Note: MDPI stays neutral with regard to jurisdictional claims in published maps and institutional affiliations.

Abstract: In ectotherm species, environmental temperature plays a key role in development, growth, and survival. Thus, determining how temperature affects fish populations is of utmost importance to accurately predict the risk of climate change over fisheries and aquaculture, critical to warrant nutrition and food security in the coming years. Here, the potential effects of abnormal thermal regimes (24, 28 and 32 °C; TR24, TR28, and TR32, respectively) exclusively applied during embryogenesis in tropical gar (*Atractosteus tropicus*) has been explored to decipher the potential consequences on hatching and growth from fertilization to 16 days post-fertilization (dpf), while effects on skeletal development and body morphology were explored at fertilization and 16 dpf. Egg incubation at higher temperatures induced an early hatching and mouth opening. A higher hatching rate was obtained in eggs incubated at 28 °C when compared to those at 24 °C. No differences were found in fish survival at 16 dpf, with values ranging from 84.89 to 88.86%, but increased wet body weight and standard length were found in larvae from TR24 and TR32 groups. Thermal regime during embryogenesis also altered the rate at which the skeletal development occurs. Larvae from the TR32 group showed an advanced skeletal development, with a higher development of cartilaginous structures at hatching but reduced at 16 dpf when compared with the TR24 and TR28 groups. Furthermore, this advanced skeletal development seemed to determine the fish body morphology. Based on biometric measures, a principal component analysis showed how along development, larvae from each thermal regime were clustered together, but with each population remaining clearly separated from each other. The current study shows how changes in temperature may induce craniofacial and morphological alterations in fish during early stages and contribute to understanding the possible effects of global warming in early development of fish and its ecological implications.

Keywords: temperature; skeletal development; ossification; morphological alterations

1. Introduction

Climatic variations through time due to anthropogenic activities and global warming have become a significant threat to ecosystems and biodiversity (Intergovernmental Panel on Climate Change [1]). Global warming of 1.5 °C is predicted to negatively impact the natural environment, including droughts, floods, increase sea level and ocean acidification [1]. Temperature fluctuations in aquatic habitats promote changes in the development, physiology, and behavior of fish species [2–8]. Exposure to different temperatures within and outside the optimal species-specific range during early development can specifically promote alterations in survival, growth performance, and metabolism [9,10], including changes in cortisol, sodium, potassium, glucose levels and osmolality [11] or changes in muscular development that affects swimming efficiency [12]. The potential effect on the fish skeletal development and the induced skeletal malformations are also of particular interest, as they affect the fish survival and growth [13].

The potential effect of increased global temperature and ocean acidification has been explored in different fish species, mainly in Teleost of commercial relevance. For example, changes in temperature during metamorphosis and juvenile stages of gilthead seabream (*Sparus aurata*) produce changes in gill cover, hemal lordosis, and anomalies in the caudal and dorsal fins [14]. Temperature above 18 °C during egg incubation induce the appearance of deformities in caudal vertebrae of *Solea senegalensis* when compared to those incubated at 15 °C [15]. Temperatures above 29 °C cause deformities in the mandible and vertebrae of *Trachinotus ovatus* [16]. Pimentel and co-workers [17] have shown as eggs of gilthead seabream and meagre (*Argyrosomus regius*) exposed to future ocean conditions (+4 °C in water temperature and −0.5 ΔpH of acidification) had lower hatching success and larval survival. However, while no differences in body length were observed at hatching, a significant interaction between pCO_2 and species for somatic growth length was found at a certain age (at 15 dph for *S. aurata* and at 10 dph for *A. regius*). Similar to what was reported by [14], the incidence of body malformations in *S. aurata* larvae was significantly increased under these future ocean conditions, which was suggested to affect larval performance and recruitment success, altering the abundance of fish stocks [17]. In contrast, the projected ocean acidification scenarios seemed to not affect the development of contemporary European sea bass (*Dicentrarchus labrax*) larvae when exposed to them from hatching onwards [18]. These results suggest that the effects of climate change predicted scenarios might be developmental and/or species-dependent.

As above reviewed, the effects of increased mean water temperature derived from a climate change scenario have been described for different fish species, but always considering contemporaneous specimens (i.e., not transiently exposed to temperature increase through different generations) and thus, somehow neglecting their capacity to adapt to the new environmental conditions. Climate change also predicts an increase on the frequency and intensity of extreme events such as heat waves [1]. Heat waves are defined as very high temperatures over a sustained period of days and is directly affecting the contemporary specimens. It can represent one of the most enduring effects of climate change [19]. In the last decade, heat waves have increased the mortality of aquatic and terrestrial organisms due to its correlation with physiological stress [20], limiting their ability to cope with environmental challenges [21]. However, there is scarce information available on how an altered temperature due to a heat wave might impact freshwater fish species, specially to Holostei fish species.

Lepisosteids are an infraclass of Actinopterygii lacking the extra whole genome duplication of Teleosts, and a key group to understand vertebrate's evolution [22]. Moreover, Lepisosteid larvae are suitable piscine models to study the early life stages of fishes due to their rapid embryonic and larval development [23,24]. In particular, the tropical gar (*Atractosteus tropicus*), one of the seven extant Lepisosteids, is found in freshwater environments such as rivers, streams, lagoons, and swamps with abundant vegetation from southeast Mexico to Costa Rica [25]. The tropical gar (known as 'pejelagarto') has an important ecological role by regulating fish and amphibian populations, but also it has a cultural and commercial significance in southeast Mexico. It is captured and cultured because of its high nutritional value, for souvenirs with handicrafts of their scales and/or whole fish, and one of the most exotic sports fishing species [26,27]. Furthermore, its environment has been under constant pressure for the past 50 years, with declining populations in Central America [28]. Indeed, habitat degradation and destruction were suggested to be responsible for the drop in the population of this species [29].

The present study aimed to evaluate the effect of a water temperature fluctuation during the embryogenesis of the tropical gar, particularly on craniofacial development and body morphology. A correct and timely precise development of the craniofacial skeleton is necessary for proper growth and survival since it is essential for efficient fish breathing and preys' capture. We hypothesize that the craniofacial development of the larvae will be influenced by the occurrence of heat waves during embryonic development. Recorded water temperature variations in the region of Tabasco (México) ranged from $25.3 \pm 0.9\ °C$ to $32.0 \pm 0.8\ °C$ [30] in the Centla wetland, and from 21 to 31 °C in the Usumacinta river (one of the main rivers in Tabasco [31]), specifically from 22.96 to 33.88 °C in the lower basin of Usumacinta [32]. Based on our results, some reflections on how heat waves might sculpt fish morphology at population level under a climate change context will be also presented at the discussion section.

2. Materials and Methods

2.1. Ethical Statement

Fish were handled in compliance with the standards for the good welfare practices of laboratory animals from the Norma Mexicana NOM-062-ZOO-1999 de la Secretaría de Agricultura, Ganadería, Desarrollo Rural, Pesca y Alimentación.

2.2. Animal Acquisition and Care

Fertilized eggs of *A. tropicus* were obtained from a broodstock held at the Tropical Aquaculture Laboratory of the Universidad Juárez Autónoma de Tabasco, Mexico. One female was anesthetized with clove oil and injected with $1\ mL\ kg^{-1}$ of gonadotropin releasing a hormone (GnRH, Sanfer) to artificially induce breeding. The female was deposited in a 2000 L tank with six males. Artificial substrate was introduced to mimic the natural vegetation used for egg adhesion [33]. The spawning occurred at 28 °C. Eggs were collected after the female finished laying eggs (approx. 6 h after spawning started). A total of 1800 eggs were collected and placed in 10-L tanks (60 eggs per tank) with non-chlorinated water, continuously aerated and under a natural photoperiod of 12 h light-12 h darkness. A quick adaptation to the experimental thermal regimes was performed in a frame of 2 h.

2.3. Experimental Design and Sampling

Collected eggs were submitted to three thermal regimes (TR) from egg fertilization to mouth opening: low temperature (24 °C; TR24), normal (control) temperature (28 °C; TR28) and high temperature (32 °C; TR32). For each thermal regime, 600 eggs were randomly distributed in 10 tanks of 70 L (60 fish per replicate). After mouth opening, rearing temperature was progressively ($1\ °C\ day^{-1}$) increased to 28 °C in tanks from TR24 group, decreased in TR32 or remained constant in those from TR28 group and maintained at 28 °C until 16 dpf (Figure 1). At each group and experimental condition, temperature was daily monitored and maintained at the set temperature with variations of $\pm 0.5\ °C$ using aquarium chillers and heaters (e.g., TECO® TK-2000 and EHEIM 200).

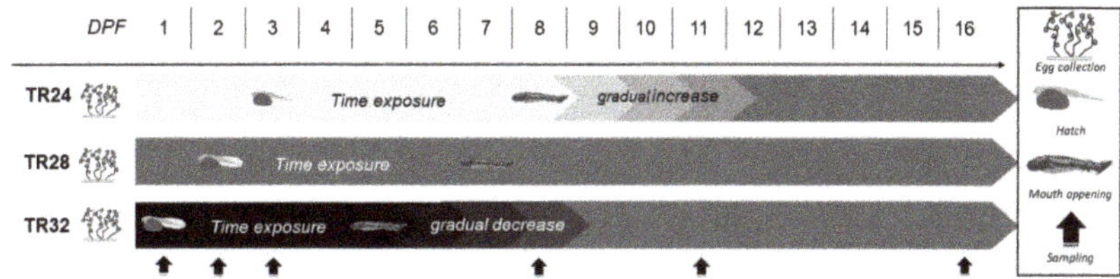

Figure 1. Experimental design to determine the effect of temperature during the early development of tropical gar (*Atractosteus tropicus*) (16 dpf). The exposure period is shown for each treatment, as well as the temperature transition and when the samplings were performed.

Larvae were fed with brine shrimp nauplii (*Artemia sp*) every 3 h from 8:00 to 17:00 h for five days. Subsequently, co-feeding of brine shrimp nauplii and trout diet (Silver Cup, 45% protein, 16% lipids) was provided to apparent satiation. One hour after every meal, dead brine shrimp and feces were removed by siphoning. Fifty percent water was renewed every 48 h. Taking into account the density of fish per tank and their airbreathing capacity, no aeration has been provided. Abiotic factors were daily monitored and recorded values were as follows: 6–8 pH and dissolved oxygen (>6 mg/L).

Fifteen larvae were randomly collected at hatching, 8, 11 and 16 dpf from each treatment for standard length (SL) and wet body weight (WBW) individual assessment. For skeletal development and body morphology analysis, 10 larvae were collected from each sampling time, 40 per thermal group. For both analytical purposes, fish were first euthanized with an overdose of MS-222. After, larvae for skeletal development were fixed in 4% paraformaldehyde (PFA) with $1\times$ phosphate-buffered saline (PBS) at pH 7.4 for 24 h. Then, larvae were rinsed in $1\times$ PBS for 15 min and progressively dehydrated in absolute ethanol-PBS solutions (25:75, 50:50 and 75:25 *v/v*) and finally preserved in 100% absolute ethanol until processing.

2.4. Survival and Growth

Survival was assessed by counting the number of dead larvae every day from every TR. Growth performance was evaluated as changes in WBW, SL and Fulton's condition factor (K). WBW (in milligrams) of the larvae was accounted using an analytical balance; SL was measured using a digital caliper to the closest millimeter. Condition factor was calculated as $K = (WBW \div SL^3) \times 100$ [34].

2.5. Skeletal Development Assessment and Body Morphology Biometrics

To analyze the degree of development of skeletal structures in the tropical gar, the acid-free double stain protocol described by [35] was previously adapted and conducted.

Processed larvae were observed under a dissecting microscope SMZ 25 to analyze osteological development and high-resolution photographs were taken. The degree of skeletal development was evaluated as the proportions of 'red pixels' (for bone) or 'blue pixels' (for cartilage) of the total surface of the fish (in pixels). Photographs and measurements were analyzed in the software Image J (Version 1.50i, https://imagej.nih.gov/ij/download.html (accessed on 1 December 2021)).

To explore how the exposure to different TRs determined the body morphology, different biometric measures (including pre-orbital length, body depth at cleithrum, prepectoral length, pre-pelvic length, pre-anal length, head length and width, jaw length and width, distance between ceratohyals at ossification front and length of ossified ceratohyal) were assessed.

Since fish metabolism is temperature dependent, as well as skeletal development is growth dependent, biometric data was normalized with SL and $°C\ day^{-1}$. Data normalization procedure consisted in dividing the corresponding values by the respective SL of the specimen and the $°C\ day^{-1}$ at each sampling point. Data on bone skeletal development degree (ratios) was normalized by the $°C\ day^{-1}$ at each sampling point. This procedure allowed us to reduce and/or avoid the potential effect of sampling individuals at different stages (temperature dependent) and effect of interindividual variability within each experimental group [36].

2.6. Statistical Analysis

Otherwise indicated, results are given as mean values $\pm$ standard deviations. All data were previously checked for normality (Kolmogorov–Smirnov test) and homoscedasticity of variance (Bartlett's test). Results were compared by means of one-way ANOVA to detect differences among experimental groups at each sampling time, and when detected, the post-hoc Tukey's multiple comparisons test was performed using GraphPad Prism 8.0 (GraphPad Software, Inc., San Diego, CA, USA). A Principal Component Analysis using

SPSS was performed with the biometric data. Since data followed normal distribution and equal variance, data was not log-transformed. The matrix of covariance was analyzed. In all analysis, statistical significance was set at $p < 0.05$.

3. Results

3.1. Hatching Rate, Survival and Growth Performance

Hatching occurred at 1-, 2- or 3-days post-fertilization (dpf) in eggs under TR32, TR28 and TR24, respectively. Hatching rate has been significantly affected by the thermal regime applied (Figure 2). Hatching rate was significantly higher when *Atractosteus tropicus* eggs were incubated at 28 °C than at 24 °C (90.17 ± 4.67% vs. 85.17 ± 5.12%; ANOVA, $p < 0.05$). Hatching rate at 32 °C (89.83 ± 2.88%) was not significantly different from both TR24 (85.16 ± 5.11%) and TR28 (90.16 ± 4.67%) groups. Increased temperature also advanced fish development, particularly regarding the timing of mouth opening, taking place at 8, 6 and 5 dpf in TR24, TR28 and TR32 larvae (results not shown), respectively. In contrast, no significant differences were observed in the survival rate of *A. tropicus* larvae at the end of the trial (16 dpf; Figure 3; ANOVA, $p > 0.05$), with values ranging from 84.89 to 88.86%.

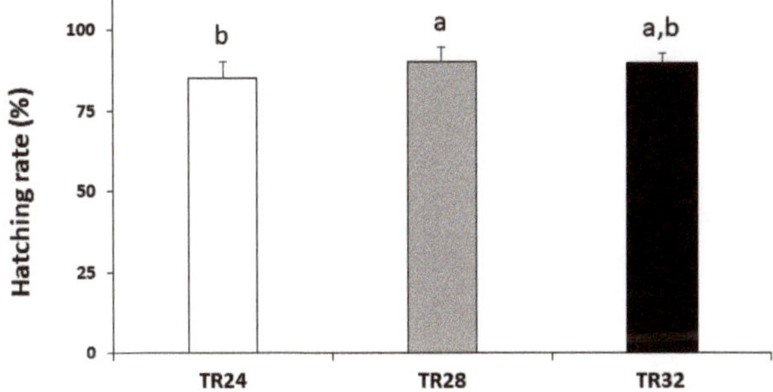

Figure 2. Hatching rate (mean ± standard deviation) of *Atractosteus tropicus* larvae when embryos were incubated at different thermal regimens. Lowercase letters at the top of each bar indicate statistical differences among experimental groups (ANOVA, $p < 0.05$; $N = 10$).

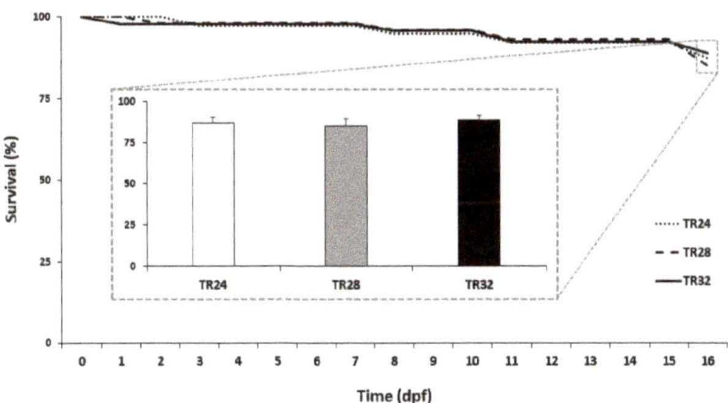

Figure 3. Mean survival of *Atractosteus tropicus* larvae along and at the end of the experimental trial (16 dpf) when embryos were incubated at different thermal regimens. The values of survival represented in the histogram bars are the mean ± standard deviation at 16 dpf.

Fish growth in terms of wet body weight (WBW; Figure 4a) and standard length (SL; Figure 4b) increased progressively during larval development, while the Fulton's condition factor (K) sharply decreased after hatching (Figure 4c). No differences in WBW were found at hatching. At 16 dpf, larvae from TR24 and TR32 reached similar WBW (108.20 ± 20.58 mg and 107.18 ± 15.85 mg, respectively) and significantly higher than that of larvae from TR28 (82.20 ± 8.91; ANOVA, $p < 0.05$). Although larvae from TR32 showed higher WBW than those from TR24 and TR28 (ANOVA, $p < 0.05$) at 8 and 11 dpf, such differences might be related to the differential yolk-sac resorption status which the larvae from these two last experimental groups might be at. At hatching, SL were significantly higher in *A. tropicus* from TR24 (9.37 ± 0.49 mm) than that of larvae from TR28 and TR32 (ranging from 8.96 ± 0.14 to 9.00 ± 0.01 mm; Figure 4b; ANOVA, $p < 0.05$). However, these differences were not maintained along larval development. Indeed, larvae from TR32 start to show greater length than the other thermal regimes, and maintained such differences with larvae from TR28 until the end of the trial (ANOVA, $p < 0.05$), reaching 31.00 ± 2.16 mm. In line with WBW results, larvae from TR24 achieved similar SL to that of larvae from TR32 (29.79 ± 1.41 mm; ANOVA, $p > 0.05$), while SL and condition factor at 8 and 11 dpf might be also influenced by different rates on yolk-sac resorption in TR24 and TR28 larvae. At 16 dpf, Fulton's condition factor in TR24 (0.41 ± 0.02) was significantly higher than the one of larvae from TR32 (0.36 ± 0.03; ANOVA, $p < 0.05$), with larvae from TR28 showing intermediate values (0.37 ± 0.03; Figure 4c).

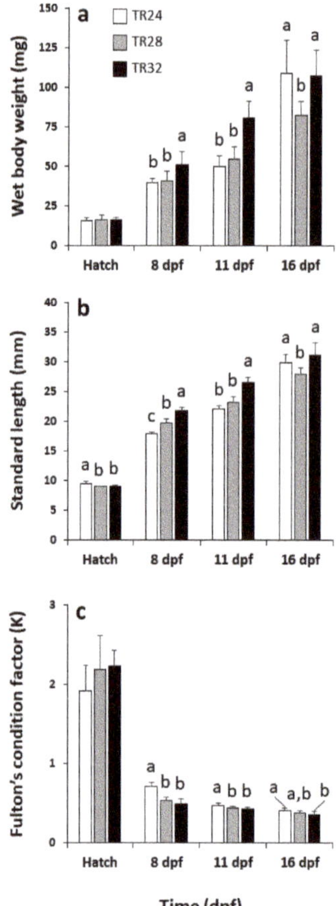

Figure 4. Growth performance (mean ± standard deviation) of *Atractosteus tropicus* larvae when embryos were incubated at different thermal regimens. Wet body weight (**a**), Standard length (**b**), and Fulton's condition factor (**c**). Lowercase letters at the top of each bar indicate statistical differences among experimental groups at the specific developmental sampling time (ANOVA, $p < 0.05$; $N = 10$). White bars, TR24; grey bars, TR28; and black bars, TR32.

3.2. Skeletal Development

Development of the skeletal structures in *A. tropicus* larvae progressed quite rapidly, particularly those composing the cranial region. Skeletal structures development (expressed as median values of the ratio of cartilage and bone surfaces over total larval surface per °C day^{-1}; Supplementary Data S1) was determined by the thermal regime during egg incubation and is presented in Figure 5. At hatching, higher quantity of cartilage was observed in larvae from TR32 group (0.51) than in larvae from TR24 (0.23) and TR28 (0.33; Figure 5a; ANOVA, $p < 0.05$). In contrast, at 16 dpf, the quantity of cartilage decreased in *A. tropicus* larvae, and statistical differences were again observed among the experimental groups. Larvae from TR32 groups exhibited the lowest cartilage quantity (median surface ratio per °C day^{-1} of 0.06), larvae from TR28 showed an intermediate value (0.07) and larvae from TR24 revealed the lowest value (0.09; Figure 5b; ANOVA, $p < 0.05$).

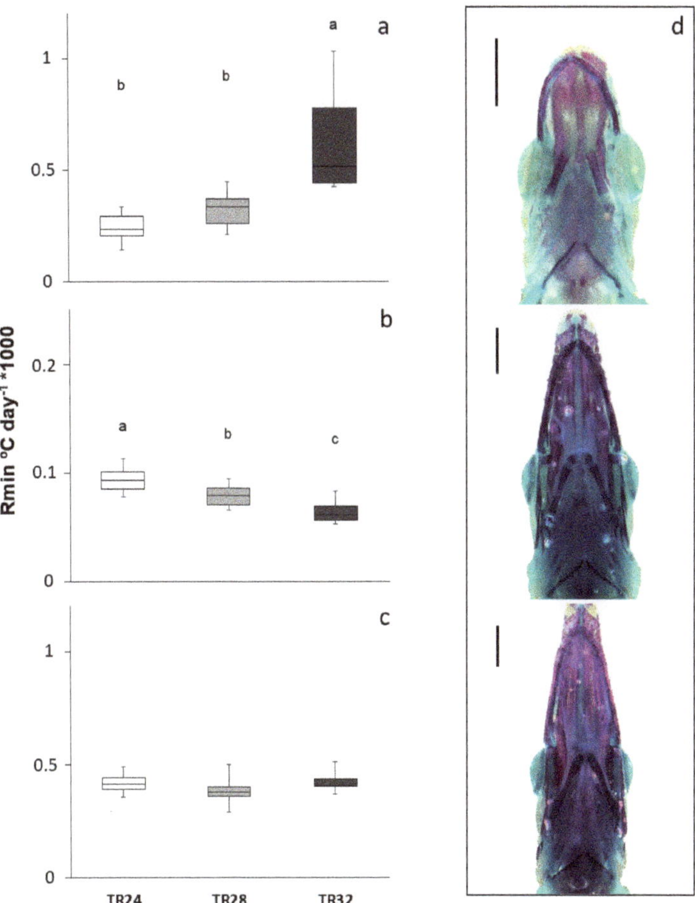

Figure 5. Skeletal development degree rates (expressed as median ± max/min values) per °C day (Rmin °C day^{-1} × 1000) in *Atractosteus tropicus* larvae when embryos were incubated at different thermal regimens. Cartilage quantity (ratio of blue stained surface over total larval surface) at hatching (**a**) and 16 (**b**) days post-fertilization (dpf); and bone structures (ratio of red stained surface over total larval surface) at 16 (**c**) dpf. Examples of tropical gar juveniles stained with alcian blue and alizarin red (ventral view) showing progressive ossification of cranial bones (**d**). Different letters at the top of the boxes denote significant differences ($p < 0.05$). Scale bar = 1 mm.

Regarding the bone ossification of the skeletal structures, it reflected the normal progression of increased ossification along larval development. At hatching, none of the skeletal elements showed bone ossification regardless the experimental group considered. First bone structures to start to be ossified were those involved in respiration and live prey capture (e.g., cleithrum and jaws; results not shown). Higher bone ossification was observed in *A. tropicus* larvae along larval development, and no differences were observed among the experimental groups at 16 dpf (median values ranging from 0.37 to 0.41, respectively; Figure 5c).

3.3. Body Morphology and Biometric Lengths

In order to decipher how advanced skeletogenesis under different thermal regimes during egg incubation might affect body growth and development, we further evaluated different body (including SL, body depth at cleithrum, and pre-orbital, pre-pectoral, pre-pelvic and pre-ana lengths) and cranial (including head length, head width, jaw length, jaw width, distance between ceratohyals at ossification front and length of ossified ceratohyal) biometric measures in *A. tropicus* (Figure 6; Supplementary Data S2).

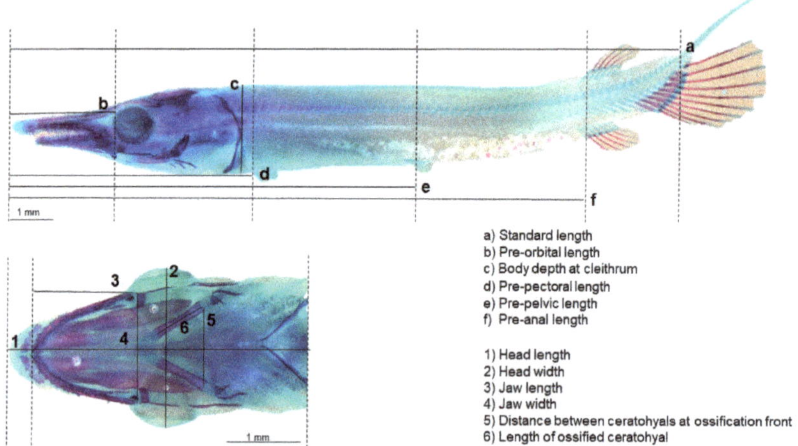

Figure 6. Different biometric measurements performed in the body and head of tropical gar (*Atractosteus tropicus*) larvae when embryos were incubated at different thermal regimens.

A principal component analysis (PCA) was conducted to identify the variables that explain the differences between larvae from the thermal groups at hatching and 16 dpf (Figure 7; Supplementary Data S3). Results showed how along the larval development, biometrics are altered depending on the TR to which *A. tropicus* embryos were exposed to. At hatching, Component 1 explained the 95.35% of the variability observed between TRs, and larvae from TR32 were clearly clustered separately from TR24 and TR28 larvae (Figure 7a). At the end of the trial (16 dpf), variance was mostly explained by Component 1 (74.37%), and includes pre-anal, pre-pectoral and head lengths as well as head width as the main contributing variables (Figure 7b).

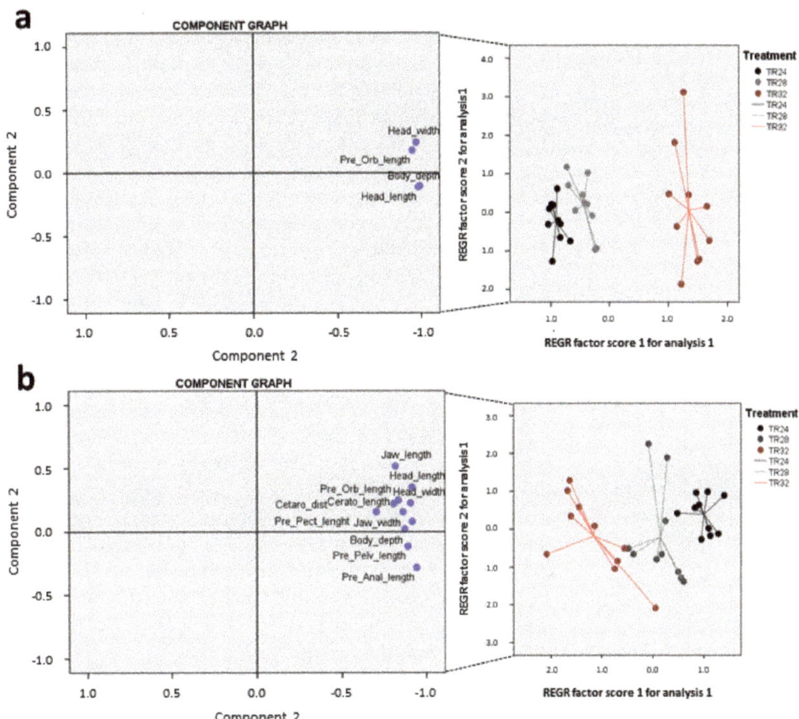

Figure 7. Scatter plot of experimental groups (with centroid distribution of the replicates) separated by the two principal components obtained from a principal component analysis (PCA), to reduce the dimensionality of the dataset, when embryos of *A. tropicus* were incubated at different thermal regimens. Representation is according to normalized biometric data at hatching (**a**) and 16 days post-fertilization (dpf) (**b**).

4. Discussion

Experimental approaches where the natural occurring alterations of the environmental conditions are mimicked more precisely are essential to unveil their potential consequences and for an accurate risk assessment. This is the case of studies focused on establishing the potential effects of the climate change over fish species. Most studies on this issue exposed contemporary populations of fishes to global warming (higher mean water temperature and ocean acidification) that progressively will take place in the coming years [14–18,37], neglecting the capacity of fish species to adapt to this environmental condition through the different generations. Here, we explored the effect of a more extreme event (although shorter in time, only lasting some days) related to climate change, the heat waves, a contemporary event which frequency and intensity are predicted to be increased [1]. This extreme climatic event has been previously suggested to represent one of the most enduring effects of climate change [19], increasing the mortality of aquatic and terrestrial organisms [20]. In this sense, the potential effects of a short alteration of the water temperature during the embryogenesis of tropical gar, an emblematic fish species in Central America, was explored to envisage their potential consequences in the short-term (16 dpf).

In the last decade, greater interest and efforts to increase aquaculture production of tropical gar have been placed due to the reduced capture of this species in the natural environments [28,29]. These efforts have been translated in a broader knowledge on the optimal rearing conditions and husbandry practices [23,24,38–42]. The optimal rearing temperature is 26–28 °C (Álvarez-González et al., unpublished results); thus, an alteration of just 4 °C seemed to be a quite realistic approach to explore the effects of a heat wave, considering that broader thermal alterations were registered in freshwater bodies [43]. Good hatching rate and survival at 16 dpf (both > 80%) from the Control group (TR28) suggests that present results are reliable and not driven by low egg quality and/or suboptimal rearing conditions. Moreover, although slight differences were found in hatching rate when different thermal regimes were applied during embryogenesis, all experimental groups exhibited good hatching rates and survival, evidencing a great thermal tolerance (Δ8 °C) of *A. tropicus* embryos. Other fish species were shown to be more sensitive to thermal alterations. For example, in European eel (*Anguilla anguilla*) larvae, increasing temperature from 18 °C (suggested to be the optimal temperature) to 22 °C (Δ4 °C) affected hatching success, survival and growth, and accelerated larval development [44]. Additionally, increased temperature, from 15 to 21 °C (Δ6 °C), during embryonic development until hatching in Senegalese sole (*Solea senegalensis*) leads to an increased incidence of skeletal deformities [15]. Therefore, these results are in line with the reported high resistance of *A. tropicus* larvae to suboptimal rearing conditions with low requirements for water quality (regarding pH, dissolved oxygen, and pollutants, and with high ammonia tolerance) [24,45–47]. Nevertheless, thermal variation during embryonic development was shown to induce sublethal effects in fish growth, skeletal development, and body morphology.

Alteration of rearing temperature has been shown to dramatically alter fish metabolism, growth potential, muscle development, immune response, and even sex differentiation, among other processes [48–54]. Here, a different thermal regime during *A. tropicus* embryogenesis induced an altered growth in terms of WBW and SL that varied along with the larval development. Although encountered differences among experimental groups at 8 and 11 dpf might be due to differences in developmental stages (e.g., differences in the rate of yolk-sac resorption) and/or differences in the current rearing temperature; this might not be the case at hatching (where all the specimens from the different experimental groups are at the same developmental stage) or at 16 dpf, where the thermal regime in all groups has been restored to a common situation (water temperature at 28 °C). In general, the higher the temperature of egg incubation, the higher growth in all developmental stages analyzed unless at 16 dpf. The general trend of higher growth in larvae from TR32 is in line with previous studies showing higher growth with increased rearing temperatures [44]. The rationale behind the equal growth reached in larvae from TR24 and TR32 at the end of the trial (16 dpf) remains unknown; although the lower hatching rate, leading to higher availability of food and less competence within congeners in the tanks and/or a kind of compensatory growth, might partially explain how these two extreme groups reached similar growth (in terms of WBW and SL).

How the embryo and its skeleton are formed in gars have been previously described, particularly the head region, and mainly in the spotted gar (*Lepisosteus oculatus*) [55–62]. Some descriptions of the early development of *A. tropicus* have been published in the last decade regarding the buccal cavity [23]. In general, the skeletal development of *A. tropicus* was similar to its closest sister species, the Cuban gar (*Atractosteus tristoechus*), which has been particularly described in [62]. In contrast to other studies, no specific skeletal deformities were found in *A. tropicus* regardless of the thermal regime applied. This might be related to the higher thermal tolerance of the species or to the early developmental stage here evaluated (larvae of only 16 dpf). Unless for severe alterations of the skeletal development, in order for any disequilibrium to be translated in such a process in specific skeletal anomalies, a longer rearing time or a longer thermal alteration exposure (not only during embryogenesis) might be needed. In fact, skeletal deformities are regularly detected at larval or juvenile stages [63,64]. The axial skeleton of our larvae was not fully ossified at 16 dpf, this occurring at later stages (e.g., at 118 days post-hatching in *A. tristoechus*; [62]). Previous studies found that abnormal thermal regimes during embryogenesis induced skeletal deformities in *S. senegalensis*, mainly in the axial and caudal complex skeletons [15]. However, this species is already known to be prone to show a high incidence of skeletal deformities [65]. Indeed, skeletal deformities were only induced with extreme temperatures during embryogenesis [66,67], or with moderated changes of thermal regimes applied during both embryogenesis and larval development [44] or larval development [14,16,17,68,69]. These effects on the skeleton have been suggested to be induced through a disruption in the armonic development of bone structures and muscle growth, with higher temperatures, faster muscle growth, and the subsequent increased mechanical load over the skeleton [70,71]. Another plausible hypothesis of increased temperature inducing skeletal deformities is that it also advanced the development of the skeletal structures, as shown in European eel [44]. Although skeletal development is a process with some plasticity [72], advanced or delayed ossification has been suggested to induce the appearance of skeletal deformities [73]. In the present study, an advanced skeletal development was observed in larvae from TR32, with increased cartilage quantity at hatching, and lowest at 16 dpf. These events might be related to an advanced endochondral ossification, the process by which several skeletal structures develop through a cartilaginous anlagen that will be finally replaced by bone tissue [74]. Other abiotic and biotic factors have been shown to advance skeletal development [13,75] that finally will induce abnormal skeletogenesis. Nevertheless, if the thermal alteration here performed during embryogenesis induced any skeletal deformity afterward remains to be deciphered.

Higher temperatures accelerate the rate of development, resulting in the appearance of some structures at smaller larval sizes [76], as it has been observed for fin formation and metamorphosis [77–79]. Higher temperature during embryogenesis induced an early hatching and mouth opening (1 or 2 days earlier than eggs incubated at 24 °C; please see Figure 1), as well as an advanced skeletal ossification at 16 dpf in *A. tropicus*. A similar effect has been recorded in the sister species *A. tristoechus* when eggs were incubated at 26–30 °C [80]. Further altered development related to the thermal regime applied in the *A. tropicus* embryos were recorded here. Biometric data showed how tropical gars from each thermal regime have a common body morphology, but distinct from each other. At 16 dpf, pre-anal, pre-pectoral, jaw, head and length of ossified ceratohyal as well as body depth at cleithrum and head width were the biometric variables that mainly explained the variability observed between larvae from different thermal regimes. Similarly, Cuban gar larvae reared at increasing temperatures (from 26 to 30 °C) also showed an accelerated inflexion points of different morphometric characters [81]. Since the survival of gars has been intrinsically related with their ability to catch prey, with a proper head elongation (longer heads with reduced heights and widths) during early stages warranting an efficient food capture [80]; present results suggest that heat wave occurrence during *A. tropicus* embryogenesis might be a risk factor for natural populations. In the long-term, alteration of the characteristic allometric growth along development of the gars due to increased temperature (specifically during embryogenesis) might alter the larval survival capacity of the wild populations.

Recently, long lasting effects of early temperature exposure have been identified in metamorphosing gilthead seabream, including decreased critical swimming speed and the incidence of caudal-fin abnormalities [82]. Although further research is needed to decipher the specific implications of the long-term heat waves during tropical gar embryogenesis, present results showed how a thermal regime alteration during embryogenesis induced lower hatching rate, increased growth, advanced skeletal development, and modified body morphology of tropical gar. These results suggest that heat waves related to climate change might be a source of biodiversity loss. Furthermore, due to the relevance of this species in Central America, the currently predicted climate change scenarios might have a detrimental economic and social impact on their communities.

5. Conclusions

In the present study, the effects of an altered thermal regime (mimicking an extreme event related to climate change, the heat waves) during *A. tropicus* embryogenesis were described. Through the comparison of larval performance, skeletal development and body morphometrics of larvae at the same developmental stages (hatching) and after the time exposure to thermal alteration (16 dpf), alterations on hatching rate, skeletal development and body morphometry have been evidenced. Fish incubated during embryogenesis at higher temperatures exhibited an advanced skeletal development that was translated in a distinct body morphometry. Although to decipher the long-term implications of these alterations over fish survival and the population dynamics further research efforts are needed, present results anticipate climate change as an additional risk for wild fisheries conservation of *A. tropicus*.

Supplementary Materials: The following are available online at https://www.mdpi.com/article/10.3390/fishes7010016/s1, Supplementary Data S1: normalized mineralization degree. Supplementary Data S2: normalized biometrics for PCA. Supplementary Data S3: Data from larvae at hatching and at 16 dpf.

Author Contributions: Conceptualization, I.F.; methodology, C.A.Á.-G. and I.F.; formal analysis, S.E.C.-d.l.C., M.F.R., G.M.-B., D.C.-R. and I.F.; writing—original draft preparation, S.E.C.-d.l.C., M.F.R. and I.F.; writing—review and editing, S.E.C.-d.l.C., M.F.R., G.M.-B., D.C.-R., T.M.-B., E.S.P.-M., C.A.Á.-G. and I.F. All authors have read and agreed to the published version of the manuscript.

Funding: This work was partially funded by "Study of the digestive physiology in larvae and juveniles of tropical gar (*Atractosteus tropicus*) based on histological, biochemical and molecular techniques" project (Ref. CB-2016-01-282765) from the National Council for Science and Technology (CONACyT) of Mexico. I.F. acknowledges the funding from the MICIU and the European Social Fund, "The European Social Fund invests in your future" through the Ramón y Cajal (Ref. RYC2018-025337-I) contract from the Plan Estatal de Investigación Científica y Técnica e Innovación 2017–2020.

Institutional Review Board Statement: The animal study protocol was approved by the Secretaría de Agricultura, Ganadería, Desarrollo Rural, Pesca y Alimentación (protocol code NOM-062-ZOO-1999, 12 May 2016).

Data Availability Statement: The data that support the findings of this study are available upon request from the authors.

Acknowledgments: Authors also thanks the support from the RED LARVAplus "Estrategias de desarrollo y mejora de la producción de larvas de peces en Iberoamérica" (117RT0521) funded by the Programa Iberoamericano de Ciencia y Tecnología para el Desarrollo (CYTED).

Conflicts of Interest: The authors declare no conflict of interest.

References

1. Arias, P.; Bellouin, N.; Coppolam, E.; Jones, R.; Krinner, G.; Marotzke, J.; Zickfeld, K. *Climate Change 2021: The Physical Science Basis. Contribution of Working Group I to the Sixth Assessment Report of the Intergovernmental Panel on Climate Change*; Technical Summary; Cambridge University Press: Cambridge, UK, 2021; In Press.
2. Enders, E.C.; Boisclair, D. Effects of environmental fluctuations on fish metabolism: Atlantic salmon *Salmo salar* a case study. *J. Fish Biol.* **2016**, *88*, 344–358. [CrossRef]
3. Jonsson, B.; Jonsson, N. A review of the likely effects of climate change on anadromous Atlantic salmon *Salmo salar* and brown trout *Salmo trutta*, with particular reference to water temperature and flow. *J. Fish Biol.* **2009**, *75*, 2381–2447. [CrossRef] [PubMed]
4. Lim, M.Y.; Manzon, R.G.; Somers, C.M.; Boreham, D.R.; Wilson, J.Y. The effects of fluctuating temperature regimes on the embryonic development of lake whitefish (*Coregonus clupeaformis*). *Comp. Biochem. Physiol. A. Mol. Integr. Physiol.* **2017**, *214*, 19–29. [CrossRef] [PubMed]
5. Marcos-López, M.; Gale, P.; Oidtmann, B.C.; Peeler, E.J. Assessing the impact of climate change on disease emergence in freshwater fish in the United Kingdom. *Transbound. Emerg. Dis.* **2010**, *57*, 293–304. [CrossRef] [PubMed]
6. Magel, J.M.T.; Dimoff, S.A.; Baum, J.K. Direct and indirect effects of climate change-amplified pulse heat stress events on coral reef fish communities. *Ecol. Appl.* **2020**, *30*, e02124. [CrossRef] [PubMed]
7. Murdoch, A.; Mantyka-Pringle, C.; Sharma, S. The interactive effects of climate change and land use on boreal stream fish communities. *Sci. Total Environ.* **2020**, *700*, 134518. [CrossRef] [PubMed]
8. Servili, A.; Canario, A.V.M.; Mouchel, O.; Muñoz-Cueto, J.A. Climate change impacts on fish reproduction are mediated at multiple levels of the brain-pituitary-gonad axis. *Gen. Comp. Endocrinol.* **2020**, *291*, 113439. [CrossRef]
9. Abdo-de la Parra, M.I.; Martínez-Rodríguez, I.E.; González-Rodríguez, B.; Rodríguez-Ibarra, L.E.; Duncan, N.; Hernández, C. Efecto de la temperatura y salinidad del agua en la incubación de huevos de botete diana *Sphoeroides annulatus*. *Rev. De Biol. Mar. Y Oceanogr.* **2012**, *47*, 147–153. [CrossRef]

10. Landsman, S.J.; Gingerich, A.J.; Philipp, D.P.; Suski, C.D. The effects of temperature change on the hatching success and larval survival of largemouth bass *Micropterus salmoides* and smallmouth bass *Micropterus dolomieu*. *J. Fish Biol.* **2011**, *78*, 1200–1212. [CrossRef]

11. Mateus, A.P.; Costa, R.; Gisbert, E.; Pinto, P.; Andree, K.B.; Estévez, A.; Power, D.M. Thermal imprinting modifies bone homeostasis in cold-challenged sea bream (*Sparus aurata*). *J. Exp. Biol.* **2017**, *220*, 3442–3454. [CrossRef]

12. Herbing, I.H.V. Effects of temperature on larval fish swimming performance: The importance of physics to physiology. *J. Fish Biol.* **2002**, *61*, 865–876. [CrossRef]

13. Boglione, C.; Gisbert, E.; Gavaia, P.; Witten, P.E.; Moren, M.; Fontagné, S.; Koumoundouros, G. Skeletal anomalies in reared European fish larvae and juveniles. Part 2: Main typologies, occurrences and causative factors. *Rev. Aquac.* **2013**, *5*, S121–S167. [CrossRef]

14. Georgakopoulou, E.; Katharios, P.; Divanach, P.; Koumoundouros, G. Effect of temperature on the development of skeletal deformities in Gilthead seabream (*Sparus aurata* Linnaeus, 1758). *Aquac. Res.* **2010**, *308*, 13–19. [CrossRef]

15. Dionísio, G.; Campos, C.; Valente, L.M.P.; Conceição, L.E.C.; Cancela, M.L.; Gavaia, P.J. Effect of egg incubation temperature on the occurrence of skeletal deformities in *Solea senegalensis*. *J. Appl. Ichthyol.* **2012**, *28*, 471–476. [CrossRef]

16. Yang, Q.; Ma, Z.; Zheng, P.; Jiang, S.; Qin, J.G.; Zhang, Q. Effect of temperature on growth, survival and occurrence of skeletal deformity in the golden pompano *Trachinotus ovatus* larvae. *Indian J. Fish.* **2016**, *63*, 74–82. [CrossRef]

17. Pimentel, M.S.; Faleiro, F.; Marques, T.; Bispo, R.; Dionísio, G.; Faria, A.M.; Rosa, R. Foraging behaviour, swimming performance and malformations of early stages of commercially important fishes under ocean acidification and warming. *Clim. Chang.* **2016**, *137*, 495–509. [CrossRef]

18. Crespel, A.; Zambonino-Infante, J.L.; Mazurais, D.; Koumoundouros, G.; Fragkoulis, S.; Quazuguel, P.; Claireaux, G. The development of contemporary European sea bass larvae (*Dicentrarchus labrax*) is not affected by projected ocean acidification scenarios. *Mar. Biol.* **2017**, *164*, 155. [CrossRef]

19. Meehl, G.A.; Tebaldi, C. More intense, more frequent, and longer lasting heat waves in the 21st century. *Science* **2004**, *305*, 994–997. [CrossRef]

20. Stillman, J.H. Heat waves, the new normal: Summertime temperature extremes will impact animals, ecosystems, and human communities. *Physiology* **2019**, *34*, 86–100. [CrossRef]

21. Alfonso, S.; Gesto, M.; Sadoul, B. Temperature increase and its effects on fish stress physiology in the context of global warming. *J. Fish Biol.* **2021**, *98*, 1496–1508. [CrossRef]

22. Braasch, I.; Gehrke, A.R.; Smith, J.J.; Kawasaki, K.; Manousaki, T.; Pasquier, J.; Amores, A.; Desvignes, T.; Batzel, P.; Postlethwait, J.H.; et al. The spotted gar genome illuminates vertebrate evolution and facilitates human-teleost comparisons. *Nat. Genet.* **2016**, *48*, 427–437. [CrossRef]

23. Burggren, W.W.; Bautista, G.M.; Coop, S.C.; Couturier, G.M.; Delgadillo, S.P.; García, R.M.; González, C.A.A. Developmental cardiorespiratory physiology of the air-breathing tropical gar, *Atractosteus tropicus*. *Am. J. Physiol. Regul. Integr. Comp. Physiol.* **2016**, *311*, R689–R701. [CrossRef]

24. Martínez, G.; Peña, E.; Martínez, R.; Camarillo, S.; Burggren, W.; Álvarez, A. Survival, Growth, and Development in the Early Stages of the Tropical Gar *Atractosteus tropicus*: Developmental Critical Windows and the Influence of Temperature, Salinity, and Oxygen Availability. *Fishes* **2021**, *6*, 5. [CrossRef]

25. Miller, R.R.; Minckley, W.L.; Norris, S.M. *Freshwater Fishes of Mexico*; University of Chicago Press: Chicago, IL, USA, 2005; p. 490.

26. Resendez Medina, A. Estudio de los peces de la Laguna de Términos, Campeche, México. II. Última parte. *Biótica* **1981**, *6*, 345–430.

27. Reséndez-Medina, A.; Salvadores, M.L. Contribución al conocimiento de la biología del pejelagarto *Lepisosteus tropicus* (Gill) y la tenguayaca *Petenia splendida* Günther, del estado de Tabasco. *Biótica* **1983**, *8*, 413–426.

28. Mora, M.; Cabrera-Peña, J.; Galeano, G. Reproducción y alimentación del gaspar *Atractosteus tropicus* (Pisces: Lepisosteidae) en el refugio nacional de vida silvestre Caño Negro, Costa Rica. *Rev. Biol. Trop.* **1997**, *45*, 861–866.

29. Barrientos-Villalobos, J.; Espinosa de los Monteros, A. Genetic variation and recent population history of the tropical gar *Atractosteus tropicus* Gill (Pisces: Lepisosteidae). *J. Fish Biol.* **2008**, *73*, 1919–1936. [CrossRef]

30. Torres, J.R.; Barba, E.; Choix, F.J. Mangrove productivity and phenology in relation to hydroperiod and physical-chemistry properties of wáter and sediment in biosphere reserve, Centla wetland, Mexico. *Trop. Cons. Sci.* **2018**, *11*, 1–14. [CrossRef]

31. Soares, D.; García, A. *La Cuenca Del Río Usumacinta Desde la Perspectiva Del Cambio Climático*; Instituto Mexicano de Tecnología del Agua: Jiutepec, Mexico, 2017; 422p. Available online: https://www.gob.mx/imta/documentos/la-cuenca-del-rio-usumacinta-desde-la-perspectiva-del-cambio-climatico (accessed on 13 December 2021).

32. Galaviz Villa, I.; Sosa Villalobos, C.A. (Eds.) Fuentes Difusas y Puntuales de Contaminación. Calidad de las Aguas Superficiales y Subterráneas. *Universidad Autónoma de Campeche*. 2019; 142p. Available online: https://drive.google.com/file/d/1E1tcVEztzkG0 6ORUT6jAYgs_Lb1-A10F/view (accessed on 13 December 2021).

33. Márquez-Couturier, G.; Vázquez-Navarrete, C.J.; Contreras-Sánchez, W.M.; Álvarez-González, C.A. *Acuicultura Tropical Sustentable: Una Estrategia Para la Producción y Conservación del Pejelagarto (Atractosteus tropicus) en Tabasco, México*, 2nd ed.; Narciso Rovirosa, C.J., Ed.; Universidad Juárez Autónoma de Tabasco: Tabasco, México, 2015; Volume 3, p. 223.

34. Ricker, W.E. Computation and interpretation of biological statistics of fish populations. *Bull. Fish. Res. Board Can.* **1975**, *191*, 1–382.

35. Walker, M.B.; Kimmel, C.B. A two-color acid-free cartilage and bone stain for zebrafish larvae. *Biotech. Histochem.* **2007**, *82*, 23–28. [CrossRef]

36. Tarasco, M.; Laizé, V.; Cardeira, J.; Cancela, M.L.; Gavaia, P.J. The zebrafish operculum: A powerful system to assess osteogenic bioactivities of molecules with pharmacological and toxicological relevance. *Comp. Biochem. Physiol. Part C* **2017**, *197*, 45–52. [CrossRef]

37. Pimentel, M.S.; Faleiro, F.; Machado, J.; Pousão-Ferreira, P.; Rosa, R. Seabream larval physiology under ocean warming and acidification. *Fishes* **2020**, *5*, 1. [CrossRef]

38. Huerta-Ortíz, M.; González, C.A.Á.; Couturier, G.M.; Sánchez, W.M.C.; Cerecedo, R.C.; Bores, E.G. Sustitución total de aceite de pescado con aceite vegetal en larvas de pejelagarto *Atractosteus tropicus. Kuxulkab'* **2009**, *15*. [CrossRef]

39. Frías-Quintana, C.A.; Álvarez-González, C.A.; Márquez-Couturier, G. Diseño de microdietas para el larvicultivo de pejelagarto *Atractosteus tropicus*, Gill 1863. *Univ. Cienc.* **2010**, *26*, 265–282.

40. Guerrero-Zárate, R.; Álvarez-González, C.A.; Olvera-Novoa, M.A.; Perales-García, N.; Frías-Quintana, C.A.; Martínez-García, R.; Contreras-Sánchez, W.M. Partial characterization of digestive proteases in tropical gar *Atractosteus tropicus* juveniles. *Fish Physiol. Biochem.* **2014**, *40*, 1021–1029. [CrossRef] [PubMed]

41. Frías-Quintana, C.A.; Domínguez-Lorenzo, J.; Álvarez-González, C.A.; Tovar-Ramírez, D.; Martínez-García, R. Using cornstarch in microparticulate diets for larvicultured tropical gar (*Atractosteus tropicus*). *Fish Physiol. Biochem.* **2016**, *42*, 517–528. [CrossRef] [PubMed]

42. Martínez-Cárdenas, L.; Hernández-Cortez, M.I.; Espinosa-Chaurand, D.; Castañeda-Chavez, M.R.; León-Fernández, A.E.; Valdez Hernández, E.F.; Álvarez-González, C.A. Effect of stocking density on growth, survival and condition factor in tropical gar (*Atractosteus tropicus* Gill, 1863) juveniles. *Lat. Am. J. Aquat. Res.* **2020**, *48*, 570–577. [CrossRef]

43. Ferreira-Rodríguez, N.; Fernandez, I.; Cancela, M.L.; Pardo, I. Multibiomarker response shows how native and non-native freshwater bivalves differentially cope with heat-wave events. *Aquatic. Conserv. Mar. Freshw. Ecosyst.* **2018**, *28*, 934–943. [CrossRef]

44. Politis, S.N.; Mazurais, D.; Servili, A.; Zambonino-Infante, J.L.; Miest, J.J.; Sorensen, S.R.; Butts, I.A. Temperature effects on gene expression and morphological development of European eel, *Anguilla anguilla* larvae. *PLoS ONE* **2017**, *12*, e0182726. [CrossRef]

45. Mendoza Alfaro, R.; González, C.A.; Ferrara, A.M. Gar biology and culture: Status and prospects. *Aquac. Res.* **2008**, *39*, 748–763. [CrossRef]

46. Aranda-Morales, S.A.; Peña-Marín, E.S.; Jiménez-Martínez, L.D.; Martínez-Burguete, T.; Martínez-Bautista, G.; Álvarez-Villagómez, C.S.; De la Rosa, S.; Camarillo-Coop, S.; Martínez-García, R.; Álvarez-González, C.A.; et al. Expression of ion transport proteins and routine metabolism in juveniles of tropical gar (*Atractosteus tropicus*) exposed to ammonia. *Comp. Biochem. Physiol. Part C: Toxicol. Pharmacol.* **2021**, *250*, 109166. [CrossRef] [PubMed]

47. Córdova-de la Cruz, S.E.; Martínez-Bautista, G.; Peña-Marín, E.S.; Martínez-García, R.; Núñez-Nogueira, G.; Adams, R.H.; Burggren, W.W.; Alvarez-González, C.A. Morphological and cardiac alterations after crude oil exposure in the early-life stages of the tropical gar (*Atractosteus tropicus*). *Environ. Sci. Pollut. Res.* **2021**, 1–12. [CrossRef]

48. Navarro-Martin, L.; Vinas, J.; Ribas, L.; Diaz, N.; Gutierrez, A.; Di Croce, L.; Piferrer, F. DNA methylation of the gonadal aromatase (cyp 19a) promoter is involved in temperatura dependent sex ratio shifts in the European sea bass. *PLoS Genet.* **2011**, *7*, e1002447. [CrossRef]

49. Campos, C.; Valente, L.M.P.; Conceicao, L.E.C.; Engrola, S.; Fernandes, J.M.O. Temperature affects methylation of the myogenin putative promoter, its expression and muscle cellularity in *Senegalese sole* larvae. *Epigenetics* **2013**, *8*, 389–397. [CrossRef]

50. Burgerhout, E.; Mommens, M.; Johnsen, H.; Aunsmo, A.; Santi, N.; Andersen, O. Genetic background and embryonic temperature affect DNA methylation and expression of myogenin and muscle development in Atlantic salmon (*Salmo salar*). *PLoS ONE* **2017**, *12*, e0179918. [CrossRef] [PubMed]

51. Ribas, L.; Liew, W.C.; Diaz, N.; Sreenivasan, R.; Orban, L.; Piferrer, F. Heat-induced masculinization in domesticated zebrafish is family-specific & yields a set of different gonadal transcriptomes. *Proc. Natl. Acad. Sci. USA* **2017**, *114*, E941–E950. [CrossRef]

52. Di Santo, V. Ocean acidification and warming affect skeletal mineralization in a marine fish. *Proc. R. Soc. B Biol. Sci.* **2019**, *286*, 20182187. [CrossRef] [PubMed]

53. Zhou, C.Q.; Zhou, P.; Ren, Y.L.; Cao, L.H.; Wang, J.L. Physiological response and miRNA-mRNA interaction analysis in the head kidney of rainbow trout exposed to acute heat stress. *J. Therm. Biol.* **2019**, *83*, 134–141. [CrossRef]

54. Valdivieso, A.; Ribas, L.; Monleon-Getino, A.; Orban, L.; Piferrer, F. Exposure of zebrafish to elevated temperature induces sex ratio shifts and alterations in the testicular epigenome of unexposed offspring. *Environ. Res.* **2020**, *186*, 109601. [CrossRef] [PubMed]

55. Arratia, G.; Schultze, H.P. Palatoquadrate and its ossifications: Development and homology within osteichthyans. *J. Morphol.* **1991**, *208*, 1–81. [CrossRef]

56. Long, W.L.; Ballard, W.W. Normal embryonic stages of the longnose gar, *Lepisosteus osseus. BMC Dev. Biol.* **2001**, *1*, 1–18. [CrossRef]

57. Diogo, R.; Ziermann, J.M.; Molnar, J.; Siomava, N.; Abdala, V. *Muscles of Chordates: Development, Homologies, and Evolution*, 1st ed.; CRC Press: Boca Raton, FL, USA, 2018; ISBN 978-1-1385-7116-7.

58. Grande, L. An Empirical Synthetic Pattern Study of Gars (Lepisosteiformes) and Closely Related Species, Based Montly on Skeletal Anatomy. The Resurrection of Holostei. Copeia. American Society of Icthyologists and Herpetolologists, 2010. ppt 6, 1187. Available online: https://www.jstor.org/stable/20787269 (accessed on 13 December 2021).

59. Eames, B.F.; Amores, A.; Yan, Y.L.; Postlethwait, J.H. Evolution of the osteoblast: Skeletogenesis in gar and zebrafish. *BMC Evol. Biol.* **2012**, *12*, 27. [CrossRef]

60. Carvalho, M.; Bockmann, F.A.; de Carvalho, M.R. Homology of the fifth epibranchial and accessory elements of the cerato-branchials among gnathostomes: Insights from the development of ostariophysans. *PLoS ONE* **2013**, *8*, e62389. [CrossRef]
61. Hilton, E.J.; Konstantinidis, P.; Schnell, N.K.; Dillman, C.B. Identity of a unique cartilage in the buccal cavity of gars (Neopterygii: Lepisosteiformes: Lepisosteidae). *Copeia* **2014**, *1*, 50–55. [CrossRef]
62. Scherrer, R.; Hurtado, A.; Machado, E.G.; Debiais-Thibaud, M. MicroCT survey of larval skeletal mineralization in the Cuban gar *Atractosteus tristoechus* (Actinopterygii; Lepisosteiformes). *MorphoMuseum* **2017**, *3*. [CrossRef]
63. Koumoundouros, G. Morpho-anatomical abnormalities in Mediterranean marine aquaculture. *Recent Adv. Aquac. Res.* **2010**, *661*, 125–148.
64. Boglione, C.; Gavaia, P.; Koumoundouros, G.; Gisbert, E.; Moren, M.; Fontagné, S.; Witten, P.E. Skeletal anomalies in reared E uropean fish larvae and juveniles. Part 1: Normal and anomalous skeletogenic processes. *Rev. Aquac.* **2013**, *5*, S99–S120. [CrossRef]
65. Fernández, I.; Granadeiro, L.; Darias, M.J.; Gavaia, P.J.; Andree, K.B.; Gisbert, E. *Solea senegalensis* skeletal ossification and gene expression patterns during metamorphosis: New clues on the onset of skeletal deformities during larval to juvenile transition. *Aquaculture* **2018**, *496*, 153–165. [CrossRef]
66. Wang, L.H.; Tsai, C.L. Effects of temperature on the deformity and sex differentiation of tilapia, *Oreochromis mossambicus*. *J. Exp. Zool.* **2000**, *286*, 534–537. [CrossRef]
67. Hansen, T.K.; Falk-Petersen, I.B. The influence of rearing temperature on early development and growth of spotted wolffish *Anarhichas minor* (Olafsen). *Aquac. Res.* **2001**, *32*, 369–378. [CrossRef]
68. Sfakianakis, D.G.; Koumoundouros, G.; Divanach, P.; Kentouri, M. Osteological development of the vertebral column and of the fins in *Pagellus erythrinus* (L. 1758). Temperature effect on the developmental plasticity and morpho-anatomical abnormalities. *Aquaculture* **2004**, *232*, 407–424. [CrossRef]
69. Georgakopoulou, E.; Angelopoulou, A.; Kaspiris, P.; Divanach, P.; Koumoundouros, G. Temperature effects on cranial deformities in European sea bass, *Dicentrarchus labrax* (L.). *J. Appl. Ichthyol.* **2007**, *23*, 99–103. [CrossRef]
70. Ytteborg, E.; Baeverfjord, G.; Torgersen, J.; Hjelde, K.; Takle, H. Molecular pathology of vertebral deformities in hyperthermic Atlantic salmon (*Salmo salar*). *BMC Physiol.* **2010**, *10*, 12. [CrossRef]
71. Sfakianakis, D.G.; Georgakopoulou, E.; Papadakis, I.E.; Divanach, P.; Kentouri, M.; Koumoundouros, G. Environmental determinants of haemal lordosis in European sea bass, *Dicentrarchus labrax* (Linnaeus, 1758). *Aquaculture* **2006**, *254*, 54–64. [CrossRef]
72. Martini, A.; Huysseune, A.; Witten, P.E.; Boglione, C. Plasticity of the skeleton and skeletal deformities in zebrafish (*Danio rerio*) linked to rearing density. *J. Fish Biol.* **2021**, *98*, 971–986. [CrossRef] [PubMed]
73. Fernández, I.; Ortiz-Delgado, J.B.; Sarasquete, C.; Gisbert, E. Vitamin A effects on vertebral bone tissue homeostasis in gilthead sea bream (*Sparus aurata*) juveniles. *J. Appl. Ichthyol.* **2012**, *28*, 419–426. [CrossRef]
74. Hall, B.K. *Bones and Cartilage: Developmental and Evolutionary Skeletal Biology*, 2nd ed.; Academic Press: Cambridge, MA, USA, 2015; ISBN 9780124166851.
75. Fernández, I.; Darias, M.; Andree, K.B.; Mazurais, D.; Zambonino-Infante, J.L.; Gisbert, E. Coordinated gene expression during gilthead sea bream skeletogenesis and its disruption by nutritional hypervitaminosis A. *BMC Dev. Biol.* **2011**, *11*, 1–20. [CrossRef]
76. Fuiman, L.A.; Poling, K.R.; Higgs, D.M. Quantifying developmental progress for comparative studies of larval fishes. *Copeia* **1998**, 602–611. [CrossRef]
77. Seikai, T.; Tanangonan, J.B.; Tanaka, M. Temperature influence on larval growth and metamorphosis of the Japanese flounder *Paralichthys olivaceus* in the laboratory. *Bull. Jpn. Soc. Sci. Fish.* **1986**, *52*, 977–982. [CrossRef]
78. Polo, A.; Yufera, M.; Pascual, E. Effects of temperature on egg and larval development of *Sparus aurata* L. *Aquaculture* **1991**, *92*, 367–375. [CrossRef]
79. Koumoundouros, G.; Divanach, P.; Anezaki, L.; Kentouri, M. Temperature-induced ontogenetic plasticity in sea bass (*Dicentrarchuslabrax*). *Mar. Biol.* **2001**, *139*, 817–830. [CrossRef]
80. Comabella, Y.; Hurtado, A.; Canabal, J.; Galano, T.G. Effect of temperature on hatching and growth of Cuban gar (*Atractosteus tristoechus*) larvae. *Ecosistemas y Recursos Agropecuarios* **2014**, *1*, 19–32.
81. Combella, Y.; Azanza, J.; Hurtado, A.; Canabal, J.; García-Galano, T. Allometric growth in Cuban gar (*Atractosteus tristoechus*) larvae. *Universidad Ciencia* **2013**, *29*, 301–315.
82. Kourkouta, C.; Printzi, A.; Geladakis, G.; Mitrizakis, N.; Papandroulakis, N.; Koumoundouros, G. Long lasting effects of early temperature exposure on the swimming performance and skeleton development of metamorphosing Gilthead seabream (*Sparus aurata* L.) larvae. *Sci. Rep.* **2021**, *11*, 8787. [CrossRef] [PubMed]

Article

Direct and Molecular Observation of Movement and Reproduction by Candy Darter, *Etheostoma osburni*, an Endangered Benthic Stream Fish in Virginia, USA

Kathryn E. McBaine [1,2], Eric M. Hallerman [1,*] and Paul L. Angermeier [1,3]

1   Department of Fish and Wildlife Conservation, Virginia Polytechnic Institute and State University, Blacksburg, VA 24061, USA; katimc9@vt.edu (K.E.M.); biota@vt.edu (P.L.A.)
2   Idaho Department of Fish and Game, 324 417 E, Jerome, ID 83338, USA
3   U.S. Geological Survey, Virginia Cooperative Fish and Wildlife Research Unit, Blacksburg, VA 24061, USA
*   Correspondence: ehallerm@vt.edu

Abstract: Direct and indirect measures of individual movement provide valuable knowledge regarding a species' resiliency to environmental change. Information on patterns of movement can inform species management and conservation but is lacking for many imperiled fishes. The Candy Darter, *Etheostoma osburni*, is an endangered stream fish with a dramatically reduced distribution in Virginia in the eastern United States, now known from only four isolated populations. We used visual implant elastomer tags and microsatellite DNA markers to directly describe movement patterns in two populations. Parentage analysis based on parent-offspring pairs was used to infer movement patterns of young-of-year and age-1 individuals, as well as the reproductive contribution of certain adults. Direct measurements of movement distances were generally similar between methods, but microsatellite markers revealed greater distances moved, commensurate with greater spatial frames sampled. Parent-offspring pairs were found throughout the species' 18.8-km distribution in Stony Creek, while most parent-offspring pairs were in 2 km of the 4.25-km distribution in Laurel Creek. Sibship reconstruction allowed us to characterize the mating system and number of spawning years for adults. Our results provide the first measures of movement patterns of Candy Darter as well as the spatial distribution of parent-offspring pairs, which may be useful for selecting collection sites in source populations to be used for translocation or reintroductions. Our results highlight the importance of documenting species movement patterns and spatial distributions of related individuals as steps toward understanding population dynamics and informing translocation strategies. We also demonstrate that the reproductive longevity of this species is greater than previously described, which may be the case for other small stream fishes.

Keywords: conservation planning; dispersal; genetic markers; mating system; parentage analysis; visual tags

Citation: McBaine, K.E.; Hallerman, E.M.; Angermeier, P.L. Direct and Molecular Observation of Movement and Reproduction by Candy Darter, *Etheostoma osburni*, an Endangered Benthic Stream Fish in Virginia, USA. *Fishes* **2022**, *7*, 30. https://doi.org/10.3390/fishes7010030

Academic Editor: Manuel Manchado

Received: 1 December 2021
Accepted: 27 December 2021
Published: 27 January 2022

Publisher's Note: MDPI stays neutral with regard to jurisdictional claims in published maps and institutional affiliations.

1. Introduction

Understanding individual movement patterns can provide information valuable for species management and conservation [1,2]. Knowledge of movement patterns is useful for evaluating a species' resilience to environmental change [3,4], gene flow, demography, and population structure [5]. Demographic connectivity relies on the survival and successful reproduction of dispersing individuals that ultimately determine the receiving population's vital rates [5]. Population connectivity is influenced by many factors controlling species-specific movement by individuals, including the capacity to navigate physical barriers (natural or anthropogenic), sex, seasonality of movement [6], life-stage [7], interspecific interactions [8], and habitat suitability [9]. Hence, assessment of movement patterns provides critical information on the spatial extent of populations that must be recognized to effectively monitor population dynamics and to inform conservation. Further, understanding the spatial distribution of individuals' relatedness within the scope of their movement can provide insight into population genetic processes such as propensity to inbreeding [10].

Fish movement can be associated with daily, seasonal, or occasional behaviors, which may manifest over short or long distances. Fishes exhibit complex life histories, often requiring multiple habitats to complete their life cycle [11]. For instance, small-scale daily movements may be associated with habitats used for feeding and/or refuge from predators. Labbe and Fausch [6] documented seasonal movements by Arkansas Darter *Etheostoma cragini* between reproductive habitat and overwintering habitat in intermittent streams in Colorado (USA). Spawning by North American darters generally coincides with high spring-time flows that facilitate passive advective transport of larvae downstream (i.e., larval drift; [12]). The extent of dispersal distances resulting from larval drift varies among species and relates in part to streamflow [13]. Conversely, adults may move upstream prior to spawning to compensate for drift.

General approaches to studying movement patterns include real-time observation (e.g., radio tags), capture-mark-capture, microchemistry analysis, and DNA markers. The selection of approach depends on study objectives and the spatial and temporal scales of movement meant to be detected [14]. Each approach has limitations to its application: some require culling individuals to retrieve hard parts, while others may be limited to specific sizes of fish appropriate for implanting tags [14]. An inherent source of bias in many movement studies is the sampling design. Albanese et al. [15] found that sampling longer recapture sections increased their ability to detect long-distance movements by stream fishes in Virginia (USA). Marking techniques such as visible implant elastomer (VIE) tags are useful for tagging small-bodied fishes but are subject to observer bias associated with tag colors and/or loss [16]. Roberts and Angermeier [17] suggested that VIE tag performance may be species-specific because of the variability in fish and tag color and in tag retention. Given the limitations of any single fish-marking approach, dual-tagging methods may provide more reliable characterizations of fish movement patterns.

Molecular methods, which indirectly measure dispersal over a range of spatial and temporal scales, are increasingly used to describe fish movement patterns. For example, Argentina et al. [18] found no genetic structuring among populations of Variegate Darter *Etheostoma variatum* presumed to be isolated by dams, suggesting that gene flow was mediated by dispersal among watersheds up to 400 km apart. Using molecular data for characterizing movement patterns poses a few disadvantages. Although costs associated with molecular analyses may exceed those of analyses based on traditional methods, recent advances in technology can make genetic markers more cost-effective overall [19]. The availability of molecular markers for non-game species (those lacking fisheries) may limit some studies, but the conservation value of information gained through molecular data may far outweigh the monetary costs.

The degree to which larval drift and sub-adult movements contribute to within-stream demographic connectivity and population structure remains to be investigated for most fishes. In the absence of immigration and emigration, isolated populations rely on survival and local recruitment for persistence. To understand how movement and recruitment influence a population's vital rates, it is important to combine genetic data with measures of movement to explain demographic connectivity [5]. Few studies of stream fishes have combined direct measures of movement with measures of demographic connectivity based on molecular data. Ruppert et al. [20] used passive integrated transponder (PIT) tags to track adult movements and genetic markers to characterize the population structure of Rocky Mountain Sculpin *Cottus* sp. in British Columbia and Alberta (Canada). They concluded that most adults are sedentary and suggested that larval or juvenile dispersal is a main contributor to genetic connectivity but did not explicitly measure movement by sub-adults.

Parentage analysis based on molecular methods has recently been used to estimate the number of spawning adults, number of spawning events [21], and extent of larval dispersal [22]. Roberts et al. [2] used pedigree reconstruction and assignment tests to identify reproductively successful long-distance movements (14–55 km) by Roanoke Logperch *Percina rex* in Virginia (USA), revealing the importance of managing the species at the watershed scale. They also assessed natal dispersal distances but were unable to distinguish dispersal direction for many sibling pairs because natal sites could not be determined. Applying parentage analyses in small streams may prove more successful than in rivers, as it is more feasible to capture related individuals and infer natal dispersal in small streams.

Parentage analysis may also advance understanding of reproductive biology and provide insight into population persistence. Throughout a fish's lifetime, it may move among a mosaic of habitat patches to grow, survive, and reproduce, which contributes to individual fitness. The number of surviving offspring, a measure of reproductive success, varies among individuals and years [23]. Estimating reproductive success can provide inference regarding effective population size, generally described as the number of individuals contributing to the population gene pool. Understanding which and how many individuals contribute genes to subsequent generations can lead to inferences of susceptibility to inbreeding and the degree of functional habitat connectivity. Kanno et al. [23] showed the importance of combining measures of movement and reproductive success of Brook Trout *Salvelinus fontinalis* to determine connectivity among headwater streams and a mainstem river. An analogous approach may prove useful for understanding relations between temporal variability in reproductive success and the loss of genetic diversity in isolated populations.

Habitat degradation and restricted geographic range are major factors associated with the imperilment of minnows and darters, the most imperiled freshwater fish families in North America [24]. The lack of knowledge for these species is striking given their degree of imperilment. Enhancing our understanding of the ecology and connectivity of isolated populations of imperiled species can provide information necessary for species management and conservation. The Candy Darter *Etheostoma osburni* (Figure S1a,b) represents a suite of other imperiled fishes in terms of their ecology, life-history traits, and vulnerability to anthropogenic threats. Candy Darter was listed as federally Endangered in response to widespread hybridization with an invasive congener, habitat degradation (increasing fine sediment, water temperature, and fragmentation), catastrophic events, and vulnerability to introduced predators [25]. Endemic to the upper Kanawha River drainage in Virginia and West Virginia (USA), Candy Darter is restricted to four isolated populations in Virginia but occurs more extensively in West Virginia. Adults are habitat specialists [26], generally restricted to areas with silt-free substrates and cool temperatures [27]. Preferred microhabitats range from low-velocity areas used by age-0 fish to swift, shallow areas with complex substrates used by adults [28]. Basic life-history traits, population demographics, and genetic structure of populations in Virginia are being assessed to inform potential conservation actions.

Despite the extensive literature on stream fish movement, we lack basic information for most species, including Candy Darter [25]. Understanding movement patterns will increase our ability to identify barriers to dispersal and define the spatiotemporal extent of populations, which can inform future surveys, designation of critical habitat, and translocations. By comparing the results of traditional mark-recapture methods with those of molecular methods, we can evaluate the relative effectiveness of these methods for characterizing adult movements. Additionally, enhancing our understanding of the spatiotemporal distribution of parent-offspring pairs can provide insight into demographic connectivity, early-life movement, and the susceptibility of populations to decreased genetic diversity and increased inbreeding. Finally, understanding aspects of the species' reproductive biology can provide insight into the mating system and reproductive success of individuals. To that end, our research involving Candy Darter addressed the following objectives: (1) measure and compare spatiotemporal patterns of adult movement based on traditional visual versus molecular marks, (2) describe the spatiotemporal distribution of related pairs to infer movement patterns of sub-adults, and (3) assess the reproductive contribution of adults to subsequent generations.

2. Materials and Methods

2.1. Study Area

We examined the movement of Candy Darter in Stony and Laurel creeks in Virginia in the southeastern United States. These represent two of the four remaining populations in Virginia and 17 populations range-wide. Virginia populations belong to a distinct genetic form (Figure S1a,b) occurring in the Valley and Ridge physiographic province relative to populations occurring in the Appalachian Plateau physiographic province in West Virginia. Range-wide, many populations are small and/or isolated by physical barriers or long reaches of unsuitable habitat [25]. Stony Creek supports one of the largest and densest Candy Darter populations in Virginia, while Laurel Creek supports a smaller, less dense population. These streams are the closest pair of populations in Virginia based on fluvial distance but are separated by approximately 50 fluvial kilometers.

Stony and Laurel creeks are third-order tributaries of the New River. Stony Creek is a cold-water, high-gradient stream in Giles County [29]. Stream widths are 8–15 m. Candy Darter occupy approximately 18.8 km of Stony Creek. Because of underground mining, surface flow in the lower 1.5 km of Stony Creek is intermittent, which may act as a seasonal barrier to fish movement. Laurel Creek is a cool-water, high-gradient stream in Bland County [28]. Stream widths are 1.2–8 m. Candy Darter occupy the lower 4.25 km, below a series of milldams.

We selected mark-recapture sites based on Candy Darter distributions reported in Dunn [30]. Two sites were selected in Stony Creek, one near the center of the distribution (Interior) and one just upstream of the intermittent-flow reach near the mouth (Lower Stony; Figure 1 and Figure S2). Initial mark sites in Stony Creek included 3–4 riffles in a 150–200-m reach. We selected one site near the center of Candy Darter distribution in Laurel Creek. The initial mark site included five riffles in a 500-m reach (Mark Site; Figure 1 and Figure S3). Recapture sites for both streams encompassed the initial mark site(s) plus 7–17 riffles within 500 m upstream and downstream. All detected movements were considered minimum movements because we did not track movements continuously.

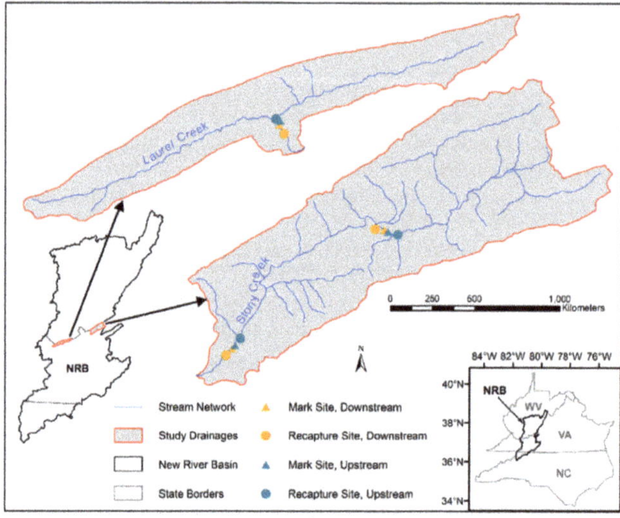

Figure 1. Mark-recapture sites in Laurel Creek and Stony Creek watersheds of Virginia. NRB = New River basin.

2.2. Sampling

We used quadrat-based electrofishing to sample fish. We used pulsed direct current from an LR-24 backpack (Smith-Root, Vancouver, WA, USA) to conduct a single electrofishing pass through all riffles in the mark and recapture sites. Beginning at the downstream end of each site and on either bank, a 1.5- × 3-m weighted seine with 5-mm mesh was held to the stream bottom by two crew members as a third person electro-fished downstream while disturbing the substrate, thereby allowing the streamflow to carry stunned fish into the seine [31]. Electrofishing was conducted in transects, where a transect consisted of contiguous, non-overlapping seine sets across the stream width. After completing a transect, the crew moved 4 m upstream and began another transect. Each seine set encompassed a quadrat measuring 3 m × 4 m; sampling continued throughout the riffles of the sites. Our sampling protocol was similar to that used for Roanoke Logperch *Percina rex* [31], an-other endangered darter that is similar in habitat use to Candy Darter. We observed low mortality (<5%) for Candy Darter during our sampling. Mark and recapture events were separated by at least four weeks during May–September of 2016–2018.

We identified recaptured fish via data on length, sex, and tags. We measured the length of each captured Candy Darter and assigned a sex if determinable. Candy Darters were anesthetized in an immersion solution of AQUI-S 20E (20 mg/L; Aqua Tactics Fish Health, Kirkland, WA, USA) and stream water. When anesthetized, fish $\geq$ 55 mm total length received a subcutaneous visual implant elastomer (VIE; Northwest Marine Technology, Inc., Anacortes, WA, USA; Figure S1c) tag. Riffle-specific batch tags were implanted in combinations of colors (red, black, orange, and yellow) and marking locations (ventral fin, ventral caudal peduncle, first dorsal fin, and second dorsal fin). During recapture events, all Candy Darter were visually inspected for previous tags using an ultraviolet light (Figure S1c), then given a unique color-locale tag regardless of the presence of a previous tag. Individuals were tagged during two initial mark surveys (spring 2016 and 2017) and two recapture surveys (fall 2016 and 2017). A final recapture survey was conducted in spring 2018, but no new tags were given.

In addition to applying VIE tags, we collected a fin clip from each Candy Darter captured in mark-recapture sites as well as in supplemental sites designated for concurrent studies (Figure 1, Figures S2 and S3). Five supplemental sites were established in Stony Creek, approximately evenly distributed along the 18.8-km distribution (Figure S2): Lower Stony, Below Vims, Pole Bridge, Cherokee Flats, and Glen Alton. Four supplemental sites were established in Laurel Creek (Figure S3), approximately evenly distributed along the 4.25-km distribution: School, Aker's Towing, Recapture Site, and Church. Names of supplemental sites refer to landmarks. Supplemental sites included 2–5 riffles within a 150–200-m reach and were sampled during May–June and August–October in 2017–2018. Supplemental sites were included to increase the chances of documenting long-distance movements. We clipped the lower lobe of the caudal fin for every individual upon every capture. Fin clips were placed in a small paper envelope for air-drying and labeled with a unique alphanumeric code corresponding to the capture location. After measuring lengths (standard and total) and weight and collecting a fin clip, we placed fish in a recovery tank with stream water until they resumed normal behaviors, then released them at the site of capture.

2.3. Molecular Analysis

Nine nuclear microsatellite loci were screened for genetic variation using primer pairs developed by Switzer et al. [32] for Candy Darter. PCR protocols for nuclear markers were adapted from those of Switzer et al. [32]. DNA was amplified using two multiplex PCR reactions (multiplex 1: *EosC208, EosC207, EosC112, EosC117*; multiplex 2: *EosD10, EosC3, EosC2, EosD108, EosD11*). The PCR amplification was performed in a final volume of 10 µL and contained 2.0 µL of DNA extract, 2.0 µL Nanopure water, 2.0 µL 5× GoTaq Flexi Buffer (Promega, Madison, WI, USA), 1.75 µL 25 mM $MgCl_2$ (Promega, Madison, WI, USA), 1.15 µL 2.5 µM dNTPs (Promega, Madison, WI, USA), 0.5 µL of each primer, and

0.1 units/µL of GoTaq DNA polymerase (Promega, Madison, WI, USA). The PCR protocol consisted of an initial denaturation at 95 °C for 15 min; 25 cycles of: 94 °C denaturation for 30 s, 57 °C annealing for 90 s, and 72 °C extension for 1 min; and a 30-min extension at 60 °C. An aliquot of the PCR product was used for confirmation of amplification of DNA in a 2% agarose gel. PCR products were sent to the Cornell University Core Laboratory (Ithaca, NY, USA) for fragment-size analysis using an ABI 3730XL DNA Analyzer (Applied Biosystems, Waltham, MA, USA). Allele calls were scored using GeneMarker [33].

We used identity analysis, which identifies matching genotypes within samples, to identify genetic recaptures. Multilocus genotypes were compared across all Candy Darter captured using Cervus v2.0 [34]. Individuals were considered recaptures if genotypes at all loci matched. Genetic recaptures were then matched with commensurate field data, including date of capture, sex, total length (TL), and site of capture. We used comparisons of TL at initial and subsequent captures as the main criteria to validate genetic recaptures. We assumed fish would exhibit minimal shrinkage during our study, and removed from further analysis any recaptures for which TL at recapture was >5 mm smaller than the TL at initial capture.

Cervus v2.0 [34] also was used for parentage analysis. The program considers multilocus genotypes among sampled offspring, then assigns putative parents of those individuals. Each Cervus run simulated 100,000 offspring for parentage of known sexes, with 0.15 as the proportion of parents sampled, 0.85 as proportion of loci typed, and 0.01 as proportion of mistyped loci, parameters that we estimated from our data. Following the simulation, the two most likely parent pairs were selected based on the logarithm of the odds (LOD) score.

Three separate parentage analysis runs were conducted. Individuals were assigned to age classes based on TL designations in Dunn [29]: adults were coded as 1 (females > 60 mm TL, males > 65 mm TL), juveniles as 2 (<60/65 mm and >45 mm TL), and young-of-year (YOY) as 3 ($\leq$45 mm TL). Cohorts for each sampling year (2016–2018) were assigned as previously described. However, for 2017 and 2018 data, juveniles sampled from the previous year were coded as adults and YOY from the previous year were coded as juveniles.

Distances between assigned mothers and fathers and their offspring were estimated from fluvial distances between their respective capture sites. The distance between an offspring found at a different site than the assigned parent was measured from the midpoint of the offspring collection site to the midpoint of the parent collection site. Assigned parents captured downstream or upstream of offspring were given negative or positive distances, respectively. Similar methods were used for measuring distances between members of inferred sib pairs, including pairs captured in the same year and different years. We used Welch's *t*-tests to assess differences in distances between mother-offspring pairs and father-offspring pairs. We stratified tests by offspring age (YOY versus juvenile), stream (Stony Creek versus Laurel Creek), and year of capture (same year versus different year). Genetic recaptures from the identity analysis were noted in order to evaluate distances between members of offspring-parent pairs through time. Multiple parent-pair assignments among runs of Cervus were assessed, noting the respective LOD scores.

Sibling relationships were established by observing multiple offspring assignments to a single candidate mother or father. Full-sibs had identical assignments for both parents, while half-sibs had identical assignments for one parent. We estimated spawning periodicity and frequency from inferred ages of offspring and reconstruction of half-sib families. YOY were assumed to be spawned the year of capture (age = 0) and juveniles were assumed to be spawned the previous year (age = 1). For instance, if a candidate mother was assigned to two YOY and one juvenile captured in 2017, the inferred spawning years would be 2017 and 2016, respectively, and we would infer that she had spawned in at least two consecutive years. In addition, if the assigned fathers were different for the YOY, we would infer that the female spawned at least twice in 2017. The number of mothers per father in a given year also was estimated this way.

We applied this analytic approach on a year-by-year basis to infer the number of spawning years for individuals and to assess the dynamics of the spatial distribution of parent-offspring pairs. That is, conducting analyses for year-specific age classes accounts for contributions of additional candidate parents in subsequent years, but does not assume parentage outside of the sampling period (2016–2018). However, we assumed that adults survived throughout the sampling period. Additionally, genotypes belonging to YOY and juveniles with indistinguishable sexes at the time of capture were assigned as appropriate to candidate mother or father pools once sexual maturation was observed (at age 2+) in subsequent years.

2.4. Analysis of Movement Data

Movement distance was characterized in reference to mesohabitat (riffle, run, pool) patches. If individuals were recaptured in the same riffle as initially marked, they were categorized as showing "no movement". Individuals recaptured in a riffle other than the initial riffle were categorized as exhibiting "movement". Movement upstream or downstream of the initial riffle was represented by positive or negative values, respectively. Lengths of mesohabitat patches in the mark and recapture sites were measured each year. For within-year recaptures, movement distances were computed from that year's patch lengths. We used mean patch lengths across recapture years to compute between-year movement distances. A given movement was calculated by assigning zero distance to the midpoint of the initial-capture patch, then adding (for upstream) or subtracting (for downstream) the lengths of all intervening patches up to the midpoint of the recapture patch. We used identical methods to calculate movement distances based on VIE versus molecular marks.

Welch's *t*-test for unequal variance and non-parametric Kruskal-Wallis tests were used to compare movement distances between streams, sexes, and recapture methods. Welch's *t*-tests were used to evaluate differences in distances moved, while differences in movement direction were evaluated using non-parametric Kruskal-Wallis tests with three options: upstream, none, and downstream. Genetic recaptures were summarized in two ways: including VIE recaptures and excluding VIE recaptures.

3. Results

3.1. Visible Inplant Elastomer Recaptures

Most fish marked with VIE tags were never recaptured. A total of 286 individuals were marked with VIE tags in Stony Creek in 2016 and 2017 (Table 1, Figure 1 and Figure S2); 16% of those were recaptured at least once. Eleven individuals were recaptured at least twice, and a single male was recaptured three times. A total of 200 individuals were marked with VIE tags in Laurel Creek in 2016 and 2017 (Table 1; Figure 1 and Figure S3); 12% of those were recaptured at least once. Nine individuals were recaptured at least twice, and four individuals were recaptured three times. We observed few recaptures (4% of recaptures in Stony Creek, 8% in Laurel Creek) two years after fish were initially tagged.

Table 1. Numbers of Candy Darters marked and recaptured, and mean minimum distances (meters) moved in two streams based on two marking methods (visual implant elastomer [VIE] and molecular markers). "Identified in field" represents the numbers of individuals observed with VIE marks during sampling. Molecular recaptures were not identified in the field ("-"). "Molecular, including VIE" represents all individuals detected with either method, as all individuals recaptured with VIEs were also identified as molecular recaptures, but not vice versa. "SE" = standard error.

Stream	Method	Marked	Recaptured (%)	Identified in Field (%)	Mean Minimum Distances (SE)		
					Female	Male	Overall
Laurel Creek	VIE	200	30 (15)	24 (12)	179 (86)	266 (97)	242 (70)
	Molecular, excluding VIE	157	28 (18)	-	131 (51)	343 (120)	230 (64)
	Molecular, including VIE	357	58 (16)	-	146 (43)	304 (77)	233 (47)
Stony Creek	VIE	286	55 (19)	47 (16)	107 (34)	109 (32)	108 (23)
	Molecular, excluding VIE	451	29 (6)	-	276 (207)	518 (319)	370 (173)
	Molecular, including VIE	737	84 (11)	-	190 (102)	245 (114)	216 (76)

Overall, fish recaptured with VIE tags exhibited limited movement, especially in Stony Creek. Individuals moved a mean distance of 108 m (range: 0–1060 m) in Stony Creek (Table 2); the maximum detectable distance was 1350 m. Twenty-three individuals (45%) stayed in the same riffle (Figure 2). Males and females moved similar distances ($p = 0.97$, $df = 46.72$). Twenty individuals moved upstream of their initial capture, while nine moved downstream. Directionality of movement was similar between sexes (Kruskal-Wallis test; $p = 0.9$, $df = 2$). Detected movement distances in Laurel Creek ranged from 0–1300 m across the 1500-m detectable distance (mean = 242 m; Table 1; Figure 2). There were no differences between male and female distances moved or directionality ($p = 0.51$, $df = 20.62$; $p = 0.27$, $df = 2$). Five individuals (2.5%) stayed in the same riffle. Seventeen of the 24 recaptured individuals (71%) moved upstream from initial capture. Mean movement distances were statistically similar between streams ($p = 0.09$, $df = 27.67$), but movements exhibited more upstream bias in Laurel Creek compared to Stony Creek ($p = 0.04$, $df = 2$).

Table 2. Numbers of offspring captured in two streams during 2016–2018 for an assignment using parentage analysis, and numbers of parent-offspring assignments made using Cervus v2.0 [Kalinowski et al. 2007 = [34]]. YOY = young-of-year. Percentages assigned are in parentheses.

Stream	Year	Offspring to Be Assigned			Parentage Assignments		
		YOY	Juveniles	Total	YOY	Juveniles	Total
Laurel Creek	2016	7	8	15	5 (71)	3 (38)	8 (53)
	2017	9	9	18	9 (100)	3 (33)	12 (67)
	2018	4	9	13	3 (75)	8 (89)	11 (85)
Stony Creek	2016	4	23	27	4 (100)	16 (70)	20 (74)
	2017	19	32	51	17 (89)	27 (84)	44 (86)
	2018	15	60	75	14 (93)	52 (87)	66 (88)

3.2. Molecular Recaptures

Most fish with clipped fins were never recaptured, but all individuals identified as VIE recaptures also were identified as molecular recaptures. We collected 838 fin clips in Stony Creek during 2016–2018 (Table 1), representing 737 fish. Eighty-four individuals (11%) were recaptured at least once; of those, 15 were recaptured twice, and one was recaptured three times. Molecular recaptures indicated movement distances of 0–4500 m (mean: 216 m) across the 17.3-km detectable distance (Table 2; Figure 3). We collected 439 fin clips in Laurel Creek during 2016–2018 (Table 1), representing 357 fish. Fifty-eight individuals (16%) were recaptured at least once; of those, seven were recaptured twice, four three times, and one four times. We observed few recaptures (5% of all recaptures) in either creek two years after a fish's initial fin-clipping. While we could often tell that an individual had a regenerated fin, we could not tell how many times we had collected it before; further, by reanalyzing its DNA markers, we could infer how far it had moved. Molecular recaptures indicated movement distances of 0–1550 m (mean: 233 m) across the 3.5-km detectable distance (Table 2; Figure 3). Many Candy Darter recaptures (37 and 34 fish in Stony Creek and Laurel Creek, respectively) were identified only via molecular methods (Table 1).

Molecular methods–but not VIE methods–allowed us to indirectly observe young-of-year (YOY) and juvenile movements. Three YOY were recaptured in Stony Creek. One was found 1 km upstream of its original capture two years prior. The other two were initially captured in 2017, then recaptured 200 m and 600 m downstream, respectively, in 2018. One juvenile in Stony Creek was recaptured in 2018, 35 m upstream of its original capture in 2017. In Laurel Creek, six YOY were recaptured. Five were found 40–130 m upstream of their original capture, and the other was recaptured 170 m downstream. Two juveniles were recaptured as adults, found 560 m and 185 m upstream of their original capture. Although our sample sizes were small, movements by Candy Darter sub-adults seemed to trend upstream in Laurel Creek but not in Stony Creek.

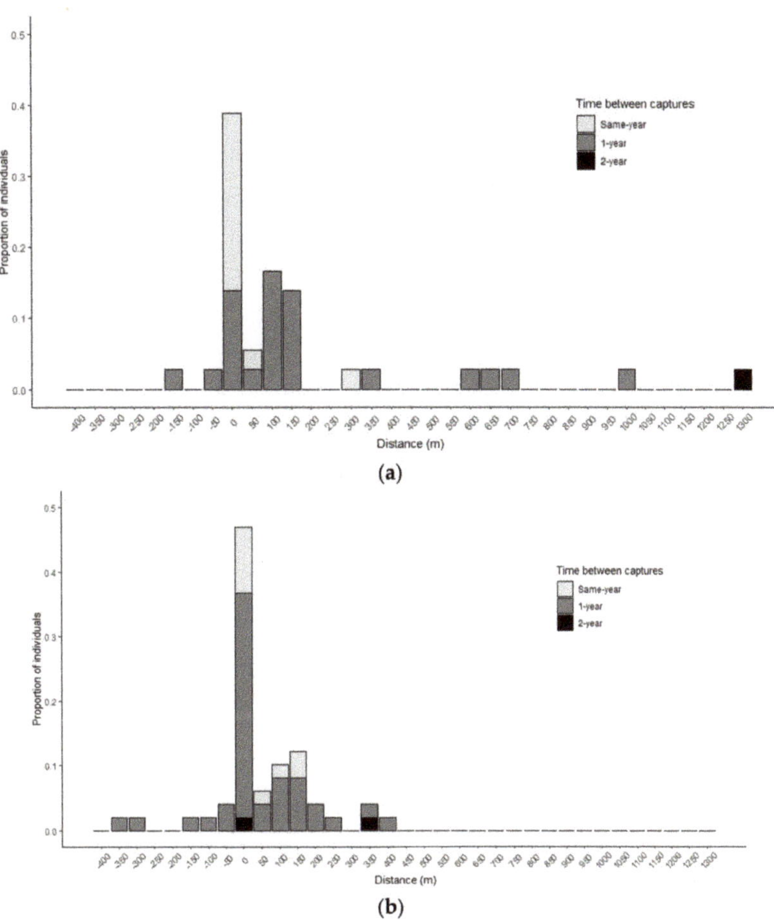

Figure 2. Distances moved by proportions of individuals recaptured during 2016–2018 using visual implant elastomer tags in (**a**) Laurel Creek and (**b**) Stony Creek. Positive numbers represent upstream movements and negative numbers represent downstream movements.

Both VIE and molecular recaptures produced similar movement patterns overall. Neither mean distances moved nor directionality of movement differed between recaptures from VIE and molecular recaptures only ($p = 0.15$, $df = 36.12$) or all molecular recaptures ($p = 0.21$, $df = 104.4$) in Stony Creek. We observed similar patterns in Laurel Creek ($p = 0.94$, $df = 52.15$; $p = 0.78$, $df = 46.01$). Greater distances were observed for molecular recaptures in Stony Creek (mean = 370 m). However, mean distances for molecular-only recaptures in Laurel Creek were similar to distances for total molecular recaptures, including VIE recaptures (mean = 230 m).

3.3. Spatial Distribution of Related Pairs

Related pairs were common in our dataset. Between 2016 and 2018 we captured 153 offspring in Stony Creek and 46 in Laurel Creek (Table 2), including juveniles and YOY in all three years. In both steams, most offspring were assigned to parents (Table 2). In Stony Creek, 371 candidate mothers and 396 candidate fathers were considered for 2018 parentage assignments (Table 3). Of those, <17% were assigned as parents. In Laurel

Creek, 189 candidate mothers and 216 candidate fathers were considered for 2018 parentage assignments, with <23% assigned as parents.

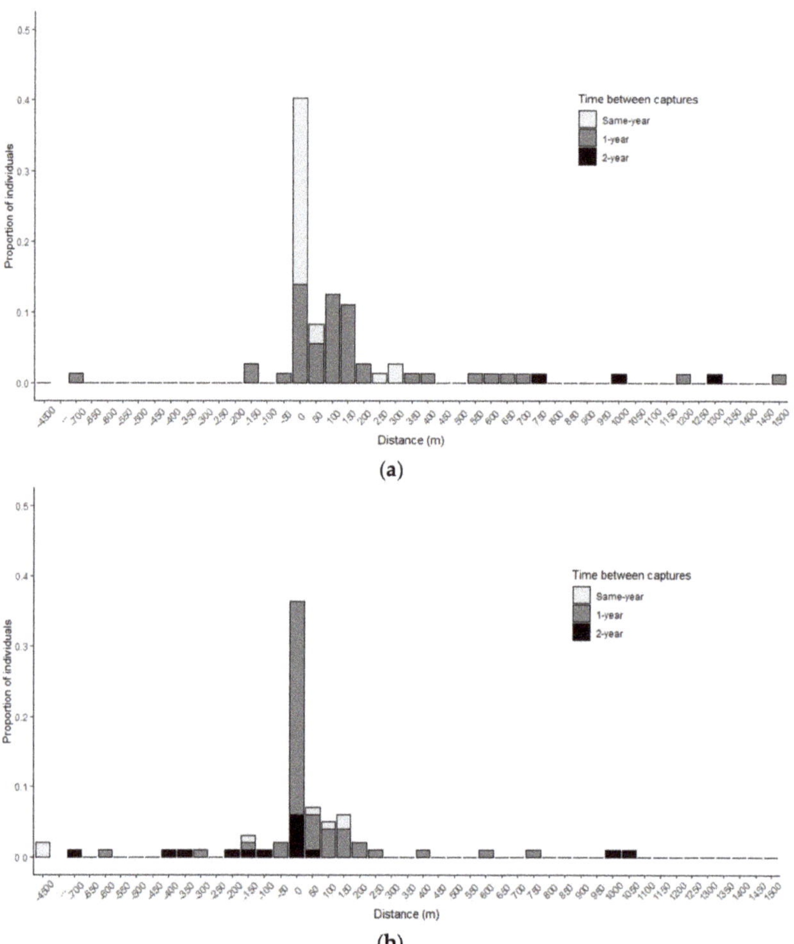

Figure 3. Distances moved by proportions of individuals recaptured during 2016–2018 using molecular markers in (**a**) Laurel Creek and (**b**) Stony Creek. Positive numbers represent upstream movements and negative numbers represent downstream movements.

Table 3. Numbers of candidate parents captured in two streams during 2016–2018 and parent-offspring pairs assigned by Cervus v2.0 [33]. Percentages assigned are in parentheses.

Stream	Year	Candidate Parents		Parentage Assignments	
		Males	**Females**	**Males**	**Females**
Laurel Creek	2016	51	44	11 (22)	10 (23)
	2017	163	138	12 (7)	11 (8)
	2018	216	189	11 (5)	13 (7)
Stony Creek	2016	129	107	18 (14)	20 (19)
	2017	259	219	40 (15)	40 (18)
	2018	396	371	49 (12)	50 (13)

Spatial distributions of adults and sub-adults differed for both streams. Adults were observed at all sampled sites, but YOY and juveniles were not detected at some sites in both streams. No YOY or juveniles were captured in the upper 3.8 km of the Candy Darter distribution in Stony Creek or the upper 1 km of their distribution in Laurel Creek.

Parentage analysis of captured related pairs of Candy Darter revealed complex patterns of post-spawning movement. We documented distances between pair-members but could not distinguish which pair-members moved nor their movement directions. For example, most offspring were not captured at the same locations as their parents. Only 19% of offspring captures occurred in the same site as both parents' captures in Stony Creek, compared to 15% in Laurel Creek. In Stony Creek, 31% of parents were captured in the same site versus 41% in Laurel Creek. Locations of parent and offspring captures were separated by up to 18.5 km and 1.7 km in Stony Creek and Laurel Creek, respectively. Although mean distances between members of assigned mother-offspring pairs were generally greater than distances between members of assigned father-offspring pairs in both streams, we found no significant differences for any year-specific capture group ($p > 0.09$). On average, mothers were captured 1.1 km and 200 m further from offspring than fathers in Stony Creek and Laurel Creek, respectively.

Observed distances between members of parent-offspring pairs suggest Candy Darters move substantial distances during early-life stages. Although we did not measure distances that larvae drift, YOY were generally further from parents than were juveniles in both streams. In Stony Creek, average separation distances were significantly greater for YOY-mother pairs compared to juvenile-mother pairs for both same-year and different-year captures ($p = 0.05$). Additionally, YOY were downstream of assigned mothers more frequently than random based on Kruskal-Wallis tests, but less frequently than random for assigned fathers in Stony Creek ($p < 0.0001$). Similarly, YOY distances from mothers in Laurel Creek were significantly greater than juvenile distances ($p = 0.01$), and the analogous pattern for fathers was nearly significant ($p = 0.06$). Juveniles and YOY were downstream of mothers more frequently than random based on Kruskal-Wallis tests, but less frequently than random for assigned fathers in Laurel Creek ($p < 0.0001$). In summary, Candy Darter offspring, parents, or both moved in ways that resulted in decreasing separation between offspring and parents during the first two years of life. Further, given that mothers and fathers necessarily co-occur during spawning, their differential juxtapositions relative to offspring indicate sexual dimorphism in post-spawning movements.

Candy Darter siblings were more widely separated in Stony Creek than in Laurel Creek. Capture locations of most sibling pairs (82%) from the same inferred spawning year in Stony Creek were separated by at least 100 m, with distances between full siblings spanning 0.6–10.25 km (Table 4). Members of three of the four half-sibling pairs from the same inferred spawning year in Laurel Creek were separated by <120 m. Separation distances between half-siblings spawned in different years ranged from 0–13.75 km in Stony Creek, compared to 0.08–0.6 km in Laurel Creek. These differences in separation distances between streams are consistent with differences in Candy Darter distributions (i.e., 18.8 km in Stony Creek versus 4.25 km in Laurel Creek).

Table 4. Numbers of sibling (sib) pairs assigned in two streams and mean number of assigned fathers known to spawn with assigned mothers per year. Mean, maximum (Max), and minimum (Min) distances (km) between pair members are shown for full-sibs and half-sibs. "-" entries indicate no data.

Stream	Full-Sib Pairs				Half-Sib Pairs				Mean Number of Males Spawning with a Given Female
	N	Mean	Max	Min	N	Mean	Max	Min	
Laurel Creek	0	-	-	-	20	0.3	0.6	0.08	1.25
Stony Creek	6	5.3	10.25	0.6	96	4	13.75	0	1.4

3.4. Reproductive Contributions

Individual Candy Darters spawned over multiple years with multiple partners. The inferred number of spawning-years for assigned mothers ranged from one to four in Stony Creek and one to two in Laurel Creek. The inferred number of spawning-years for assigned fathers ranged from one to three in both streams. On average, 1.4 and 1.25 males spawned with each female per year in Stony Creek and Laurel Creek, respectively (Table 4). We detected up to four males spawning with a given female in a year.

4. Discussion

Improving understanding of fish movement is critical for advancing conservation, for example informing protection of habitats critical for completion of a target species' life cycle. Our understanding of the ontogenetic movement of stream fishes, is, however, limited. We examined patterns of movement for two populations of Candy Darter using both physical and genetic marking methods. Our results present novel insights into movement and mating patterns for isolated populations of Candy Darter in Virginia, including variation in movement across streams, years, sexes, and life stages. Additionally, we provide evidence that the reproductive lifetime of adult Candy Darter is longer than previously described.

4.1. Spatiotemporal Patterns of Candy Darter Movement

Although the average distances detected for VIE and molecular genetic markers were similar, longer distance movements were detected using molecular methods, likely resulting from the greater spatial frame effectively sampled. However, some VIE markings were misidentified in the field as a result of fading or lost tags. VIE marks were less effective than molecular methods at identifying recaptures. VIE tags have various retention rates among colors and species [35]. Our study found similar results, with individuals bearing orange, yellow, and red colors mis-assigned to specific riffles. Further, marking with VIE poses stresses to the fish due to anesthesia, injection, extra handling, and more time out of water [36,37]. Depending on the objectives of the study, molecular methods for mark-recapture work may be more reliable than VIE methods. Given that there are highly polymorphic loci and existing PCR primer sets for a species, molecular methods may be a more reliable way to "mark" individuals, as marks cannot be lost and are individual-specific. In addition, a great deal of information can be gleaned from molecular methods. For the endangered Candy Darter, individual assignment tests allowed us to infer movements of young and adult life stages. Molecular methods are not confined to use in large fish and are non-lethal, which is of great importance when dealing with species of conservation concern. However, it is important to note that both methods had low recapture rates, which is likely a reflection of our study design. Low recapture rate may have results from the technical impossibility of sampling entire stream lengths and some fish may have died after sampling from handling stress or natural causes.

Our results revealed complex, previously undocumented variations in movement and habitat use between Candy Darter sexes and life stages. Recaptured Candy Darters were generally found in the same riffle during capture and recapture events in one of the streams, while recaptured individuals in the other stream tended to move upstream from the point of initial capture. In both streams, adults were observed in the upper reaches of their known distributions, but neither YOY nor juveniles were observed there. YOY were generally farther from parents compared to juveniles in both streams. Females moved farther than males during the spawning season, perhaps reflecting female searches for and spawning with multiple males, in contrast with males remaining near and defending a particular spawning site. The overall spatiotemporal pattern of behavior emerging from the genetic marking data suggests that spawners avoid the most-upstream reaches, YOY drift or swim downstream of spawning sites, post-spawning, females move further upstream than males, including into non-spawning areas, and as YOY mature, they tend to move upstream, which brings them closer to parents.

We showed that some patterns of Candy Darter movement are similar to those exhibited by other small-bodied stream fishes, but other patterns are new. Studies of other darters have found that most marked fish are never recaptured, most recaptured adults are found within 200 m of where they were marked, and a few marked fish may be recaptured great distances (>2 km) from where they were marked [38]. Roberts and Angermeier et al. [35] found movements among microhabitats within riffle-run complexes were common for darter species during summer and fall months.

Our study design offered new insight into how spatial variation in habitat templates, as reflected by stream size and/or separation of suitable habitat patches, may affect fish movement. For example, for Candy Darter, long-distance movements were more common in Laurel Creek (smaller) than in Stony Creek (larger). On average, riffles and intervening runs and pools were larger (in length and width) in Stony Creek than in Laurel Creek, which may have promoted more Candy Darter movement in Laurel Creek. Roberts and Angermeier [35] showed that the isolation of riffles by intervening in an unsuitable habitat inhibited movements by Fantail Darter *Etheostoma flaballare*, Riverweed Darter *Etheostoma podostemone*, and Roanoke Darter *Percina roanoka* but they did not examine effects of stream size. Beyond the darters living there, movements by individuals of five species in two Arkansas streams were three times greater from pools bounded by "short" riffles than from pools bounded by "long" riffles, and movements from "long-riffle" pools were directed downstream in a higher proportion than from "short-riffle" pools [39]. Hodges and Magoulik [40] showed that movement by Creek Chub *Semotilus atromaculatus* into pools of an Ozark stream in Arkansas, USA increased their survival and abundance during seasonal drying, supporting the hypothesis that pools act as a refuge habitat for the species. Population dynamics of darters may be driven by processes and habitat juxtapositions that transcend specific stream reaches. For example, Roberts et al. [2] documented the effects of catchment-wide processes on local habitat use and abundance of Roanoke Logperch *Percina rex*, another small benthic fish. Examining ecological correlates of movement for eight fish species in the upper James River watershed in Virginia, USA, Albanese et al. [3] found that the probability of emigrating from a reach was positively related to stream intermittency (one species) and fish body size (one species), and negatively related to distance from the mainstem creek (two species) and local habitat complexity (one species). Habitats with greater structural complexity may have a higher probability of supplying requirements, thereby obviating the need to move often [3]; however, exceptions exist [41]. Further research might explore other influences of the stream habitat template on fish movement to better understand the colonization-extinction dynamics of populations. Reviewing the literature on movement by individual stream fish, Rasmussen and Belk [42] concluded that future research should focus on interactions among extrinsic ecological factors and intrinsic individual factors to advance understanding of the ecological and evolutionary causes and implications of movement.

Characterization of fish movement is sensitive to the spatiotemporal extent of the study and the detectability of the focal species. Tagged individuals may not be recaptured as a result of emigration from the sampling area, mortality, low sampling effort, or low sampling efficiency. Albanese et al. [15] found increasing detectability of movements with greater extents of recapture sites, but long-distance movements are inherently less detectable than short-distance movements, and species are not equally detectable. While a majority of VIE and genetic recaptures were within our intended recapture sites, the extent of those sites probably limited our ability to detect long-distance movements. Although our study design provided opportunistic observations of long-distance movements outside of the core recapture areas and the extent of our sampling effort was great than those of most movement studies, the unsampled areas between the additional sites may have contributed to our low recapture rates. As is typically the case, our sampling effort was constrained by time and funding, which forced tradeoffs between sampling intensity (number of electrofishing passes) and sampling extent (length of stream sampled).

Genetic marks were useful as a double-tagging method to understand movement patterns and assess the long-term effectiveness of VIE marks. All individuals identified as recaptures with VIE also were identified as recaptures via genetic analysis, but not all VIE-tagged individuals were identified as recaptures in the field, perhaps due to faded or lost VIE marks. Had these individuals not been identified as recaptures using molecular methods, several long-distance movements would have been missed. Although mean differences in movements detected by VIE versus molecular methods were not statistically significant, the use of a single marking method would have yielded different results. Additionally, sub-adult movements would not have been described, as VIE marks cannot be applied to very small fishes. Given our study objectives, VIE methods offered no advantages over molecular methods.

4.2. Spatiotemporal Distribution of Related Pairs

Although we did not determine spawning locations to directly measure the dispersal of offspring, our data on the spatial distributions of parent-offspring pairs offer new insights into fish movement during and after spawning. In both streams, adults were captured in the upper reaches of known Candy Darter distributions, but sub-adults were not detected there. This pattern suggests that larvae drifted or swam downstream or that spawning does not occur in the upper reaches, or both. For example, we collected 56% of the offspring at our most downstream site in Stony Creek, compared to 37% of assigned mothers and 34% of assigned fathers. This finding supports the general model of downstream dispersal of larvae via drift, followed by upstream movement of post-larval stages. Additionally, this site is just upstream of the intermittent portion of the stream, which may, at different times, serve as a nursery habitat or population sink. Hooley-Underwood et al. [43] documented larval drift of suckers (Family Catostomidae) in a drying stream, which resulted in stranding and high mortality in pools. Alternatively, Davey and Kelly [44] observed fish movements to avoid drying reaches.

Spatial patterns in Laurel Creek were similar to those in Stony Creek. We captured 80% of the offspring in the lower 2 km, while 76% of assigned mothers and 50% of assigned fathers were captured in the upper 2 km of the known Candy Darter distribution. The lower number of offspring captured in Laurel Creek compared to Stony Creek probably reflects the smaller Candy Darter population in Laurel Creek but may also reflect offspring dispersal into Wolf Creek downstream. While a few Candy Darter have been observed in Wolf Creek [30], undocumented movement may occur between these streams. Overall, these patterns reveal high levels of demographic connectivity throughout both streams. Although data on the spatial juxtaposition of offspring (YOY and juveniles) and adults could have been acquired without parentage analysis, the results of this analysis revealed new patterns of gene flow and dispersal distances.

4.3. Reproductive Contribution

Understanding the reproductive biology of a species can inform conservation and management strategies. Kreiser et al. [45] used sibship analysis to estimate the number of spawning adults in populations of Alligator Gar *Atractosteus spatula* in Choke Canyon and the Trinity River, Texas. Ultimately, their results revealed the relative contributions of spawning adults and the spatial distribution of spawning events, which were useful in assessing the risk of population declines due to harvest. We used parentage analysis to characterize parent-offspring and sibling-pair relationships and to infer the periodicity and frequency of spawning. Our results suggest that <25% of captured adults contributed to each year-class. However, these results are limited to the offspring-parent assignments we made and may be regarded as a minimum number of spawning adults within each population; clearly, we did not collect every offspring in the respective systems. Our results suggest that families may use an entire stream network to complete their life history, with spatiotemporal variation in occupancy among life stages. Therefore, parentage analysis may aid in assessing risks of stochastic events or anthropogenic alterations, designating critical habitats, and selecting individuals for translocation.

While many studies have addressed the initiation and duration of the reproductive season [46,47], the number of clutches [47], and fecundity [48] of darter species, few have focused on the potential reproductive contribution throughout an individual's lifespan. Jenkins and Burkhead [49] described Candy Darter as sexually mature at age-2, with a lifespan of three years, thereby implying a maximum of two spawning years. McBaine and Hallerman [50] recently reported that individuals can live to at least age-5, based on otolith readings. This finding aligns with our evidence that females can reproduce in four consecutive years with multiple males each year, which coincides with sexual maturity at age 2. This was the first study to show that many Candy Darters spawn in two or three years. In addition, parentage assignment provided evidence that both sexes mate with multiple partners. These results are helpful in describing the species' life history and suggest implications for conservation that would not have been realized without the use of genetic methods. Understanding the number of spawning partners within and across years may provide valuable insight into population persistence and the ability to transmit genetic variation to new generations. For instance, promiscuous species may be less vulnerable to inbreeding, especially if individuals spawn with multiple partners within and throughout the years. Hunter et al. [22] used sibship analysis and pedigree reconstruction to describe the number of spawning years, events, and locations of Lake Sturgeon *Acipenser fulvescens*. They suggested that the reproductive strategy of mating with multiple partners will safeguard against recruitment failure from site-specific mortality and could reduce the risk of losing genetic diversity. Future studies designed to estimate larval survival for Candy Darter could provide valuable information on recruitment dynamics and population persistence.

4.4. Management Implications

Our findings advance knowledge of the spatial structure across Virginia populations of Candy Darter. Analysis of our movement data shows that individuals use multiple complementary and supplementary habitats through their life cycle, involving most of the length of the streams studied. Other studies of darters also have shown that their population dynamics are driven by processes that transcend specific stream reaches. For example, Roberts et al. [31] doc-umented effects of catchment-wide processes on local habitat use and abundance of Roanoke Logperch (*Percina rex*), another small benthic fish. Noting that habitat critical for all life stages must be protected in order to support recruitment and persistence, a key implication for managing imperiled fishes is that focusing management on locations or habitats used by juveniles or adults only may not prove effective for maintaining viable populations or recovering species. Results of our study could prove useful in designating critical habitat for Candy Darter in these two stream systems. We note, however, that additional segments essential for Candy Darter persistence, but still undocumented, may occur elsewhere in our study streams. Unfortunately, we seldom know all the crucial habitats for stream fishes [11], a knowledge gap that could be targeted by future research. Candy darter offspring were assigned only to parents within their respective streams, which supports the current presumption that each stream is an independent demographic unit. This pattern was corroborated by population genetic structure analyses [51]. We are unaware of any studies that have compared species movements across isolated populations or metapopulations of stream fishes.

Understanding the distribution of parents and offspring within populations may provide guidance for practical aspects of translocating Candy Darters, such as suitable source and recipient sites and age composition of translocated individuals. This knowledge is timely because managers in Virginia (Mike Pinder, Virginia Department of Wildlife Resources, personal communication, 2021) and West Virginia (Nate Owens, West Virginia Department of Natural Resources, personal communication, 2021) are considering translocations as a central tactic to promote recovery of the species. We showed that Candy Darter parents and offspring may be separated by >10 km and that sub-adult siblings may be separated by >4 km. If a management goal is to maximize genetic diversity in recipient populations while maintaining their genetic signatures, individuals should be collected from throughout the within-stream distribution of source populations. This could reduce the risk of collecting individuals from a limited number of family groups [52]. Todd and Lintermans [53] assessed the composition of age-classes to use for translocations and the timeframe for establishing new populations of Macquarie Perch *Macquaria australasica*, a long-lived (>25 years) endangered freshwater species. Translocating juveniles and adults over a period of five years was the most successful strategy, with little impact on the source population. However, the risk of sampling from a limited number of families may increase if collections include a range of life stages [54]. Understanding individual dispersal distances can inform choices of release sites. George et al. [55] suggested releasing small numbers of individuals in multiple nearby sites for species that are poor dispersers, but more individuals in a single site for species that are good dispersers (i.e., whose larvae drift).

Future studies may be warranted to describe movement patterns in larger, more connected populations of Candy Darter in West Virginia, where we expect dispersal and movement to be more extensive than is possible for Virginia populations. Further, documenting movement patterns in more connected ecosystems may prove useful in identifying recipient streams for the reintroduction of Candy Darter in Virginia, with an eye toward recreating a functional metapopulation maintained via natural movement among subpopulations. That is, reintroductions may prove more successful if recipient streams are near existing populations and the habitat is similar to the source system(s). Reintroductions might be prioritized to restore connectivity among the four Virginia populations, which would likely enhance the long-term viability of the species.

5. Conclusions

Knowledge of individual movement patterns would inform conservation planning but is lacking for most stream fishes. We employed both physical (visible implant elastomer) and genetic (microsatellite DNA) marks to, directly and indirectly, infer movements of Candy Darters in two-stream ecosystems in Virginia in the eastern United States. The DNA markers revealed somewhat greater distances moved and allowed reconstruction of family relationships, showing parent-offspring pairs throughout the 18.8-km distribution in Stony Creek and within 2 km in the 4.25-km distribution in Laurel Creek. Molecular markers also showed that Candy Darters of both sexes had multiple mates and a longer reproductive lifetime than previously recognized. Our results will inform conservation planning for this Endangered species.

Supplementary Materials: The following supporting information can be downloaded at: https://www.mdpi.com/article/10.3390/fishes7010030/s1: Figure S1. Adult Candy Darter male (a) and female (b). Red visual implant elastomer tag on dorsal surface of Candy Darter (c). Figure S2. Stony Creek watershed in Giles County, Virginia. The locations of six study sites are also shown. "candydarter_points" = presence of study species at that study site. Figure S3. Laurel Creek watershed in Bland County, Virginia. The locations of all study sites are shown.

Author Contributions: Conceptualization, P.L.A. and E.M.H.; funding acquisition, P.L.A. and E.M.H.; methodology, K.E.M., E.M.H. and P.L.A.; field and laboratory work, K.E.M.; data analysis, K.E.M., E.M.H. and P.L.A.; writing of original draft, K.E.M.; review and editing of manuscript, P.L.A. and E.M.H. All authors have read and agreed to the published version of the manuscript.

Funding: This research was supported by the Virginia Department of Wildlife Resources (VDWR) under grant no. 449509. We thank Mike Pinder and Craig Roghair for their in-kind support in conducting fieldwork, Logan Sleezer for developing the maps, and Brett Albanese for reviewing the manuscript. The participation of author EMH was supported in part by the U.S. Department of Agriculture National Institute of Food and Agriculture Hatch Program. The Virginia Cooperative Fish and Wildlife Research Unit is jointly sponsored by U.S. Geological Survey, Virginia Tech,

Virginia Department of Wildlife Resources, and Wildlife Management Institute. Any use of trade, firm, or product names is for descriptive purposes only and does not imply endorsement by the U.S. Government.

Institutional Review Board Statement: This work was carried out under the auspices of the Virginia Tech Institutional Animal Care and Use Committee Protocol 16-095.

Data Availability Statement: The data that support the findings of this study will be available on request to the corresponding author.

Conflicts of Interest: The authors declare no conflict of interest.

References

1. Albanese, B.; Angermeier, P.L.; Peterson, J.T. Does mobility explain variation in colonisation and population recovery among stream fishes? *Freshw. Biol.* **2009**, *54*, 1444–1460. [CrossRef]
2. Roberts, J.H.; Angermeier, P.L.; Hallerman, E.M. Extensive dispersal of Roanoke Logperch (*Percina rex*) inferred from genetic marker data. *Ecol. Freshw. Fish* **2016**, *25*, 1–16. [CrossRef]
3. Albanese, B.; Angermeier, P.L.; Dorai-Raj, S. Ecological correlates of fish movement in a network of Virginia streams. *Can. J. Fish Aquat. Sci.* **2004**, *61*, 857–869. [CrossRef]
4. Railsback, S.F.; Lamberson, R.H.; Harvey, B.C.; Duffy, W.E. Movement rules for individual-based models of stream fish. *Ecol. Model.* **1999**, *123*, 73–89. [CrossRef]
5. Lowe, W.H.; Allendorf, F.W. What can genetics tell us about population connectivity? *Mol. Ecol.* **2010**, *19*, 3038–3051. [CrossRef] [PubMed]
6. Labbe, T.R.; Fausch, K.D. Dynamics of intermittent stream habitat regulate persistence of a threatened fish at multiple scales. *Ecol. Appl.* **2000**, *10*, 1774–1791. [CrossRef]
7. Fraser, D.F.; Sise, T.E. Observations of stream minnows in a patchy environment: A test of a theory of habitat distribution. *Ecology* **1980**, *61*, 790–797. [CrossRef]
8. Schaefer, J. Riffles as barriers to interpool movement by three cyprinids (*Notropis boops, Campostoma anomalum* and *Cyprinella venusta*). *Freshw. Biol.* **2001**, *46*, 379–388. [CrossRef]
9. Eros, T.; Campbell Grant, E.H. Unifying research on the fragmentation of terrestrial and aquatic habitats: Patches, connectivity and the matrix in riverscapes. *Freshw. Biol.* **2015**, *60*, 1487–1501. [CrossRef]
10. Piertney, S.B.; MacColl, A.D.C.; Lambin, X.; Moss, R.; Dallas, J.F. Spatial distribution of genetic relatedness in a moorland population of red grouse (*Lagopus lagopus scoticus*). *Biol. J. Linn. Soc.* **1999**, *68*, 317–331. [CrossRef]
11. Schlosser, I.J.; Angermeier, P.L. Spatial variation in demographic processes of lotic fishes: Conceptual models, empirical evidence, and implications for conservation. *Am. Fish Soc. Symp.* **1995**, *17*, 392–401.
12. Turner, T.F. Comparative study of larval transport and gene flow in darters. *Copeia* **2001**, *2001*, 66–774. [CrossRef]
13. Brown, A.V.; Armstrong, M.L. Propensity to drift downstream among various species of fish. *J. Freshw. Ecol.* **1985**, *3*, 3–17. [CrossRef]
14. Pine, W.E.; Hightower, J.E.; Coggins, L.G.; Lauretta, M.V.; Pollock, K.H. Design and analysis of tagging studies. In *Fisheries Techniques*, 3rd ed.; Zale, A.V., Parrish, D.L., Sutton, T.M., Eds.; American Fisheries Society: Bethesda, MD, USA, 2012; pp. 521–572.
15. Albanese, B.; Angermeier, P.L.; Gowan, C. Designing mark-recapture studies to reduce effects of distance weighting on movement distance distributions of stream fishes. *Trans. Am. Fish. Soc.* **2003**, *132*, 925–939. [CrossRef]
16. Jungwirth, A.; Balzarini, V.; Zöttl, M.; Salzmann, A.; Taborsky, M.; Frommen, J.G. Long-term individual marking of small freshwater fish: The utility of Visual Implant Elastomer tags. *Behav. Ecol. Sociobiol.* **2019**, *73*, 49. [CrossRef]
17. Roberts, J.H.; Angermeier, P.L. A comparison of injectable fluorescent marks in two genera of darters: Effects on survival and retention rates. *N. Am. J. Fish. Manag.* **2004**, *24*, 1017–1024. [CrossRef]
18. Argentina, J.E.; Angermeier, P.L.; Hallerman, E.M.; Welsh, S.A. Spatial extent of analysis influences observed patterns of population genetic structure in a widespread darter species (*Percidae*). *Freshw. Biol.* **2018**, *63*, 1185–1198. [CrossRef]
19. Hudy, M.; Coombs, J.A.; Nislow, K.H.; Letcher, B.H. Dispersal and within-stream spatial population structure of Brook Trout revealed by pedigree reconstruction analysis. *Trans. Am. Fish. Soc.* **2010**, *139*, 276–1287. [CrossRef]
20. Ruppert, J.L.W.; James, P.M.A.; Taylor, E.B.; Rudolfsen, T.; Veillard, M.; Davis, C.S.; Watkinson, D.; Poesch, M.S. Riverscape genetic structure of a threatened and dispersal limited freshwater species, the Rocky Mountain Sculpin (*Cottus* sp.). *Conserv. Genet.* **2017**, *18*, 925–937. [CrossRef]
21. Blankenship, S.M.; Schumer, G.; Van Eenennaam, J.P.; Jackson, Z.J. Estimating number of spawning white sturgeon adults from embryo relatedness. *Fish. Manag. Ecol.* **2017**, *24*, 163–172. [CrossRef]
22. Hunter, R.D.; Roseman, E.F.; Sard, N.M.; DeBruyne, R.L.; Wang, J.; Scribner, K.T. Genetic family reconstruction characterizes Lake Sturgeon use of newly constructed spawning habitat and larval dispersal. *Trans. Am. Fish. Soc.* **2020**, *149*, 266–283. [CrossRef]
23. Kanno, Y.; Letcher, B.H.; Coombs, J.A.; Nislow, K.H.; Whiteley, A.R. Linking movement and reproductive history of brook trout to assess habitat connectivity in a heterogeneous stream network. *Freshw. Biol.* **2014**, *59*, 142–154. [CrossRef]

24. Jelks, H.L.; Walsh, S.J.; Burkhead, N.M.; Contreras-Balderas, S.; Diaz-Pardo, E.; Hendrickson, D.A.; Lyons, J.; Mandrak, N.E.; McCormick, F.; Nelson, J.S.; et al. Conservation status of imperiled North American freshwater and diadromous fishes. *Fisheries* **2008**, *33*, 372–407. [CrossRef]
25. USFWS (U.S. Fish and Wildlife Service). *Species Status Assessment (SSA), Report for the Candy Darter (Etheostoma osburni)*; USFWS: West Hadley, MA, USA, 2017.
26. Chipps, S.R.; Perry, W.B.; Perry, S.A. Patterns of microhabitat use among four species of darters in three Appalachian streams. *Am. Midl. Nat.* **1994**, *131*, 175–180. [CrossRef]
27. Jenkins, R.E.; Kopia, B.L. *Population Status of the Candy Darter, Etheostoma Osburni, in Virginia 1994–95, with Historical Review*; Department of Biology, Roanoke College: Salem, VA, USA, 1995.
28. Dunn, C.G.; Angermeier, P.L. Development of habitat suitability indices for the Candy Darter, with cross-scale validation across representative populations. *Trans. Am. Fish. Soc.* **2016**, *145*, 1266–1281. [CrossRef]
29. Leftwich, K.N.; Dolloff, C.A.; Underwood, M.K. The Candy Darter (*Etheostoma osburni*) in Stony Creek, George Washington-Jefferson National Forest, Virginia. In *Annual Report to U.S. Forest Service*; Center for Aquatic Technology Transfer: Blacksburg, VA, USA, 1996.
30. Dunn, C.G. Comparison of Habitat Suitability among Sites Supporting Strong, Localized, and Extirpated Populations of Candy Darter *Etheostoma osburni*. In *Final Report*; Virginia Game and Inland Fisheries: Richmond, VA, USA, 2013.
31. Roberts, J.H.; Anderson, G.B.; Angermeier, P.L. A long-term study of ecological impacts of a flood reduction project to an endangered riverine fish: Lessons learned for assessment and restoration. *Water* **2013**, *8*, 240.
32. Switzer, J.F.; Welsh, S.A.; King, T.L. Microsatellite DNA primers for the Candy Darter, *Etheostoma osburni* and Variegate Darter, *Etheostoma variatum*, and cross-species amplification in other darters (Percidae). *Molec. Ecol. Res.* **2008**, *8*, 335–338. [CrossRef]
33. Hulce, D.; Li, X.; Snyder-Leiby, T.; Liu, C.S.J. GeneMarker® genotyping software: Tools to increase the statistical power of DNA fragment analysis. *J. Biomol. Techiques* **2011**, *22*, S35–S36.
34. Kalinowski, S.T.; Taper, M.L.; Marshall, T.C. Revising how the computer program CERVUS accommodates genotyping error increases success in paternity assignment. *Mol. Ecol.* **2007**, *16*, 1099–1106. [CrossRef]
35. Roberts, J.H.; Angermeier, P.L. Spatiotemporal variability of stream habitat and movement of three species of fish. *Oecologia* **2007**, *151*, 417–430. [CrossRef]
36. Fürtbauer, I.; King, A.J.; Heistermann, M. Visible implant elastomer (VIE) tagging and simulated predation risk elicit similar physiological stress responses in three-spined stickleback *Gasterosteus aculeatus*. *J. Fish Biol.* **2015**, *86*, 1644–1649. [CrossRef]
37. Neufeld, K.; Blair, S.; Poesch, M. Retention and stress effects of visible implant tags when marking Western Silvery Minnow and its application to other cyprinids (Family Cyprinidae). *N. Am. J. Fish. Manag.* **2015**, *35*, 1070–1076. [CrossRef]
38. Roberts, J.H.; Rosenberger, A.E.; Albanese, B.W.; Angermeier, P.L. Movement patterns of endangered Roanoke logperch (*Percina rex*). *Ecol. Freshw. Fish* **2008**, *17*, 374–381. [CrossRef]
39. Lonzarich, D.G.; Lonzarich, M.R.; Warren, M.L. Effects of riffle length on the short-term movement of fishes among stream pools. *Can. J. Fish. Aquat. Sci.* **2000**, *57*, 1508–1514. [CrossRef]
40. Hodges, S.W.; Magoulick, D.D. Refuge habitats for fishes during seasonal drying in an intermittent stream: Movement, survival and abundance of three minnow species. *Aquat. Sci.* **2011**, *73*, 513–522. [CrossRef]
41. Roni, P.; Quinn, T.P. Effects of wood placement on movements of trout and juvenile coho salmon in natural and artificial stream channels. *Trans. Am. Fish. Soc.* **2001**, *130*, 675–685. [CrossRef]
42. Rasmussen, J.E.; Belk, M.C. Individual movement of stream fishes: Linking ecological drivers with evolutionary processes. *Rev. Fish. Sci. Aquacult.* **2017**, *25*, 70–83. [CrossRef]
43. Hooley-Underwood, Z.E.; Stevens, S.B.; Salinas, N.R.; Thompson, K.G. An intermittent stream supports extensive spawning of large-river native fishes. *Trans. Am. Fish. Soc.* **2019**, *148*, 426–441. [CrossRef]
44. Davey, A.J.H.; Kelly, D.J. Fish community responses to drying disturbances in an intermittent stream: A landscape perspective. *Freshw. Biol.* **2007**, *52*, 1719–1733. [CrossRef]
45. Kreiser, B.R.; Daugherty, D.J.; Buckmeier, D.L.; Smith, N.G.; Newsome, E.B. Sibship analysis to characterize Alligator Gar reproductive contributions in two Texas systems. *N. Am. J. Fish. Manag.* **2020**, *40*, 555–565. [CrossRef]
46. Hubbs, C. Darter reproductive seasons. *Copeia* **1985**, *1*, 56–68. [CrossRef]
47. Ruble, C.L.; Rakes, P.L.; Shute, J.R.; Welsh, S.A. Captive propagation, reproductive biology, and early life history of *Etheostoma wapiti* (Boulder Darter), *E. vulneratum* (Wounded Darter), and *E. maculatum* (Spotted Darter). *Southeast Nat.* **2016**, *15*, 115–126. [CrossRef]
48. Heins, D.C.; Baker, J.A.; Guill, J.M. Seasonal and interannual components of intrapopulation variation in clutch size and egg size of a darter. *Ecol. Freshw. Fish* **2004**, *13*, 258–265. [CrossRef]
49. Jenkins, R.E.; Burkhead, N.M. *Freshwater Fishes of Virginia*; American Fisheries Society: Bethesda, MD, USA, 1994.
50. McBaine, K.M.; Hallerman, E.M. Demographic Status and Population Genetic Differentiation of Candy Darter Populations in Virginia. In *Final Report*; Virginia Department of Game and Inland Fisheries: Richmond, VA, USA, 2020.
51. McBaine, K.M. Detectability, Movement, and Population Genetic Structure of the Endangered Candy Darter in Virginia. Master's Thesis, Virginia Polytechnic Institute and State University, Blacksburg, VI, USA, May 2021.
52. Hansen, M.M.; Nielsen, E.E.; Mensberg, K.L.D. The problem of sampling families rather than populations: Relatedness among individuals in samples of juvenile brown trout *Salmo trutta* L. *Molec. Ecol.* **1997**, *6*, 469–474. [CrossRef]

53. Todd, C.R.; Lintermans, M. Who do you move? A stochastic population model to guide translocation strategies for an endangered freshwater fish in south-eastern Australia. *Ecol. Model.* **2015**, *311*, 63–72. [CrossRef]
54. Miller, L.M.; Kapuscinski, A.R. Genetic Guidelines for Hatchery Supplementation Programs. In *Population Genetics: Principles and Applications for Fisheries Scientists*; Hallerman, E.M., Ed.; American Fisheries Society: Bethesda, MD, USA, 2003; pp. 329–356.
55. George, A.L.; Kuhajda, B.R.; Williams, J.D.; Cantrell, M.A.; Rakes, P.L.; Shute, J.R. Guidelines for propagation and translocation for freshwater fish conservation. *Fisheries* **2009**, *34*, 529–545. [CrossRef]

Article

Demographic and Life History Characteristics of Black Bullheads *Ameiurus melas* in a North Temperate USA Lake

Logan W. Sikora [1,2,*], Joseph T. Mrnak [2,3], Rebecca Henningsen [1], Justin A. VanDeHey [1] and Greg G. Sass [3]

[1]   College of Natural Resources, University of Wisconsin-Stevens Point, 800 Reserve Street, Stevens Point, WI 54481, USA; rehennin@uwsp.edu (R.H.); jvandehe@uwsp.edu (J.A.V.)
[2]   Wisconsin Department of Natural Resources, Office of Applied Science, Escanaba Lake Research Station, 3110 Trout Lake Station Drive, Boulder Junction, WI 54512, USA; mrnak@wisc.edu
[3]   Center for Limnology, University of Wisconsin–Madison, 680 North Park Street, Madison, WI 53706, USA; gregory.sass@wisconsin.gov
*   Correspondence: lsiko340@uwsp.edu

Abstract: Black bullheads *Ameiurus melas* are an environmentally tolerant omnivorous fish species that are found throughout much of North America and parts of Europe. Despite their prevalence, black bullheads are an infrequently studied species making their biology, ecology, and life history poorly understood. Although limited information has been published on black bullheads, evidence suggests that bullheads can dominate the fish biomass and have profound influences on the fish community in some north temperate USA lakes. The goal of our study was to provide additional information on black bullhead population demographics, growth rates, life history characteristics, and seasonal diet preferences in a northern Wisconsin lake. Using common fish collection gears (fyke netting, electrofishing), fish aging protocols, fecundity assessments, and diet indices, our results suggested that black bullheads exhibited relatively fast growth rates, early ages at maturity, moderate fecundity, and a diverse omnivorous diet. Due to these demographic and life history characteristics, black bullheads have the potential to dominate fish community biomass in their native and introduced range. Results from our study may inform the management of black bullhead as native and invasive species.

Keywords: bullhead; black bullhead; ameiurus; *Ameiurus melas*; life history; growth; reproductive potential; fecundity; maturity; sex ratio

Citation: Sikora, L.W.; Mrnak, J.T.; Henningsen, R.; VanDeHey, J.A.; Sass, G.G. Demographic and Life History Characteristics of Black Bullheads *Ameiurus melas* in a North Temperate USA Lake. *Fishes* **2022**, *7*, 21. https://doi.org/10.3390/fishes7010021

Academic Editor: Maria Angeles Esteban

Received: 13 December 2021
Accepted: 11 January 2022
Published: 14 January 2022

Publisher's Note: MDPI stays neutral with regard to jurisdictional claims in published maps and institutional affiliations.

1. Introduction

Black bullhead *Ameiurus melas* are an ictalurid species that are common but often overlooked throughout their native range. Black bullheads are endemic to the Mississippi River drainages of North America, with populations extending from southern Saskatchewan to the Gulf of Mexico drainage near northern Mexico [1–3]. Black bullhead populations have become prevalent and widespread outside of their native range, notably across the western United States and Europe [4]. In these non-native systems, black bullheads are frequently considered invasive and (or) a nuisance [5–9] primarily due to driving undesired ecological effects [10,11]. In non-native systems, most introductions have been unintentional [5,6,10]. Following colonization, bullheads can dominate fish community biomasses, alter fish community composition, and have been shown to increase turbidity in small impoundments [10–12]. Colonization and invasive potential are often attributed to their environmental tolerance and omnivorous diet [4,5,10].

The prevalence of black bullheads can be attributed to their plastic habitat requirements and tolerance to suboptimal abiotic conditions. For example, black bullheads are tolerant to high water temperatures ($\leq$35 °C; [13]), low dissolved oxygen concentrations (>3.0 mg L−1; [14]), and degraded water quality [12,13,15]. This plasticity in habitat requirements allows black bullheads to persist in most aquatic systems such as lakes (oligotrophic and eutrophic), ponds, impoundments, diked or flooded wetlands, low gradient streams, and backwaters across their native and non-native range [10,12,14]. Despite black bullhead populations being common and widespread throughout many systems across the world, they are infrequently studied, and there is a paucity of information on their demographics, life history, and ecological role in north-temperate lakes [4,16,17].

Over the past two decades, warm-water fishes such as largemouth bass *Micropterus salmoides*, bluegill *Lepomis macrochirus*, and likely black bullheads have been increasing in abundance, while cool-water species such as walleye *Sander vitreus* and yellow perch *Perca flavescens* have been decreasing [18–21]. The mechanism(s) behind the decline of these cool-water species are largely unknown but are likely variable among systems and related to climate change, habitat loss, production overharvest, invasive species, imbalances in fish communities, species-specific angler behaviors, and/or interactions among the aforementioned [11,21–26]. In concert or independently, these mechanisms are driving abiotic and biotic change [27]. Due to the black bullhead's plasticity in habitat requirements and tolerance, it seems plausible that they will thrive in these new environmental conditions and possibly fill devoid cool-water species niche space. Therefore, expanding knowledge and understanding of this understudied species is of critical importance.

The goal of our study was to provide additional information on black bullhead population demographics, growth rates, life history characteristics, and seasonal diet preferences in a north-temperate lake. The specific objectives of our study were to: (1) determine growth rates, fecundity, sex ratio, and age at maturity of black bullheads; and (2) determine seasonal diet composition of black bullheads. This type of information will increase our management and (if needed) control capabilities of this widely infrequently studied, highly tolerant species.

2. Materials and Methods

Black bullheads were sampled monthly from Howell Lake, Forest County, Wisconsin (45.9469436°, −88.9338069°). Howell Lake is an ideal system to study black bullhead demographics and life history due to the abundant nature of the population and diverse native fish community. Howell Lake is considered eutrophic with a surface area of 69 ha and maximum depth of 4 m. The fish assemblage in Howell Lake is representative of many Northern Wisconsin lakes, including black and yellow bullhead *A. natalis*, northern pike *Esox lucius*, walleye, largemouth bass, smallmouth bass *Micropterus dolomieu*, yellow perch, bluegill, black crappie *Pomoxis nigromaculatus*, pumpkinseed *Lepomis gibbosus*, rock bass *Ambloplites rupestris*, white sucker *Catostomus commersonii*, golden shiner *Notemigonus crysoleucas*, common shiner *Luxilus cornutus*, common creek chub *Semotilus atromaculatus*, and bluntnose minnow *Pimephales notatus*. Additionally, invasive rusty crayfish *Faxonius rusticus* and Chinese mystery snails *Cipangopaludina chinensis* are present in Howell Lake.

2.1. Black Bullhead Population Characteristics

Black bullhead sampling took place during May–October 2020 with standard fisheries gear including 6 fyke nets (13-mm mesh, 1.2-m tall, 1.8-m wide, and 15–23-m lead lines), 4 mini fyke nets (4.7-mm mesh, 0.9-m tall, 0.9-m wide, and 4.5–12-m lead lines), and boat-mounted AC electrofishing equipment. Fyke nets (standard and mini) were set weekly (24-h set, picked daily) at fixed locations in the littoral zone during mid-May–mid-August. Beginning in mid-April (after ice-out) and continuing through mid-October, the entire shoreline of Howell Lake was electrofished once monthly at night. As part of an ongoing removal study, all black bullheads that were captured on Howell Lake in 2020 were removed and the population of black bullheads >100 mm was estimated using a k-pass depletion estimate [28]. Up to 30 individuals per net per day and a minimum of 50 individuals from each electrofishing survey were measured for total length (TL; mm) and weight (g). From June 10–June 25, up to 10 fish per 13 mm length bin were retained for age estimation [29]. Retained fish were placed in individually labeled Ziploc bags with their respective lengths and weights, placed on ice in the field, and brought to the laboratory at the University of

Wisconsin-Stevens Point. In the laboratory, lapilli otoliths (commonly mistaken for sagittal otoliths in siluriformes; [30]) were extracted for age estimation [31]. Sex and maturity status (i.e., immature, mature) was determined by making a small incision near the vent and visually inspecting the gonads [32].

In the laboratory, otoliths were placed in distinctly numbered vials and allowed to dry for a minimum of two weeks prior to processing [33]. Individual otoliths were placed in the wells of a 24-cavity silicone baking tray and fully submerged in West System 105/206B slow hardening two-part epoxy. After curing for a minimum of 48 h, otoliths were removed and cut through the focus on a Beuhler low-speed isomet saw, were lightly hand-sanded with wetted 1000-grit carborundum sandpaper and covered with a drop of immersion oil [33]. The epoxy that contained the halved otolith was placed under a Nikon® SMZ1500N dissecting microscope ($30\times$ magnification; Nikon, Tokyo, Japan) and illuminated from the side with a 1-mm diameter, single-strand fiber optic filament connected to a light source (Fiber-Lite model 180; Dolan-Jenner Industries, Inc., St. Lawrence, MA, USA) [34]. When all annuli were illuminated, the otolith was photographed. Ages were estimated double-blind by two independent readers by enumerating the observed annuli on otoliths. Consensus age estimates were then used for subsequent analyses.

Age estimates were used to construct an age–length key for black bullheads in Howell Lake, which was used to assign ages to un-aged fish for maturity and age and growth models (Table 1; [35]). Black bullhead growth on Howell Lake was evaluated using length at age data to inform the von Bertalanffy growth relationship [36], and by estimating mean length-at-age. Bullhead size structure was determined using proportional size distribution (PSD-X) indices [37].

Table 1. Age-length key based on 197 black bullheads *Ameiurus melas* sampled and aged in June 2020 from Howell Lake, Wisconsin. The proportion of fish in each age-length combination is followed by the number fish that were sampled in that combination in ().

Length (mm)	\multicolumn{9}{c}{Age (year)}								
	1	2	3	4	5	6	7	8	9
50	1.00 (3)								
60	1.00 (10)								
70	1.00 (6)								
80	1.00 (3)								
90	0.67 (2)	0.33 (1)							
100	0.33 (1)	0.67 (2)							
120		0.60 (3)	0.40 (2)						
130		0.40 (2)	0.60 (3)						
140		0.14 (1)	0.71 (5)	0.14 (1)					
150		0.40 (2)	0.40 (2)	0.20 (1)					
160			0.62 (5)	0.38 (3)					
170			0.73 (8)	0.27 (3)					
180			0.75 (6)	0.25 (2)					
190			0.33 (4)	0.67 (8)					
200			0.25 (2)	0.75 (6)					
210			0.10 (1)	0.70 (7)	0.20 (2)				
220			0.18 (2)	0.36 (4)	0.27 (3)	0.09 (1)	0.09 (1)		
230			0.11 (1)	0.33 (3)	0.22 (2)	0.33 (3)			
240				0.17 (2)	0.25 (3)	0.50 (6)	0.80 (1)		
250					0.11 (1)	0.89 (8)			
260					0.08 (1)	0.62 (8)	0.31 (4)		
270						0.60 (6)	0.20 (2)	0.10 (1)	0.10 (1)
280						0.18 (2)	0.55 (6)	0.18 (2)	0.09 (1)
290						0.14 (1)	0.29 (2)	0.29 (2)	0.29 (2)
300							0.17 (1)	0.50 (3)	0.33 (2)
310								1.00 (2)	

Weighted catch-curve regressions were used to estimate the total annual mortality of black bullheads in Howell Lake [38]. Based on the catch curve, we assumed that fish $\geq$ age-3 were fully recruited to the fyke nets; therefore, fish <age-3 were not included in regression analysis. The catch curve was developed from fish sampled from June 2–June 25, and we used fishes that had age estimated directly from otoliths and fish that were assigned ages based on the developed age-length key. Catch curves were developed by regressing the natural log catch of fish against age and the slope-estimated instantaneous total mortality (Z) and total annual mortality ($A = 1 - e^{-z}$; [35]). Because there is relatively little fishing pressure for bullheads in Wisconsin and Howell Lake's remote location, fishing mortality was assumed to be negligible, and any measured mortality was assumed to be natural mortality.

2.2. Spawning, Maturity, and Fecundity

The sex ratio of black bullheads in Howell Lake was estimated based on the number of males to females that were observed during sampling. The estimated sex ratio was expanded to the entire sample in a manner consistent with [29]. Sex-specific length at 50% maturity (L_{50}) and length at 90% maturity (L_{90}) was determined using logistic regression where 0 denoted immature or unknown sex fish and 1 denoted mature fish.

Ripe female black bullheads were used for fecundity and reproductive potential estimation. Sampling of these fish coincided with those that were sampled for age structures, which took place from June 10–June 25. Up to 5 females per 13 mm length bin were placed in individual Ziploc bags, labeled with their respective total lengths and weights, placed on ice in the field, and brought to the laboratory. In the laboratory, both ovaries were removed and weighed to the nearest 0.1 g. Both ovaries from each fish were agitated and rinsed to remove remnant ovarian tissue. A sub-sample of eggs were taken as a 0.5–1.5 cm cross-section from the middle of each ovary. The subsample weights varied from 10–100% of the weight of the whole ovary. The subsample of eggs was weighed and photographed for enumeration. The total number of eggs in each ovary was estimated with the equation:

$$\text{Ovary}_1 = \left(\frac{\text{Subsample Count}}{\text{Subsample Weight}} \right) \times \text{Total Ovary Weight} \tag{1}$$

and the total fecundity for each fish was estimated by adding the estimates of both ovaries together.

Fecundity estimates were paired with our population estimate, size structure, and sex ratio data to produce reproductive potential estimates for black bullheads in Howell Lake. Mean fecundity estimates were multiplied by the estimated number of females in each Gabelhouse length category [37] to determine the cumulative reproductive potential of the population.

2.3. Seasonal Diet Analysis

Up to 300 black bullheads, captured via electrofishing, were retained each month for diet analysis. Upon capture, fish were placed on ice and then brought to the WIDNR, Escanaba Lake Field Station, where they were frozen. Frozen bullheads were later brought to and processed at the University of Wisconsin-Stevens Point. After thawing, bullheads were measured for total length (mm) and weighed (g). Stomachs were extracted ventrally. Prey items were then removed, identified to the lowest possible taxonomic category (species for fishes, order or family for invertebrates), enumerated, and individual prey items were weighed wet (nearest 0.1 g).

Diets were quantified by the percentage stomachs with contents, frequency of occurrence (O_i), mean percent composition by number (MN_i), mean percent composition by weight (MW_i), and index of relative importance (IRI) for each prey taxa for each month and stratified by bullhead length (length categories from [37]) [39–42]. For analysis, each fish species in the diets were individual categories except for minnow species (common shiner and bluntnose minnow), which were pooled to create the "Cyprinid" category. Items that fell within the following categories were pooled into the appropriate category: aquatic invertebrates, terrestrial insects, and other diet items. A multivariate analysis of variance

(MANOVA) was used to test whether differences in diet were present among month of collection (May and October were excluded from the statistical analysis due to the low sample size leading to violation of the MANOVA's assumption of equal variance). If results from the MANOVA were statistically significant ($\alpha \geq 0.05$), a Tukey's HSD test for multiple comparisons was used to determine which comparisons were significantly different.

3. Results

3.1. Black Bullhead Population Characteristics

The black bullhead population (>100 mm) in Howell Lake was estimated to be 24,479 fish (95% CI = 24,171 −24,787) with an estimated density of 355/fish ha. A total of 20,952 black bullheads were sampled over the study period, ranging in length from 25 to 325 mm (mean = 197 mm; SE = 0.70; Figure 1), weight from 1 g to 474 g (mean = 121; SE = 1.34), and age from 1–9 (Figure 2). The population exhibited high natural mortality (A = 40.5%; Z = 0.52; Figure 3) and relatively fast growth rates ($L\infty$ = 381 mm; K = 0.17; t_0 = −0.64; Table 2; Figure 4). Mean lengths at ages 1–9 were 80, 132, 180, 196, 235, 260, 274, 297, and 292 mm, respectively (Table 2). Of 4966 fish that were sampled and measured in 2020, 90% of the fish were of stock length (>150 mm), 22% of the fish were of quality length (>230 mm), and only 1% of the fish were preferred length (Figure 1).

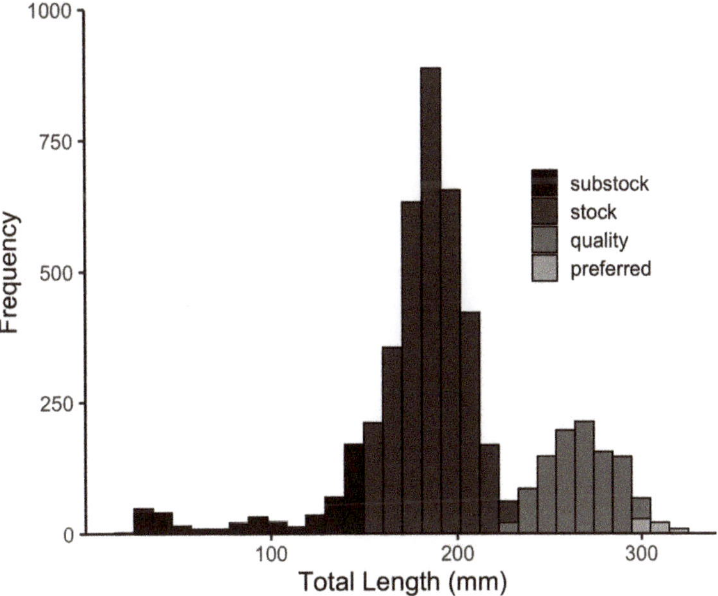

Figure 1. Length-frequency histogram for black bullheads *Ameiurus melas* (n = 4966) sampled in 2020 from Howell Lake, Wisconsin. Colors are representative of the respective Gabelhouse (1984) length categories.

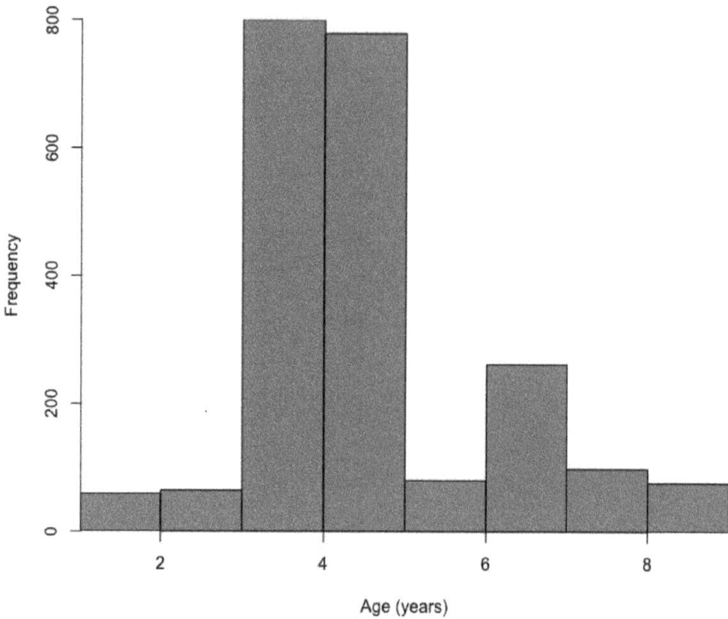

Figure 2. Age-frequency histogram for black bullheads *Ameiurus melas* (n = 4966) sampled in 2020 from Howell Lake, Wisconsin.

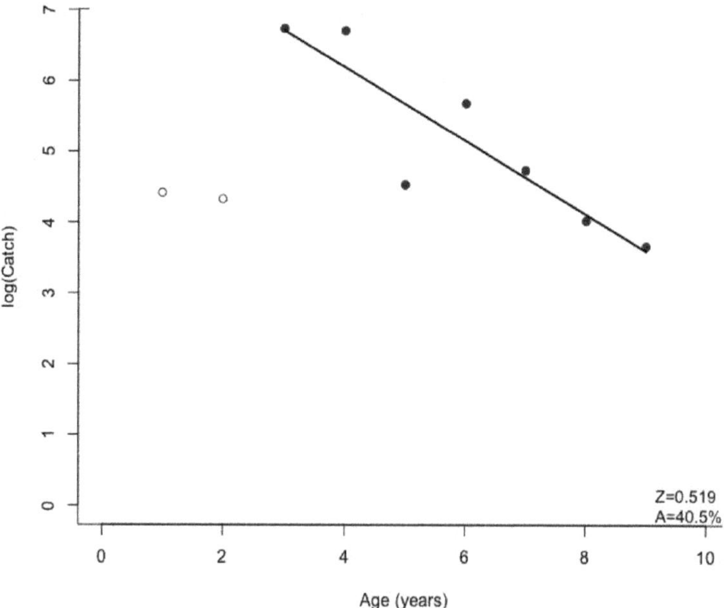

Figure 3. Catch-curve for black bullheads *Ameiurus melas* ≥age-3 sampled in June 2020 from Howell Lake, Wisconsin. Z is representative of the slope of the descending limb of the catch curve and represents instantaneous mortality and A represents the total annual mortality of the population.

Table 2. Mean length at age with standard deviation (SD) and predicted length at age from the von Bertalanffy growth function with 95% confidence intervals from black bullheads *Ameiurus melas* sampled from Howell Lake, Wisconsin.

Age (year)	N	Mean Length (mm)	Standard Deviation (SD)	Predicted Length (mm)	Lower 95% CI	Upper 95% CI
1	84	80	15.0	90	87	93
2	76	132	22.0	135	133	136
3	842	180	18.3	172	171	173
4	816	196	18.0	204	203	205
5	92	235	17.3	231	230	233
6	295	260	14.6	254	253	256
7	117	274	18.5	274	272	276
8	57	297	14.0	290	288	283
9	35	292	11.1	304	301	307

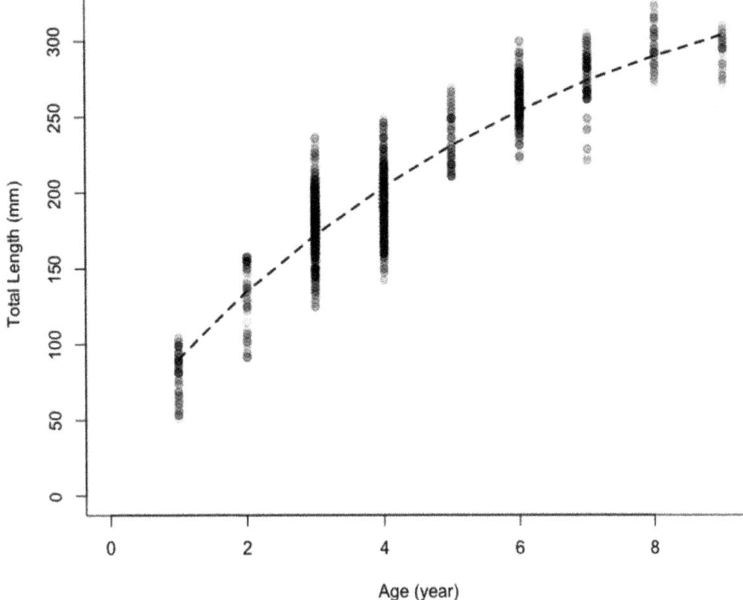

Figure 4. Best fit von Bertalanffy growth model for black bullheads *Ameiurus melas* sampled in June 2020 from Howell Lake, Wisconsin.

3.2. Spawning, Maturity, and Fecundity

In 2020, black bullhead spawning activity (nest building) was first observed on June 17 (21 °C), and fish were last observed on nests on August 3 (24 °C). Of the 959 fish sampled, the male to female ratio was nearly equal and was estimated to be 0.93:1.00 males to females. The shortest sexually mature male and female black bullheads were sampled on June 16, 2020, and were 124 mm and 127 mm, respectively. Estimated length at maturity was similar between male and female bullheads, with L_{50} = 133 mm (95% CI = 124–143 mm) for males and 134 mm (95% CI = 125–143 mm) for females. Additionally, L_{90} was 144 mm (95% CI = 126–151 mm) for males and was 144 mm (95% CI = 126–151 mm) for females (Figure 5).

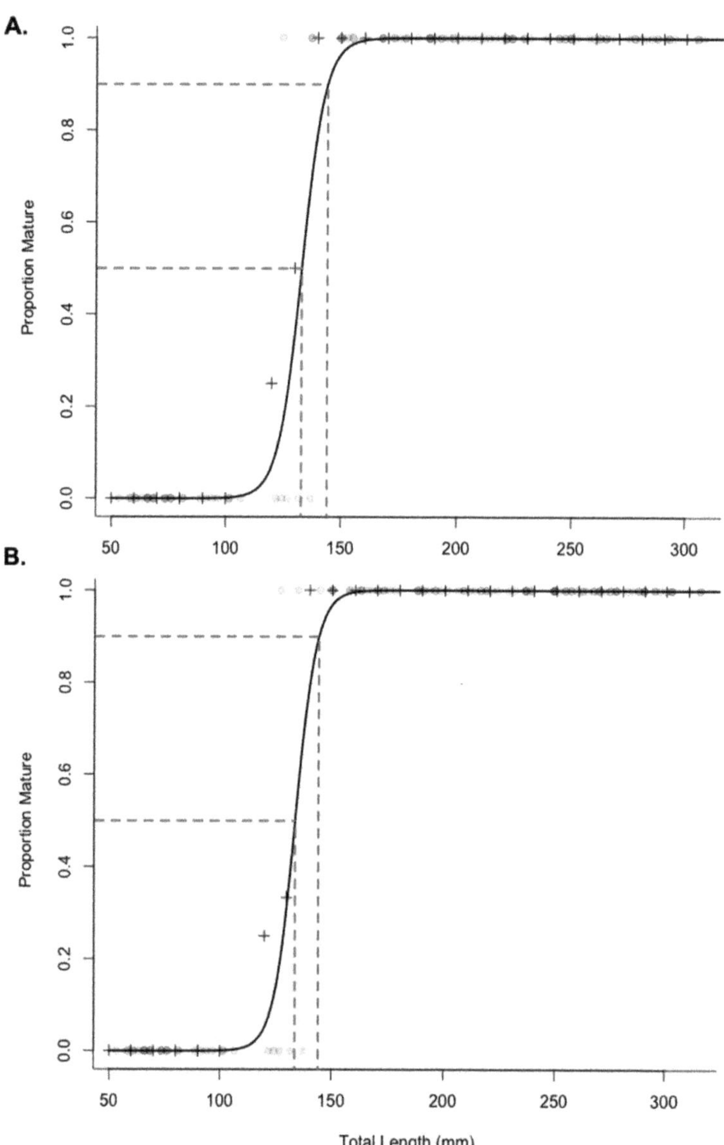

Figure 5. Logistic regression results showing length at maturity for male (**A**) and female (**B**) black bullheads *Ameiurus melas* sampled in June 2020 from Howell Lake, Wisconsin. Value 0 indicates sexually immature or unknown fish, while 1 represents sexually mature fish. The red dashed lines indicate the length at 50% maturity (L_{50}) and the blue dashed lines indicate the length at 90% maturity (L_{90}).

Ovaries of 62 sexually mature female black bullheads ranging in length from 134–308 mm (mean = 227 mm; SE = 5.81) were examined to estimate fecundity. Substock length fish (<150 mm) had a mean fecundity of 1518 eggs/female (n = 2; range = 1050–1986); stock length fish (150–230 mm) had a mean fecundity of 2,133 eggs/female (n = 30; SE = 166.72; range = 513–4128); quality length (230–300 mm) had a mean fecundity of 4319 eggs/female

(n = 27; SE = 459.29; range = 1365–12,337); and preferred length (300–390 mm) had a mean fecundity of 5,485 eggs/female (n = 3; SE = 671.78; range = 4193–6449; Figure 6). An estimated 7% of the female bullheads sampled were deemed immature. Substock length fish were estimated to comprise 3% of the sexually mature female black bullheads in Howell Lake potentially resulting in the production of an estimated 488,247 eggs (95% CI = 339,014–641,221); stock length fish were estimated to comprise 73% of the sexually mature female population, potentially resulting in the production of an estimated 17,821,634 eggs (95% CI = 4,286,216–34,490,250); quality length fish were estimated to make up 22% of the sexually mature female population, potentially resulting in the production of an estimated 10,985,501 eggs (95% CI = 3,471,917–31,379,515); and preferred length fish were estimated to make up 2% of the of the sexually mature female population, potentially resulting in the production of an estimated 784,082 eggs (95% CI = 599,390–921,886). The total reproductive potential of mature female black bullheads in Howell Lake was estimated to be 30,081,334 eggs (95% CI = 8,696,537–67,432,872).

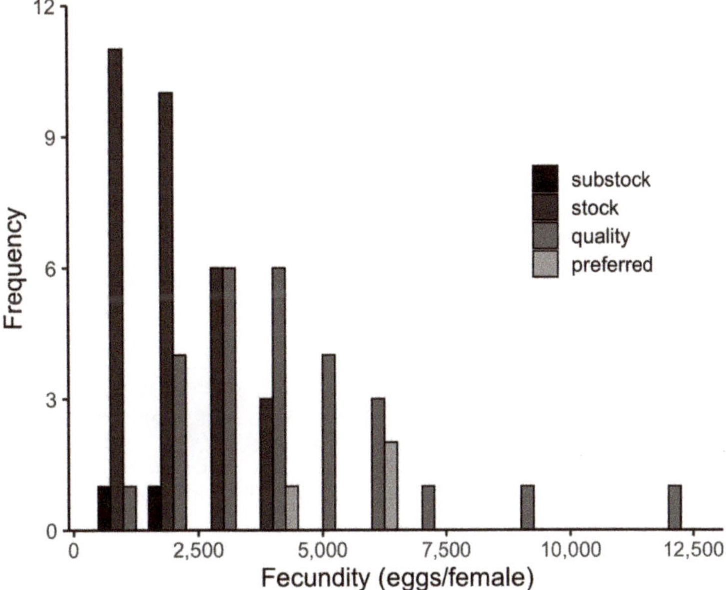

Figure 6. Frequency histogram showing black bullhead *Ameiurus melas* fecundity (eggs/female; n = 62) by Gabelhouse (1984) length category for Howell Lake, Wisconsin in June 2020.

3.3. Seasonal Diet Analysis

Of the 853 bullheads dissected for diet analysis during May–October 2020, 57% had empty stomachs (n = 452). The highest proportion of fish with empty stomachs were sampled in October at 70%, while the lowest proportion of empty stomachs were sampled in August at 47% (n = 387). Black bullheads that were dissected ranged in length from 116–308 mm (mean = 200 mm; SE = 1.03) and ranged in weight from 11–308 g (mean = 119 g; SE = 1.59). The diets of the bullheads were diverse, including seven different fish species, fish eggs, nine aquatic invertebrate taxa, three terrestrial insects, aquatic plant material, detritus, and unknown contents (Table 3). Of the diet items, snails were the most common diet item overall and yellow perch were the most common fish species for nearly every length category of bullhead during every month of sampling (Table 3; Figure 7).

Table 3. Frequency of occurrence (O_i) for seasonal diet composition of black bullheads *Ameiurus melas* collected from May–October 2020 from Howell Lake, Wisconsin. The number of fish preyed on each diet item is indicated in ().

Diet Item	May	June	July	August	October
	Oi	Oi	Oi	Oi	Oi
Fish					
Bluegill (*Lepomis macrochirus*)			0.7 (1)	1.5 (2)	16.7 (1)
Yellow Perch (*Perca flavescens*)		1.0 (1)	8.3 (12)	3.8 (5)	
Black Crappie (*Pomoxis nigromaculatus*)				0.8 (1)	
Black Bullhead (*Ameiurus melas*)					16.7 (1)
Walleye (*Sander vitreus*)	7.1 (1)				16.7 (1)
Common Shiner (*Luxilus cornutus*)		1.0 (1)		1.5 (2)	
Bluntnose Minnow (*Pimephales notatus*)				2.3 (3)	
Eggs	14.3 (2)				
Aquatic Invertebrates					
Gastropod (snails)	35.7 (5)	18.5 (19)	67.6 (98)	63.9 (85)	50 (3)
Arthropoda (crayfish)		3.9 (4)	1.4 (2)	1.5 (2)	16.7 (1)
Ephemeroptera (larvae)	7.1 (1)	4.9 (5)		0.8 (1)	
Odonota (larvae)	21.4 (3)	4.9 (5)	0.7 (1)	1.5 (2)	
Dytiscidae (beetle)		1.9 (2)			
Diptera (larvae and pupae)		1 (1)	2.6 (3)	12.8 (17)	
Trichoptera (larvae)				1.5 (2)	
Hirundea (leeches)	28.6 (4)	2.9 (3)		1.5 (2)	
Terrestrial Insects					
Odonota (adult)		1.9 (2)	0.7 (1)		
Tipulidae (adult)		1.9 (2)		0.8 (1)	
Lepidoptera (adult)		1.0 (1)			
Other					
Aquatic Vegetation		17.5 (18)	17.2 (25)	14.3 (19)	
Detritus	21.4 (3)	38.8 (40)	8.3 (12)	3.8 (5)	
Unknown		1.0 (1)			

In May, seven stock length and seven quality length bullheads with stomach contents were sampled. The diets sampled consisted of fish eggs, a walleye, and aquatic invertebrates (snails, leeches, Ephemeroptera, and Odonata; Table 3). Detritus occurred in 43%, snails in 36%, aquatic insects (Odonata) 4%, and fish eggs in 24% of the stomachs from stock length fish. Snails occurred in 29%, leeches in 5%, aquatic insects in 2% (Ephemeroptera, Odonata) in 28%, walleye in 14%, and fish eggs in 14% of quality length fish sampled in May.

In June, two sub-stock lengths, 70 stock length, 27 quality length, and four preferred length bullheads with stomach contents were sampled. The diets sampled consisted of yellow perch, common shiner, terrestrial insects (Tipulidae, Odonata, and Lepidoptera), aquatic invertebrates (snails, crayfish, leeches, Ephemeroptera, Dytiscidae, and Chironomidae), aquatic plant material, detritus, and some unknown contents (Table 3). Detritus occurred in 50% and aquatic insects in 50% of the stomachs from sub-stock length fish. Detritus occurred in 49%, aquatic plants in 14%, snails in 14%, aquatic insects in 17%, terrestrial insects in 1%, leeches in 3%, crayfish in 3%, and unknown contents in 3% of the stomachs from stock length fish. Detritus occurred in 30%, aquatic plants in 30%, snails in 33%, aquatic insects in 4%, terrestrial insects in 4%, leeches in 4%, crayfish in 7%, and yellow perch in 4% of the stomachs from quality length fish. Aquatic plants occurred in 50%, terrestrial insects in 25%, and common shiner in 25% of the stomachs from preferred length fish.

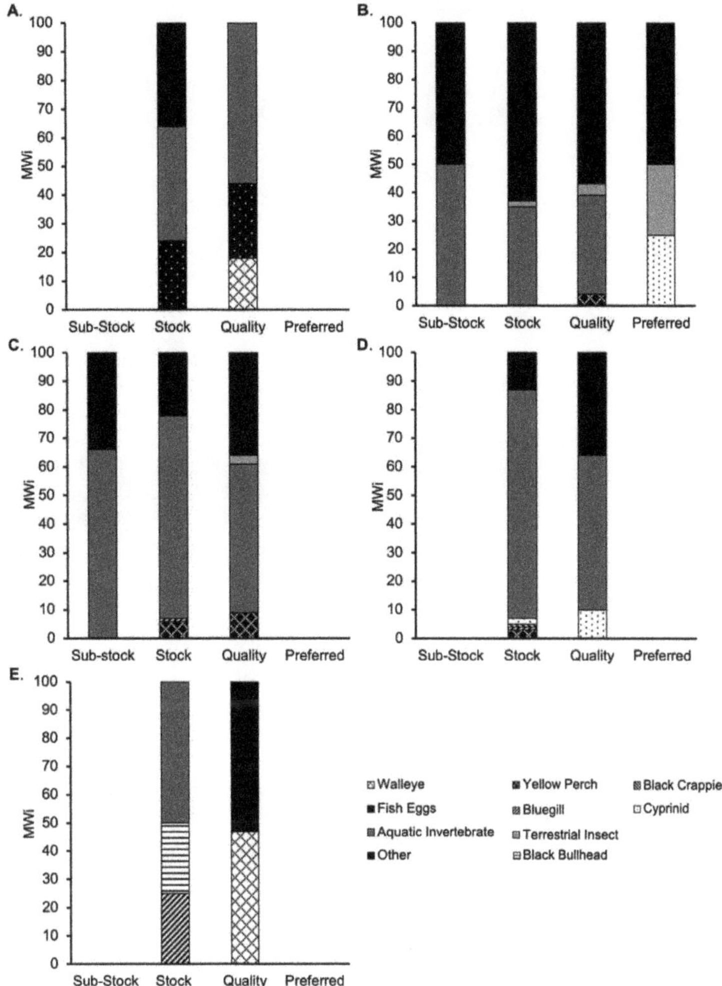

Figure 7. Mean percent composition by weight (*MWi*) of diet items from black bullheads *Ameiurus melas* separated into appropriate Gabelhouse (1984) length categories sampled from Howell Lake, Wisconsin in May (**A**), June (**B**), July (**C**), August (**D**), and October (**E**) of 2020.

In July, four sub-stock lengths, 130 stock length, and 11 quality length bullheads with stomach contents were sampled. The diets sampled consisted of yellow perch, bluegill, snails, crayfish, terrestrial insects (Odonata), aquatic insects (Odonata and Chironomidae), aquatic plant material, and detritus (Table 3). Detritus occurred in 50% and snails in 50% of the stomachs from sub-stocked length fish. Detritus occurred in 8%, aquatic plants in 16%, snails in 71%, aquatic insects in 2%, yellow perch in 8%, and bluegill in 1% of the stomachs from stock length fish. Aquatic plants occurred in 36%, snails in 45%, terrestrial insects in 9%, crayfish in 18%, and yellow perch in 9% of the stomachs from quality length fish.

In August, 115 stock length, 17 quality length, and one preferred length bullhead with stomach contents were sampled. The diets sampled consisted of yellow perch, bluegill, black crappie, common shiner, bluntnose minnow, snails, crayfish, leeches, terrestrial insects (Tipulidae), aquatic insects (Ephemeroptera, Odonata, Diptera, Trichoptera, and Chironomidae), aquatic plant material, and detritus (Table 3). Detritus occurred in 3%, aquatic plants in 11%, snails in 70%, aquatic insects in 15%, yellow perch in 3%, bluegill in 2%, cyprinids in 3%, and black crappie in 1% of the stomachs from stock length fish. Detritus occurred in 6%, aquatic plants in 35%, snails in 29%, aquatic insects in 24%, terrestrial insects in 6%, leeches in 12%, crayfish in 12%, and cyprinids in 12% of the stomachs from quality length fish. Yellow perch occurred in the one stomach from the preferred length fish.

In October, four stock length and two quality length bullheads with stomach contents were sampled. The diets sampled consisted of bluegill, a walleye, black bullhead, snails, and one crayfish (Table 3). Snails occurred in 25%, crayfish in 25%, bluegill in 25%, and black bullhead in 25% of the stomachs from stock length fish. Snails occurred in 100%, and walleye in 50% of the stomachs from the quality length fish.

Across all lengths, months and diet items, snails had the highest *IRI*, with trichopterans having the lowest *IRI* (Table 4; Figure 8). Of all prey items, snails had the highest O_i, MN_i, and MW_i throughout the study (Tables 3, 5 and 6; Figures 9 and 10). MW_i of diet items significantly differed over June, July, and August ($p < 0.001$, $df = 2$, $f = 6.82$; Table 4; Figure 7). No differences were found for the MW_i of diet items over June, July, and August for aquatic vegetation ($p = 0.45$), crayfish ($p = 0.58$), leeches ($p = 0.18$), yellow perch ($p = 0.08$), cyprinids ($p = 0.07$), bluegills ($p = 0.55$), black crappie ($p = 0.39$), and unknown diet items ($p = 0.16$; Table 4; Figure 5). Significant differences in the MW_i were observed among months for snails, terrestrial insects, aquatic insects, and detritus (Table 6; Figure 10). Snails accounted for less of the MW_i in June than in July ($p \leq 0.001$) and August ($p < 0.001$), but the MW_i of snails did not differ between July and August ($p = 0.60$; Table 6; Figure 10). Similarly, detritus accounted for significantly less of the MW_i in June compared to July ($p < 0.001$); and August ($p < 0.001$); July and August did not differ ($p = 0.52$; Table 6; Figure 10). The MWi of terrestrial insects followed the same trend being significantly less in June ($p = 0.03$; 0.02, respectively) and July and August not being different ($p = 0.98$; Table 6; Figure 10). Aquatic insects were similar and most common in diets from June and August ($p = 0.80$), while July significantly differed from both months ($p = 0.02$; 0.001, respectively; Table 6; Figure 10).

Table 4. Index of relative importance (*IRI*) for seasonal diet composition of black bullheads *Ameiurus melas* collected from May–October 2020 from Howell Lake, Wisconsin.

Diet Item	May	June	July	August	October
	IRI	IRI	IRI	IRI	IRI
Fish					
Bluegill (*Lepomis macrochirus*)			0.01	0.03	5.57
Yellow Perch (*Perca flavescens*)		0.02	1.13	0.25	
Black Crappie (*Pomoxis nigromaculatus*)				0.01	
Black Bullhead (*Ameiurus melas*)					5.57
Walleye (*Sander vitreus*)	0.64				4.01
Common Shiner (*Luxilus cornutus*)		0.02		0.03	
Bluntnose Minnow (*Pimephales notatus*)				0.07	
Eggs	15.55				
Aquatic Invertebrates					
Gastropod (snails)	10.67	6.68	89.49	78.28	37.98
Arthropoda (crayfish)		0.22	0.03	0.04	5.57
Ephemeroptera (larvae)	0.02	0.46		0.00	
Odonota (larvae)	4.10	0.48	0.01	0.00	
Dytiscidae (beetle)		0.07			
Diptera (larvae and pupae)		0.01	0.10	3.19	
Trichoptera (larvae)				0.01	
Hirundea (leeches)	4.85	0.15		0.02	

Table 4. *Cont.*

Diet Item	May	June	July	August	October
	IRI	IRI	IRI	IRI	IRI
Terrestrial Insects					
Odonota (adult)		0.05	0.00		
Tipulidae (adult)		0.04		0.00	
Lepidoptera (adult)		0.02			
Other					
Aquatic Vegetation		3.21	2.77	1.76	
Detritus	3.42	15.95	0.65	0.13	
Unknown		0.02			

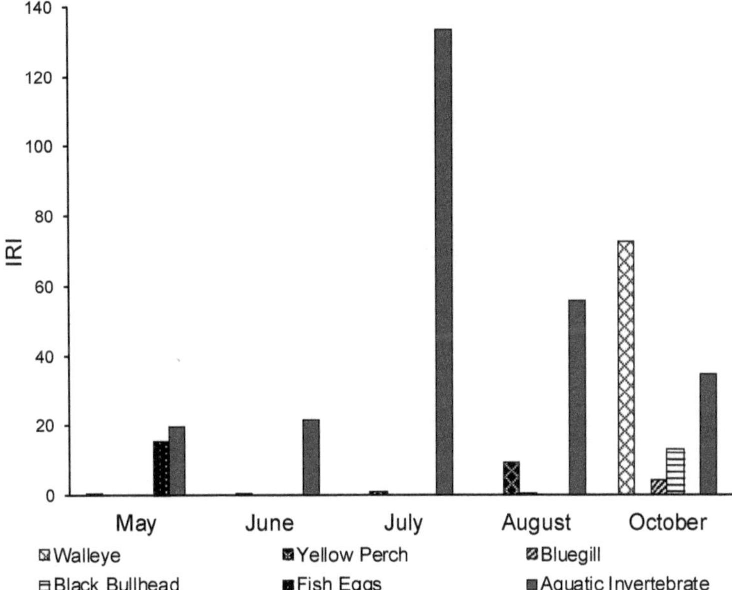

Figure 8. Index of relative importance (*IRI*) of diet items from black bullheads *Ameiurus melas* sampled from Howell Lake, Wisconsin in May, June, July, August, and October 2020.

Table 5. Mean % composition by number (MN_i) for seasonal diet composition of black bullheads *Ameiurus melas* collected from May–October 2020 from Howell Lake, Wisconsin.

Diet Item	May	June	July	August	October
	Mni	Mni	Mni	Mni	Mni
Fish					
Bluegill (*Lepomis macrochirus*)			0.69	1.13	16.67
Yellow Perch (*Perca flavescens*)		0.97	7.47	3.38	
Black Crappie (*Pomoxis nigromaculatus*)				0.38	
Black Bullhead (*Ameiurus melas*)					16.67
Walleye (*Sander vitreus*)	0.25				8.33
Common Shiner (*Luxilus cornutus*)		0.97		0.75	
Bluntnose Minnow (*Pimephales notatus*)				1.45	
Eggs	85.86				

Table 5. *Cont.*

Diet Item	May	June	July	August	October
	Mni	Mni	Mni	Mni	Mni
Aquatic Invertebrates					
Gastropod (snails)	7.00	18.01	66.55	61.92	41.67
Arthropoda (crayfish)		3.11	0.92	1.13	16.67
Ephemeroptera (larvae)	0.25	4.94		0.30	
Odonota (larvae)	4.8	4.85	0.69	1.13	
Dytiscidae (beetle)		1.94			
Diptera (larvae and pupae)		0.97	2.3	12.71	
Trichoptera (larvae)				0.45	
Hirundea (leeches)	1.77	2.91		0.83	
Terrestrial Insects					
Odonota (adult)		1.46	0.23		
Tipulidae (adult)		1.46		0.08	
Lepidoptera (adult)		0.97			
Other					
Aquatic Vegetation					
Detritus					
Unknown		1.13			

Table 6. Mean % composition by weight (MW_i) for seasonal diet composition of black bullheads *Ameiurus melas* collected from May–October 2020 from Howell Lake, Wisconsin.

Diet Item	May	June	July	August	October
	Mwi	Mwi	Mwi	Mwi	Mwi
Fish					
Bluegill (*Lepomis macrochirus*)			0.69	1.09	16.67
Yellow Perch (*Perca flavescens*)		0.97	6.10	3.12	
Black Crappie (*Pomoxis nigromaculatus*)				0.41	
Black Bullhead (*Ameiurus melas*)					16.67
Walleye (*Sander vitreus*)	8.72				15.70
Common Shiner (*Luxilus cornutus*)		0.97		1.33	
Bluntnose Minnow (*Pimephales notatus*)				1.47	
Eggs	22.87				
Aquatic Invertebrates					
Gastropod (snails)	22.90	18.07	65.8	60.58	34.30
Arthropoda (crayfish)		2.57	1.16	1.23	16.67
Ephemeroptera (larvae)	0.05	4.46		0.30	
Odonota (larvae)	14.36	4.85	0.69	1.20	
Dytiscidae (beetle)		1.94			
Diptera (larvae and pupae)		0.32	1.45	12.22	
Trichoptera (larvae)				0.12	
Hirundea (leeches)	15.20	2.14		0.77	
Terrestrial Insects					
Odonota (adult)		1.25	0.20		
Tipulidae (adult)		0.90		0.02	
Lepidoptera (adult)		0.97			
Other					
Aquatic Vegetation		18.32	16.10	12.30	
Detritus	16.00	41.10	7.80	3.42	
Unknown		1.15			

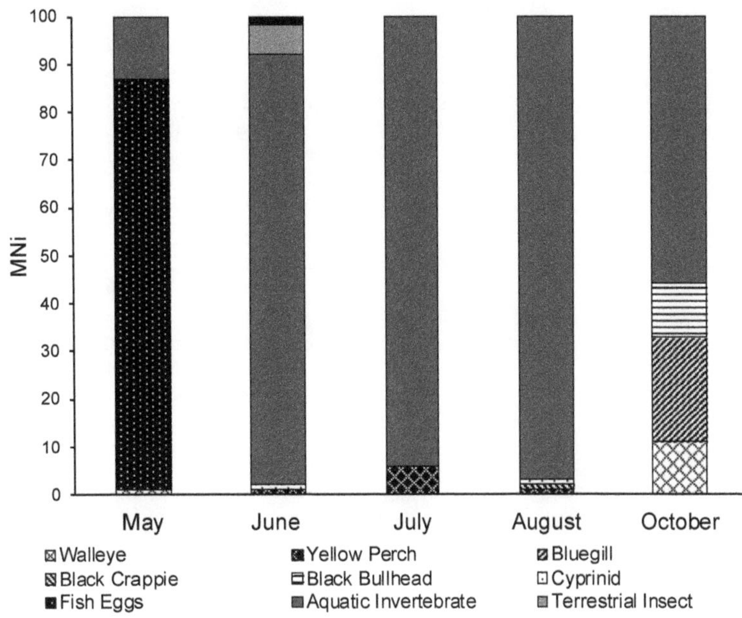

Figure 9. Mean percent composition by weight (MN_i) of diet items from black bullheads *Ameiurus melas* sampled from Howell Lake, Wisconsin in May, June, July, August, and October 2020.

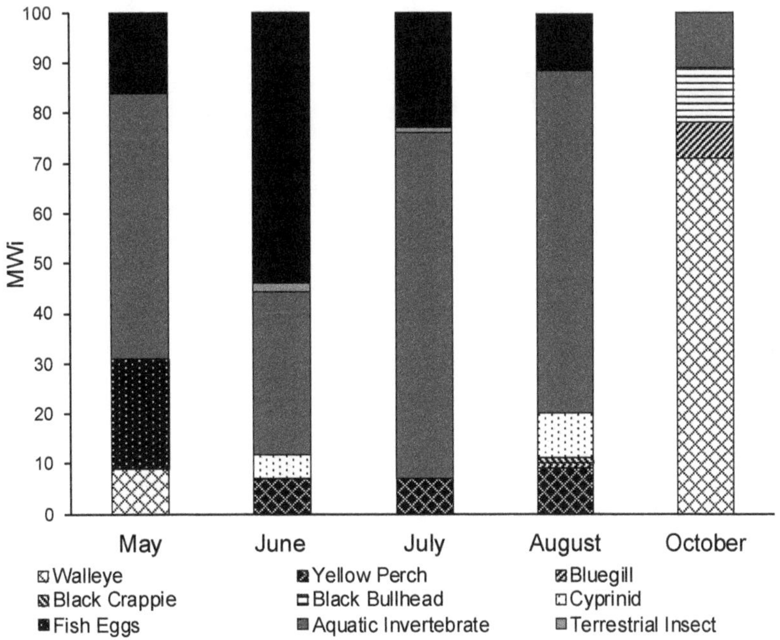

Figure 10. Mean percent composition by weight (MW_i) of diet items from black bullheads *Ameiurus melas* sampled from Howell Lake, Wisconsin in May, June, July, August, and October 2020.

4. Discussion

Black bullheads exhibit relatively fast growth rates, reach an age-at-maturity earlier than many other native north-temperate fish species, and are highly omnivorous. Further, black bullhead can reach high densities in north-temperate lakes. In concert, it appears that black bullheads have the potential to interact extensively throughout the food web. Empirical evidence suggests that high densities of bullheads can lead to undesired effects on popular sportfish, including walleye and yellow perch [11]. Despite the changing environment [27], black bullhead's plastic life history, high degree of fecundity, and extensive food web interaction capabilities will likely allow the species to remain a major component of some north-temperate lake food webs in the future, while more desirable cool-water species (e.g., walleye and yellow perch) decline and(or) become extirpated from some systems. Information realized through our research will allow for better management and mitigation of these 'nuisance species driven' negative effects.

A study of 35 black bullhead populations in South Dakota found highly variable but rapid individual growth rates using back-calculated growth increments and maximum age of eight years, with most fish being younger than age-6 [43]. The Hanchin et al. [43] age distribution generally agreed with that of our study, where the oldest identified age was nine, and most fish were estimated to be < age-5. Previous estimates of black bullhead adult length were varied. Becker [44] suggested an average adult length of 165–229 mm. Several fish in our sample far exceeded this length range, with the largest fish measuring 308 mm. Copp et al. [4] reviewed available literature related to the growth of black bullheads across their native and non-native ranges and found that body lengths and weights varied greatly, though the overall length of bullheads from the native range was greater than observed in the European populations.

Female age at maturity of bullheads in Howell Lake was young and similar to previous observations [16]. Copp et al. [4] mean sex ratio was virtually identical in the non-native and native populations for which data were available and were similar to our estimate near 1:1. Female age at maturity in the native range has been reported to span 2–5 years [13,16,45] and from 1–3.5 years in European non-native populations [4]. The only available detailed information on female maturity in the native range comes from the Mississippi River in Illinois, where bullhead females were reported to mature at 254 mm TL and age-3 [46], and Clear Lake, Iowa, where females matured at 200 mm TL and males at 216 mm TL [47]. Although limited information is available on the fecundity and reproductive potential of black bullhead, Forney [47], Carlander and Sprugel [48], and Dennison and Bulkley [49] estimated the fecundity of black bullheads in Clear Lake, Iowa. Similar to our study, these researchers found variable fecundity of fish of the same length but found that longer fish had higher fecundities. Despite lower average fecundity, stock length fish contributed an estimated 68% of the eggs to the cumulative reproductive potential of the black bullheads removed from Howell Lake in 2020. This is due to stock length black bullheads being much more abundant than quality and preferred length fish. Assuming egg and larval mortalities of 95% each [50], an estimated 75,203 age-1 black bullheads were precluded from recruiting to the Howell Lake fish community. This single whole-lake study provided valuable information on black bullhead fecundity and reproductive potential; however, our conclusions are limited to north-temperate lakes with similar limnological and ecological characteristics.

Black bullhead in Howell Lake preyed on a diverse range of diet items and exhibited an omnivorous and likely opportunistic feeding strategy. Studies have shown that in most cases, bullheads do not exhibit a preference outside of the most abundant prey species and have been shown to prey on insects, snails, clams, crayfish, frogs, plant material, detritus, and fish with varying degrees of piscivory [16,51–55]. Snow et al. [55] suggested that in Lake Carl, Oklahoma, black bullheads showed more piscivory than previous studies, while several others suggested that piscivory was not as large a part of the diets, especially in smaller fish [16,51–54]. In Howell Lake, Chinese mystery snails were found to be the most common prey item in the diets; however, fish were also a common prey item in the stomachs of lack bullheads. Seven fish species were present in the diets, all of which were generally fusiform in body shape at the life stage that they were consumed. Howell Lake has a fish assemblage representative of many northern Wisconsin lakes, which includes several small and abundant minnow species, but the most common prey fish for black bullhead were yellow perch. It is plausible that in certain systems and at certain times of the year, piscivory by black bullheads may negatively influence sport fish populations; however, our initial analyses do not strongly support the notion that black bullheads are suppressing sport fish populations in Howell Lake through direct predation. Black bullheads in Howell Lake may potentially compete with other top predators (walleye, northern pike, largemouth bass, smallmouth bass) to some degree based on the presence of fish in their diets, as yellow perch and bluegills are a common prey item for these piscivores [56–58], but this is likely not to any detectable degree in Howell Lake because of the abundance in forage. Piscivory by bullheads could be problematic if dietary overlap and resource availability is not considered in systems where black bullheads are established, and the goal of the fishery is to promote sportfishes for angling opportunities.

Results from our study show the importance of understanding the role of black bullheads in the ecosystems of north-temperate lakes, particularly when found in high abundances. Furthermore, it introduces the question of how to manage black bullhead populations in situations where they dominate fish community biomass, are highly piscivorous, and function similarly to a top predator in the system. Considerations of diet overlap and fish forage availability are critical when fisheries managers are considering management strategies for other top predators or when contemplating the introduction of a new species into an aquatic system. Further research is needed on a broader scale (multiple systems) to determine the full influences of black bullheads on sportfish populations. Furthering our knowledge and understanding of the demographics and life history of black bullheads across their range will provide managers with valuable information that is at the root of critical management decisions. Although many managers do not specifically manage bullheads directly, bullheads are present in many systems and likely play a role in structuring fish communities [11].

Black bullheads are often considered invasive species and can become overabundant within and outside of their native range. Due to their environmental tolerance (e.g., water quality, thermal) and ecological plasticity, black bullheads have the potential to disrupt native food webs and alternative fish community composition [4,5,10–12]. Our results showed that black bullhead exhibited rapid growth rates, early age at maturity, moderate fecundities, parental care of young, and a diverse omnivorous diet, which increases their invasive potential and prospective effects on native and invaded food webs and fish communities. Ongoing bullhead removal research will provide insights into the mechanistic role of bullheads in food webs to better manage these species as invasives and in situations where they may create fish community imbalances in their native range.

Author Contributions: Conceptualization, L.W.S., J.A.V. and G.G.S.; formal analysis, L.W.S. and J.T.M.; data curation, L.W.S., J.T.M., R.H., J.A.V. and G.G.S.; writing—original draft preparation, L.W.S.; writing—review and editing, L.W.S., J.T.M., R.H., J.A.V. and G.G.S.; funding acquisition, J.A.V. and G.G.S. All authors have read and agreed to the published version of the manuscript.

Funding: This research was funded by the University of Wisconsin-Stevens Point and the United States Fish and Wildlife Service, Federal Aid in Sportfish Restoration, Project F-95-P, and the Wisconsin Department of Natural Resources.

Institutional Review Board Statement: The animal study protocol was approved by the Institutional Animal Care and Use Committee (IACUC) at the University of Wisconsin-Stevens Point (Protocol: 2020.01.05 "Effects of Bullhead removals on the fish community in Northern Wisconsin lakes").

Informed Consent Statement: Not applicable.

Data Availability Statement: Data are available upon request.

Acknowledgments: We would like to thank Brenden Elwer, Levi Feucht, Glenn Schumacher, Quinn Smith, Levi Simmons, Stephanie Shaw, Katti Renik, Sam LaMarche, Steven Broda, Kayla Reed, Eric

Chase, and Sophia Callaci for their assistance with field and laboratory data collection. The University of Wisconsin-Madison, Trout Lake Station provided housing during the study.

Conflicts of Interest: The authors declare no conflict of interest.

References

1. Wheeler, A.C. Ictalurus melas (Rafinesque, 1820) and I. nebulosus (Lesueur, 1819): The North American catfishes in Europe. *J. Fish. Biol.* **1978**, *12*, 435–439. [CrossRef]
2. Page, L.M.; Burr, B.M. *A Field Guide to Freshwater Fishes: North America North of Mexico*; Houghton Mifflin: Boston, MA, USA, 1991.
3. Copp, G.H.; Bianco, P.G.; Bogutskaya, N.; Erős, T.; Falka, I.; Ferreira, M.T.; Fox, M.G.; Freyhof, J.; Gozlan, R.E.; Grabowska, J.; et al. To be, or not to be, a non-native freshwater fish? *J. Appl. Ichthyol.* **2005**, *21*, 242–262. [CrossRef]
4. Copp, G.H.; Tarkan, A.S.; Masson, G.; Godard, M.J.; Koščo, J.; Novomeská, A.; Miranda, R.; Cucherousset, J.; Pedicillo, G.; Blackwell, B. A review of growth and life-history traits of native and non-native European populations of black bullhead *Ameiurus melas*. *Rev. Fish Biol. Fish.* **2016**, *26*, 441–469. [CrossRef]
5. Vooren, C. Ecological aspects of the introduction of fish species into natural habitats in Europe, with special reference to the Netherlands. *J. Fish. Biol.* **1971**, *4*, 565–583. [CrossRef]
6. Cucherousset, J.; Paillisson, J.M.; Carpentier, A.; Eybert, M.C.; Olden, J.D. Habitat use of an artificial wetland by the invasive catfish *Ameiurus melas*. *Ecol. Freshw. Fish.* **2006**, *15*, 589–596. [CrossRef]
7. Copp, G.H. The habitat diversity and fish reproductive function of floodplain ecosystems. *Environ. Biol. Fish.* **1989**, *26*, 1–26. [CrossRef]
8. Copp, G.H. The upper River Rhône revisited: An empirical model of microhabitat use by 0+ juvenile fishes. *Folia. Zool.* **1993**, *42*, 329–340.
9. Gozlan, R.E.; Mastrorillo, S.; Dauba, F.; Tourenq, J.N.; Copp, G.H. Multi-scale analysis of habitat use by 0+ fishes during late summer in the River Garonne (France). *Aquat. Sci.* **1998**, *60*, 99–117. [CrossRef]
10. Cucherousset, J.; Paillisson, J.M.; Carpentier, A. Is mass removal an efficient measure to regulate the North American catfish *Ameiurus melas* outside of its native range? *J. Freshw. Ecol.* **2008**, *21*, 699–704. [CrossRef]
11. Sikora, L.W.; VanDeHey, J.A.; Sass, G.G.; Matzke, G.; Preul, M. Fish Community Changes Associated with Bullhead Removals in Four Northern Wisconsin Lakes. *N. Am. J. Fish. Manag.* **2021**, *41*, S71–S81. [CrossRef]
12. Braig, E.C.; Johnson, D.L. Impact of black bullhead (*Ameiurus melas*) on turbidity in a diked wetland. *Hydrobiologia* **2003**, *490*, 11–21. [CrossRef]
13. Scott, W.B.; Crossman, E.G. *Freshwater Fishes of Canada*; Fisheries Resource Board Canada: Ottawa, ON, Canada, 1973; Volume 184, p. 966.
14. Stuber, R.J. *Habitat Suitability Index Models: Black Bullhead*; United States Department of Interior, Fish and Wildlife Service: Fairfax, VA, USA, 1982.
15. Ribeiro, F.; Elvira, B.; Collares-Pereira, M.J.; Moyle, P.B. Life-history traits of non-native fishes in Iberian watersheds across several invasion stages: A first approach. *Biol. Invasions* **2008**, *10*, 89–102. [CrossRef]
16. Carlander, K.D. *Handbook of Freshwater Fishery Biology*; Iowa State University Press: Ames, IA, USA, 1996; Volume 1.
17. Bister, T.J.; Willis, D.W.; Brown, M.L.; Jordan, S.M.; Neumann, R.M.; Quist, M.C.; Guy, C.S. Proposed standard weight (Ws) equations and standard length categories for 18 warmwater nongame and riverine fish species. *N. Am. J. Fish. Manag.* **2000**, *20*, 570–574. [CrossRef]
18. Sharma, S.; Jackson, D.A.; Minns, C.K.; Shuter, B.J. Will northern fish populations be in hot water because of climate change? *Glob. Chang. Biol.* **2007**, *13*, 2052–2064. [CrossRef]
19. Hansen, G.J.A.; Gaeta, J.W.; Hansen, J.F.; Carpenter, S.R. Learning to manage and managing to learn: Sustaining freshwater recreational fisheries in a changing environment. *Fisheries* **2015**, *40*, 56–64. [CrossRef]
20. Lynch, A.J.; Myers, B.J.; Chu, C.; Eby, L.A.; Falke, J.A.; Kovach, R.P.; Krabbenhoft, T.J.; Kwak, T.J.; Lyons, J.; Paukert, C.P.; et al. Climate change effects on North American inland fish populations and assemblages. *Fisheries* **2016**, *41*, 346–361. [CrossRef]
21. Hansen, G.J.A.; Read, J.S.; Hansen, J.F.; Winslow, L.A. Projected shifts in fish species dominance in Wisconsin lakes under climate change. *Glob. Chang. Biol.* **2017**, *23*, 1463–1476. [CrossRef] [PubMed]
22. Sass, G.G.; Rypel, A.L.; Stafford, J.D. Fisheries habitat management: Lessons learned from wildlife ecology and a proposal for change. *Fisheries* **2017**, *42*, 197–209. [CrossRef]
23. Mrnak, J.T.; Shaw, S.L.; Eslinger, L.D.; Cichosz, T.A.; Sass, G.G. Characterizing the joint tribal spearing and angling walleye fisheries in the Ceded Territory of Wisconsin. *N. Am. J. Fish. Manag.* **2018**, *38*, 1381–1393. [CrossRef]
24. Embke, H.S.; Rypel, A.L.; Carpenter, S.R.; Sass, G.G.; Ogle, D.; Cichosz, T.; Hennessy, J.; Essington, T.E.; Zanden, M.J.V. Production dynamics reveal hidden overharvest of inland recreational fisheries. *Proc. Natl. Acad. Sci. USA* **2019**, *116*, 24676–24681. [CrossRef]
25. Sass, G.G.; Shaw, S.L. Walleye population responses to experimental exploitation in a northern Wisconsin lake. *Trans. Am. Fish. Soc.* **2018**, *147*, 869–878. [CrossRef]
26. Sass, G.G.; Shaw, S.L. Catch-and-release influences on inland recreational fisheries. *Rev. Fish. Sci. Aquacult.* **2020**, *28*, 211–227. [CrossRef]

27. Carpenter, S.R.; Cole, J.J.; Pace, M.L.; Batt, R.; Brock, W.A.; Cline, T.; Coloso, J.; Hodgson, J.R.; Kitchell, J.F.; Seekell, D.A.; et al. Early warnings of regime shifts: A whole-ecosystem experiment. *Science* **2011**, *332*, 1079–1082. [CrossRef] [PubMed]

28. Carle, F.L.; Strub, M.R. A new method for estimating population size from removal data. *Biometrics* **1978**, *34*, 621–630. [CrossRef]

29. Bettoli, P.W.; Miranda, L.E. A cautionary note about estimating mean length at age with sub-sampled data. *N. Am. J. Fish. Manag.* **2001**, *21*, 425–428. [CrossRef]

30. Stewart, D.R.; Long, J.M. Verification of otolith identity used by fisheries scientists for aging Channel Catfish. *Trans. Am. Fish. Soc.* **2010**, *139*, 1775–1779. [CrossRef]

31. Buckmeier, D.L.; Howells, R.G. Validation of otoliths for estimating ages of Largemouth Bass to 16 years. *N. Am. J. Fish. Manag.* **2003**, *23*, 590–593. [CrossRef]

32. Crim, L.W.; Glebe, B.D. *Reproduction. Methods for Fish Biology*; Schreck, C.B., Moyle, P.B., Eds.; American Fisheries Society: Bethesda, MA, USA, 1990; pp. 529–547.

33. Quist, M.C.; Pegg, M.A.; De Vries, D.R. *Age and Growth. Fisheries Techniques*, 3rd ed.; Alexander, A.V., Parrish, D.L., Sutton, T.M., Eds.; American Fisheries Society: Bethesda, MA, USA, 2012; pp. 677–731.

34. Buckmeier, D.L.; Irwin, E.R.; Betsill, R.K.; Prentice, J.A. Validity of otoliths and pectoral spines for estimating ages of Channel Catfish. *N. Am. J. Fish. Manag.* **2002**, *22*, 934–942. [CrossRef]

35. Ricker, W.E. *Computation and Interpretation of Biological Statistics of Fish Populations*; Fisheries Research Board of Canada Bulletin: Ottawa, ON, Canada, 1975; p. 191.

36. Von Bertalanffy, L. A quantitative theory of organic growth (Inquiries on growth laws. II). *Hum. Biol.* **1938**, *10*, 181–213.

37. Gabelhouse, D.W., Jr. A length-categorization system to assess fish stocks. *N. Am. J. Fish. Manag.* **1984**, *4*, 273–285. [CrossRef]

38. Maceina, M.J.; Bettoli, P.W. Variation in largemouth bass recruitment in four mainstream impoundments of the Tennessee River. *N. Am. J. Fish. Manag.* **1998**, *18*, 998–1003. [CrossRef]

39. Bowen, S.H. *Quantitative Description of the Diet*; Fisheries Techniques; Nielsen, L.A., Johnson, D.L., Eds.; American Fisheries Society: Bethesda, MA, USA, 1983; pp. 325–336.

40. Kline, J.L.; Wood, B.M. Food Habits and Diet Selectivity of the Brown Bullhead. *J. Freshw. Ecol.* **1996**, *11*, 145–151. [CrossRef]

41. Pope, K.L.; Brown, M.L.; Duffy, W.G.; Michaletz, P.H. A caloric-based evaluation of diet indices for largemouth bass. *Environ. Biol. Fish.* **2001**, *61*, 329–339. [CrossRef]

42. Chipps, S.R.; Garvey, J.E. *Assessment of Diets and Feeding Patterns. Analysis and Interpretation of Freshwater Fisheries Data*; Guy, C.S., Brown, M.L., Eds.; American Fisheries Society: Bethesda, MA, USA, 2007; pp. 473–514.

43. Hanchin, P.A.; Willis, D.W.; Hubers, M.J. Black Bullhead Growth in South Dakota Waters: Limnological and Community Influences. *J. Freshw. Ecol.* **2002**, *17*, 65–73. [CrossRef]

44. Becker, G. *Fishes of Wisconsin*; University of Wisconsin Press: Madison, WI, USA, 1983; pp. 693–732.

45. Jenkins, R.E.; Burkhead, N.M. *Freshwater Fishes of Virginia*; American Fisheries Society: Bethesda, MA, USA, 1993.

46. Barnickol, P.G.; Starrett, W.C. Commercial and sport fishery of the Mississippi River between Caruthersville, Missouri, and Dubuque, Iowa. *Bull. Ill. Nat. Hist. Surv.* **1951**, *23*, 267–350. [CrossRef]

47. Forney, J.L. Life history of the Black Bullhead, *Ameiurus melas* (Rafinesque), of Clear Lake, Iowa. *Iowa State Coll. J. Sci.* **1955**, *30*, 145–162.

48. Carlander, K.D.; Sprugel, G. Project 42. Bullhead management. *Quart. Rep. Iowa Coop. Wildl. Res. Units* **1951**, *15*, 44–45.

49. Dennison, S.G.; Bulkley, R.V. Reproductive potential of the Black Bullhead, Ictalurus melas, in Clear Lake, Iowa. *Trans. Am. Fish. Soc.* **1972**, *101*, 483–486. [CrossRef]

50. Houde, E.D. Mortality. In *Fishery Science. The Unique Contributions of Early Life Stages*; Fuiman, L.A., Werner, R.G., Eds.; Blackwell Publishing: Oxford, UK, 2002; pp. 64–87.

51. Harrison, H.M. The foods used by some common fish of the Des Moines River drainage. *Iowa Conserv. Comm.* **1950**, *1*, 31–44.

52. Welker, B.D. Summer food habits of Yellow Bass and Black Bullheads in Clear Lake. *Proc. Iowa Acad. Sci.* **1962**, *69*, 286–295.

53. Bauer, R.J. Digestion rate of the Clear Lake Black Bullhead. *Proc. Iowa Acad. Sci.* **1970**, *77*, 112–121.

54. Williams, W.E. Summer food of juvenile Black Bullheads (Ictalurus melas) of Mitchell Lake, Wexford Country, Michigan. *Trans. Am. Fish. Soc.* **1970**, *99*, 597–599. [CrossRef]

55. Snow, R.A.; Porta, M.J.; Robison, A.L. Seasonal Diet Composition of Black Bullhead (*Ameiurus melas*) in Lake Carl Etling, Oklahoma. *Proc. Okla. Acad. Sci.* **2018**, *97*, 54–60.

56. Clady, M.D. Food Habits of Yellow Perch, Smallmouth Bass and Largemouth Bass in Two Unproductive Lakes in Northern Michigan. *Am. Midl. Nat.* **1974**, *91*, 453–459. [CrossRef]

57. Lyons, J.; Magnuson, J.J. Effects of Walleye Predation on the Population Dynamics of Small Littoral-Zone Fishes in a Northern Wisconsin Lake. *Trans. Am. Fish. Soc.* **1987**, *116*, 29–39. [CrossRef]

58. Margenau, T.L.; Rasmussen, P.W.; Kampa, J.M. Factors Affecting Growth of Northern Pike in Small Northern Wisconsin Lakes. *N. Am. J. Fish. Manag.* **1998**, *18*, 625–639. [CrossRef]

Article

Nondestructive Monitoring of Soft Bottom Fish and Habitats Using a Standardized, Remote and Unbaited 360° Video Sampling Method

Delphine Mallet [1], Marion Olivry [2], Sophia Ighiouer [3], Michel Kulbicki [4] and Laurent Wantiez [5,*]

[1] VISIOON, 6 Rue du Docteur Fruitet, 98800 Nouméa, New Caledonia; d.mallet@visioon.nc
[2] CNAM Intechmer, Boulevard de Collignon, 50110 Tourlaville, France; marionolivry4@gmail.com
[3] École Pratique des Hautes Études, PSL University, Les Patios Saint-Jacques, 4-14 Rue Ferrus, 75014 Paris, France; sophia.ighiouer@hotmail.fr
[4] Institut de Recherche pour le Développement (IRD), UMR Entropie, Université de Perpignan, 66000 Perpignan, France; michel.kulbicki@ird.fr
[5] Entropie, University of New Caledonia, 98851 Nouméa, New Caledonia
* Correspondence: laurent.wantiez@unc.nc

Citation: Mallet, D.; Olivry, M.; Ighiouer, S.; Kulbicki, M.; Wantiez, L. Nondestructive Monitoring of Soft Bottom Fish and Habitats Using a Standardized, Remote and Unbaited 360° Video Sampling Method. *Fishes* **2021**, *6*, 50. https://doi.org/10.3390/fishes6040050

Academic Editors: Maria Angeles Esteban, Bernardo Baldisserotto and Eric Hallerman

Received: 23 September 2021
Accepted: 13 October 2021
Published: 15 October 2021

Publisher's Note: MDPI stays neutral with regard to jurisdictional claims in published maps and institutional affiliations.

Abstract: Lagoon soft-bottoms are key habitats within coral reef seascapes. Coral reef fish use these habitats as nurseries, feeding grounds and transit areas. At present, most soft-bottom sampling methods are destructive (trawling, longlining, hook and line). We developed a remote, unbaited 360° video sampling method (RUV360) to monitor fish species assemblages in soft bottoms. A low-cost, high-definition camera enclosed in a waterproof housing and fixed on a tripod was set on the sea floor in New Caledonia from a boat. Then, 534 videos were recorded to assess the efficiency of the RUV360. The technique was successful in sampling bare soft-bottoms, seagrass beds, macroalgae meadows and mixed soft-bottoms. It is easy to use and particularly efficient, i.e., 88% of the stations were sampled successfully. We observed 10,007 fish belonging to 172 species, including 45 species targeted by fishermen in New Caledonia, as well as many key species. The results are consistent with the known characteristics of the lagoon soft bottom fish assemblages of New Caledonia. We provide future users with general recommendations and reference plots to estimate the proportion of the theoretical total species richness sampled, according to the number of stations or the duration of the footage.

Keywords: underwater video; ichthyofauna; seagrass bed; macroalgae soft substrate; perireefal

1. Introduction

Soft bottom habitats constitute a major part of the coral reef seascape. These habitats make up extensive areas of mud, sand or rubble that marine plants can colonize [1,2]. In a lagoon environment, they are the key corridors between coral reefs, playing an essential role in ensuring connectivity and energy transfer within a mosaic of reef and perireefal habitats [3,4]. Many fish species, along with several emblematic species such as sea turtles or dugongs, use these habitats. Fish use such habitats as nurseries, feeding grounds or transit areas [5,6]. This very complex seascape is under increasing anthropogenic pressure, in particular due to the growing population and increased impacts such as fishing, coastal development, tourism, inputs from watersheds, the transformation of coastal landscapes and marine aquaculture.

Few studies have been devoted to soft bottom habitats compared to the other ecosystems of this seascape such as coral reefs or mangroves [7]. One of the main reasons for this is that soft bottom fish assemblages are difficult to sample, as individuals are scattered over very large areas and are often associated with significant depths. Most of the available data come from experimental fishing (essentially hook and line or trawl) or fish landings (e.g., [7–11]), which are extractive methods and present the typical problems of representativeness, sensitivity and repeatability. While standardized and nondestructive sampling methods such as underwater visual censuses (UVC) are extensively used on coral reefs, these approaches are not adapted to soft bottom habitats because of the low occurrence of fish, specific fish behavior as well as the extent or the depth of these habitats. In New Caledonia, soft bottom fish assemblages are, at present, known only from earlier programs based on experimental catch data [11–16] and fisheries survey data [17–19].

The recent development of underwater video systems [20,21] provides an opportunity to develop a standardized method to monitor fish assemblages over large areas such as soft bottom habitats. This tool has the advantage of being nondestructive for the environment, has little influence on fish behavior, and can record for long periods at various frequencies. Different video techniques exist to sample fish such as remote underwater video, whether baited or not, diver-operated video or towed video (see [21] for a review on video techniques). At present, the most widely used approach in perireefal habitats is the 'BRUV' technique (Baited Remote Underwater Video) that attracts fish around the camera with bait (see [22–25] for applications on seagrass beds). In New Caledonia, video systems have been mainly used in censuses of coral reef fish [26–28] or sharks [29,30]. Pelletier et al. (2012) used video techniques on soft bottom habitats, but the performance of the method (required number and duration of videos) was not tested.

Pilot studies on method efficiency are important to validate and optimize sampling methods as part of developing cost effective and statistically robust monitoring programs. However, most sampling designs based on video techniques are used without such pilot studies, which may compromise their results [31–33]. Considerable variability in sampling times and number of replicates characterize published studies [21]. Recently, Garcia et al. (2021) studied the possible trade-off between the number and the length of remote videos used in a rapid assessments of reef fish assemblages. With 46 videos on five sites, they indicated that increasing the sampling coverage in the reef area may be more effective than just extending the video length.

The objective of this study was to perform a pilot study to present a standardized sampling protocol to monitor the diversity, abundance and structure of perireefal fish assemblages during daytime, in relation to the environment. We used a remote and unbaited 360° video system (RUV360). The 360° camera records simultaneously all the area around each sampling point. The aim of this pilot study was to assess: (1) the limits of the RUV360 sampling method (cost, visibility, current, bottom topography); (2) the fish species targeted by the technique; and (3) the optimal recording time per station and the number of stations required to obtain representative, stable and reproducible data on perireefal fish communities.

2. Material and Methods

2.1. Study Area and Sampling Design

The main island of New Caledonia is one of the largest coral reef lagoons in the world (19.385 km^2). It includes 16,874 km^2 of nonreef habitat, with certain areas listed as a UNESCO World Heritage site [34]. This very complex seascape is under increasing anthropogenic pressure, in particular due to the growing population (268,767 inhabitants in 2014 compared to 230,789 in 2004; www.isee.nc (accessed in 2019)), and increased impacts such as fishing, coastal development, tourism, mining and marine aquaculture. The study was conducted from the 3 May until the 18 July 2018, in the Southwest Lagoon of New Caledonia. The study area is an 18.5 km long and 4 km wide transect from the coastline to the barrier reef (Figure 1). This area is representative of the coral reef seascape of the Main Island, near Nouméa, the capital city. The lagoon includes 67.5 km^2 of soft bottom habitats and two rows of coral reefs and coralline islets along a shore-barrier reef gradient. Coral heads are scattered on the lagoon bottom. Habitats with more than 50% hard substrate were excluded from the sampling.

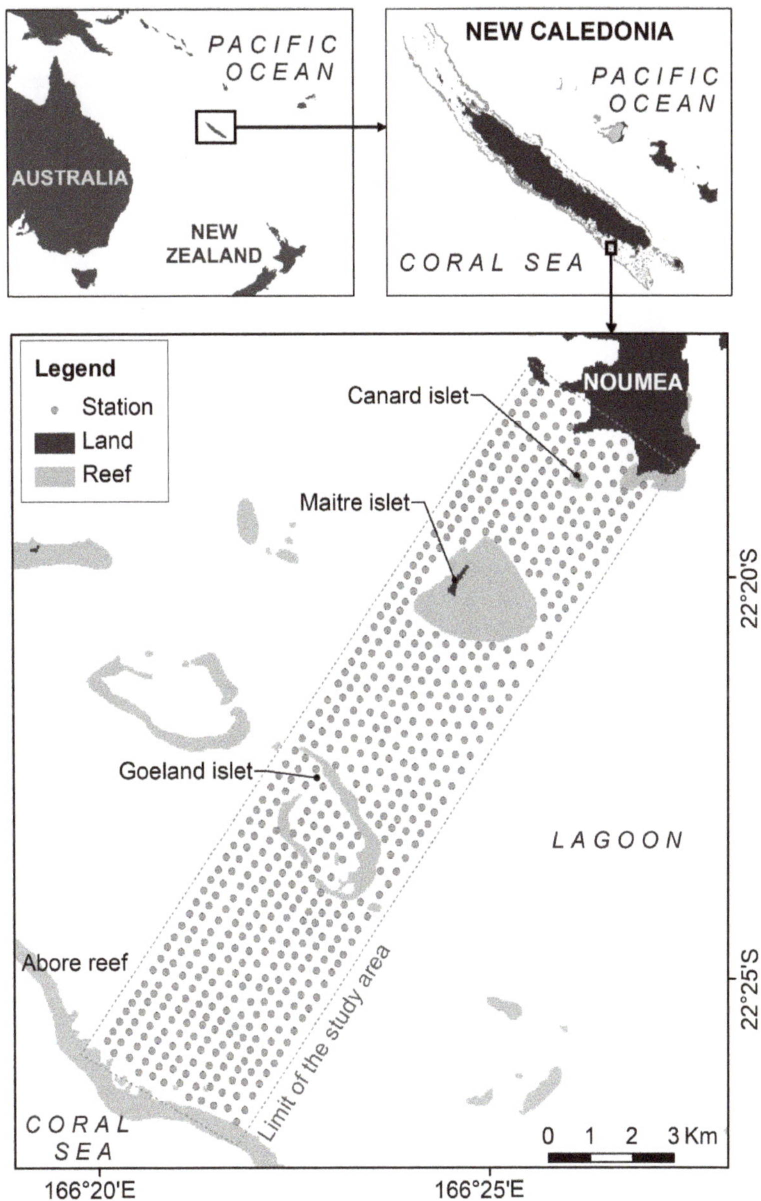

Figure 1. Studied area and sampling design. Each dot represents a station.

In order to assess the optimal recording time for each station and the number of stations required to get representative and reproducible data on soft bottom fish assemblages, we had to oversample the area. A systematic sampling protocol including 609 stations within a grid of 300 m wide cells was used. The distance between stations was sufficient to avoid overlap due to fish swimming from one station to another. The stations were sampled during daylight, at least one hour after sunrise and one hour before sunset, to avoid possible crepuscular variation in fish assemblages [35].

2.2. Sampling Technique and Images Analysis

This study used an autonomous, remote and unbaited video technique named "RUV360" (Figure 2). The camera was a low-cost camera (250€) from KODAK (model PIXPRO SP360 4K) which can record videos in very high definition (1440 × 1440 pixels, 30 fps), featuring a spherical lens with a 360° horizontal and a 235° vertical view, pointed directly upward (Figure 2). The camera was enclosed in a waterproof housing (limited to 60 m depth, 50€), fixed to an aluminum tube 17 cm above the seafloor. A tripod system was used to position and stabilize the camera on the sea floor (Figure 2). The video system was deployed from a boat without the need for the crew to enter the water. This method allowed us to maximize the number of observations while minimizing disturbance to the environment (no boat and no human were present near the video system during the recordings). To evaluate the minimum recording duration required to have representative observations, we fixed the recording duration at 25 min. This time was sufficient to observe sedentary fish and then assess the amount of additional information (passing fish) obtained over time.

Figure 2. Picture of the remote underwater video system on the seabed.

To optimize sampling at sea, we used four video systems deployed by two people aboard a small boat (<8 m). After each sampling day at sea, all videos were checked to assess (1) an appropriate field of view (visibility > 5 m), (2) an appropriate orientation of the camera allowing for a clear view of the seabed, (3) a stable camera during filming, and (4) that the habitat sampled was mainly soft (<50% of hard bottom). When a video was found to be invalid, a second attempt was made the following day.

All videos were analyzed by the same experienced observer using the camera software (Kodak Pixpro SP360 PC software, v1.7.0). The habitat was characterized by estimating the percentage of abiotic and biotic coverage over the 360° images using the "MSA" protocol [36]. The abiotic cover was classified as bare sediment (mud, sand, gravel and small boulders < 30 cm) or nonliving hard substrate (dead corals, coral slab, blocks > 30 cm). The biotic cover (live substrate) was classified as live corals (carbonated edifices that were still in place and present a coral shape) or "marine plants" (seagrass and macroalgae). The videos did not allow us to differentiate systematically between seagrass (*Cymodocea* sp., *Halophila* sp., *Halodule* sp., *Syringodium* sp., *Thalassia* sp.) and macroalgae (*Caulerpa* sp., *Halimeda* sp., *Lobophora* sp., *Sargassum* sp., *Turbinaria* sp.). All fish were counted and identified at the lowest possible taxonomic level. To avoid counting the same fish several times, we used a conservative measure of relative abundance: "MaxN" [37]. This measure of abundance is the maximum number of individuals of the same species appearing at the same time throughout the entire video. To study the influence of camera soak time on species composition and abundances, we calculated MaxN (by species) every 30 sec. This protocol made it possible to study the number of new species and new individuals observed within each time interval. Some fish species from the same genus are similar and only differ by small details (eye color, small color dots, etc.). These species are therefore difficult to distinguish on videos unless they are close enough to the camera. For our video analyses, we aggregated these species into groups: (i) *Amphiprion* gp for *Amphiprion akindynos* and *Amphiprion clarkii*; (ii) *Lethrinus* gp for *Lethrinus variegatus* and *Lethrinus genivittatus*; (iii) *Nemipterus* gp for *Nemipterus peronii*, *Nemipterus furcosus* and *Nemipterus zysron*; (iv) *Parapercis* gp for *Parapercis australis* and *Parapercis millepunctata*, and (v) *Pomacentrus* gp for *Pomacentrus amboinensis* and *Pomacentrus moluccensis*.

2.3. Sampling Cost

We estimated the cost of sampling by the time required for fieldwork and video analysis. The total time required for fieldwork each day included preparing the boat, the trip to the sampling area and the time spent within the sampling area (setup of the video systems, deployment and retrieval of video systems, travel between stations). The time required to characterize the habitat, then identify and count the macrofauna on the videos, was noted for each video during the video analysis.

2.4. Data Analysis

2.4.1. Typology of the Habitat and Fish Assemblages

We selected all stations composed of less than 50% hard bottoms for our study on soft bottom habitats. To identify the typology of the habitat, we performed a principal component analysis (PCA) on raw data and a hierarchic ascending classification (HAC) on the first three axes of the PCA (100% of the inertia), using the squared Euclidean distance and Ward's aggregation method [38].

In order to verify the discriminating nature of the type of soft bottom on the fish assemblages, a CAP (canonical analysis of principal coordinates) was carried out on the Bray Curtis similarity matrix between the stations according to species abundance, using habitat type as a classifier. We applied a square root transformation on the dataset prior to analysis to down-weight the importance of the outlier species [39]. The results of the CAP were validated by a PERMANOVA (999 permutations).

2.4.2. Influence of Soak Time and Number of Stations Sampled on Fish Assemblages

The relationship between soak time and the number of species or individuals recorded was modelled using species accumulation curves and cumulative abundance curves. Species richness and abundance were calculated at 30 s intervals until the 25 min of soak time elapsed, for the entire area and per habitat.

The species accumulation models used a rarefaction method based on raw data added in ascending order. The rarefaction model known as Mao Tau's estimate [40] is a powerful tool for detecting species richness [41]. Abundance accumulation models used the time required to reach MaxN at each station added in ascending order. The estimate of the theoretical total number of species or individuals in the area studied was calculated by fitting a nonlinear Michaelis-Menten model [42] (the most accurate of the models tested) to the accumulation data: y = (Vm × t)/(K + t), where y is the number of species or individuals after t min of recording, "Vm" is the theoretical total number of species or individuals in the study area, and "K" is the number of stations where half of the theoretical total number of species or individuals have been detected in the videos.

We calculated the proportion of the theoretical species richness (SR) according to the number of stations and the duration of the footage. These proportions were calculated as the average of the SR obtained by 180 s intervals using 999 draws (without replacement) of the required number of stations in the overall data set (534 stations).

3. Results

3.1. Sampling Cost

We validated 534 stations out of the 609 stations of the sampling protocol in the area, between 1 and 25 m depth (mean ± SE = 12.9 ± 0.3 m). Fifty stations, located in a coral habitat (more than 50% of live coral), were excluded from the study. It was not possible to position the camera correctly at 58 stations due to the relief of the seabed. The visibility of the water was too low for 8 stations and the current was too high for 52 stations (especially in the channels). Depending on wind, wave and depth conditions, the preparation of the boat and the trips took between 19 min and 113 min (mean ± SE = 46 min ± 3 min) (Table 1). Each 25 min of video required an additional of 10 min to set up, deploy and retrieve the video system. This time was reduced by using four RUV360 systems simultaneously, resulting in a total time of 40 min to 92 min to sample a set of four stations (mean ± SE = 40 min ± 3 min). The variations in time are mostly due to the requirement of correct positioning of the system on the seabed (depending on the percentage of hard corals, the depth and the relief) and the distance between two stations. The analyses of the 534 videos took 425 h in all. The time to analyze one video was between 24 min and 78 min (mean ± SE = 49 min ± 3 min), depending on the complexity (number of species and abundances) of the biodiversity in the video.

Table 1. Sampling cost. Min, max and mean (± SE) correspond to the time required per day in minutes for fieldwork preparation, per set of four stations and per station for video analysis. Totals correspond to the time required to sample and analyze the 534 videos of the study.

Time Required (Min)	Fieldwork		Analysis of One Video
	Daily Preparation of Boat and Material + Trips to the Sampling Area	Sampling a Set of 4 Stations	
Min	19	40	24
Max	113	92	78
Mean ± SE	46 ± 3	40 ± 3	49 ± 3
Total	1123	7839	25,494

3.2. Typology of the Habitat

The stations were mainly composed of bare sediment and marine plants. Overall, 31 stations were almost exclusively composed of bare sediment (more than 90% of the habitat), and 66 were almost exclusively composed of marine plants (more than 90% of the habitat); 119 stations had living corals, which never exceeded 35%, and nonliving hard substrate (max 20%) was present at 52 stations.

It was possible to identify three habitats in the studied area (Figure 3). The "vegetated soft bottom habitat" (317 stations) was dominated by marine plants (from 52% to 100%) and very little hard substrate (from 0% to 10% of living corals and from 0% to 5% of nonliving hard substrates). The "bare soft bottom habitat" (160 stations) was dominated by bare sediments (from 50% to 100%), very little hard substrate (from 0% to 10% of living corals and from 0% to 5% of nonliving hard substrates) and a lower percentage of marine plants (from 0% to 50%). The "mixed soft bottom habitat" (57 stations) was characterized by hard substrate (from 10% to 40%), including nonliving hard substrate (from 0% to 20%) and/or scattered living corals (from 0% to 35%).

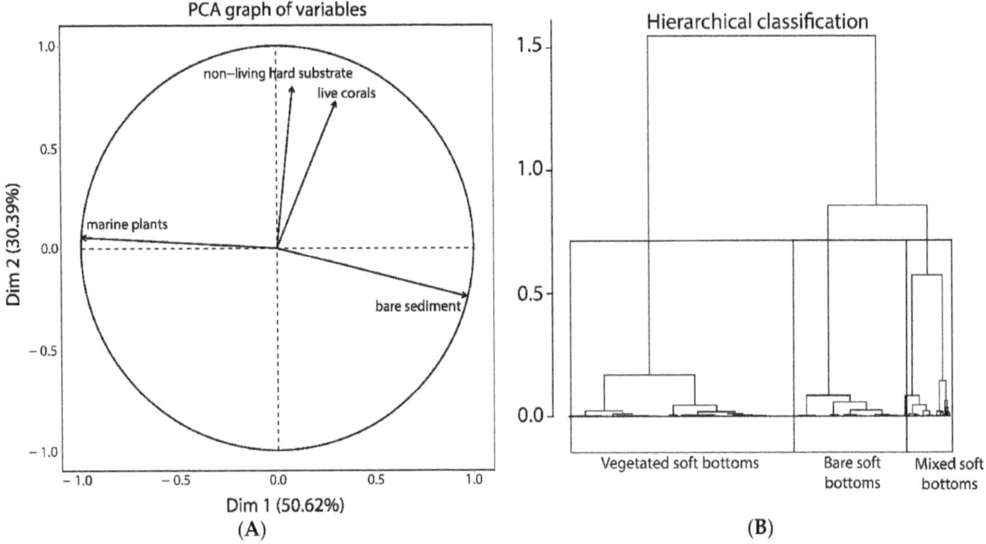

Figure 3. PCA of the habitats characteristics per station (**A**) and typology of the habitat (**B**).

3.3. Fish Assemblages

In all, 10,007 fish belonging to 172 species (98 genera and 37 families) were observed (Supplementary Materials Table S1); 3534 fish (26% of the fish) observed at 330 stations (62% of the stations) could not be identified, because they were too small (1774-50%), were located in the upper water column (607-17%) or were at the limit of detectability (506-14%). The rest of the unidentified fish showed no distinctive signs (361-10%), were blurred (260-8%) or swam too quickly (26-1%) to be identified.

Among the fish identified, the most frequent and abundant families were the Lethrinidae (frequency of occurrence (freq) = 33%, MaxN summed across all deployments (total MaxN) = 992), Pomacentridae (freq = 26.8%, total MaxN = 3390), Labridae (freq = 26.4%, total MaxN = 1175) and Mullidae (freq = 25.8%, total MaxN = 811). Most of the species were carnivores (99 species, 4184 fish). Plankton feeders were second in terms of MaxN (3866 fish), but were also the least diverse (18 species) (Table 2).

Table 2. Number of families, genera, species and abundance of fish (MaxN) per trophic group.

Trophic Group	Families	Genera	Species	MaxN
Carnivores	22	58	99	4184
Herbivores-detritus	7	14	29	1507
Piscivores	7	18	26	450
Plankton feeders	7	12	18	3866

On average, the video recorded 4.1 species and 19 fish per station for the full 25 min of deployment (Table 3). There were important variations between stations, from no fish at 119 stations to a maximum of 28 species and 269 fish at one station. Commercial species made up 29% of the fish species per station and 33% of the individuals per station. The most diverse (34% of the commercial species) and abundant (30% of the MaxN of the commercial fish) commercial fish were Lethrinidae. Scaridae (21% of the species and 23% of the MaxN of commercial fish), Carangidae (11% of the species and 15% of the MaxN of commercial fish) and Acanthuridae (11% of the species and 8% of the MaxN of commercial fish) followed in order of importance.

Table 3. Mean specific richness and abundance per station ($\pm$SE) for all the ichthyofauna, for the commercial species and for the 4 more frequent commercial families.

	Species Richness per Station	Abundance per Station (MaxN)
Total ichthyofauna	4.1 $\pm$ 0.2	19.0 $\pm$ 1.4
Commercial species	1.2 $\pm$ 0.1	6.3 $\pm$ 0.6
Lethrinidae	0.41 $\pm$ 0.03	1.86 $\pm$ 0.20
Scaridae	0.25 $\pm$ 0.03	1.44 $\pm$ 0.24
Carangidae	0.13 $\pm$ 0.02	0.95 $\pm$ 0.34
Acanthuridae	0.13 $\pm$ 0.02	0.53 $\pm$ 0.15

The fish species richness and MaxN were significantly influenced by habitat (PERMANOVA, $p = 0.001$). Species richness and MaxN were higher in the mixed soft bottom habitat than in the bare or vegetated soft bottom habitats (paired comparisons, $p < 0.001$). The fish assemblages were different on the three soft bottom habitats (PERMANOVA, $p = 0.001$). A canonical analysis was carried out on the first 42 axes of the analysis in principal coordinates (98.54% of the total inertia) (Figure 4). The CAP was validated ($p = 0.001$) and indicated an overall percentage of correct and stable classification of 63%. First, the model discriminated mixed soft bottoms communities (88% correct classification). The discrimination of the assemblages in the two other habitats was lower, i.e., 59% on the vegetated soft bottoms and 59% on the bare soft bottoms. These assemblages shared more similarities (75% misclassification between them). The mixed soft bottom fish assemblage was the most diverse. This assemblage was characterized by the presence of hard bottom species associated with corals, such as damselfish (*Dascyllus aruanus* and unidentified damselfishes), butterfly fish (*Chaetodon mertensii*), angelfish (*Centropyge tibicen*), parrotfish (*Chlorurus sordidus*, *Scarus schlegeli* and unidentified parrotfish), one wrasse (*Thalassoma lunare*), one triggerfish (*Sufflamen chrysopterum*) and coral trout (*Plectropomus leopardus*). Several ubiquitous species also characterized this community, such as goatfish (*Parupeneus barberinoides*, *Parupeneus multifasciatus*) and sea bream (*Gymnocranius* sp.). The presence of species associated with seagrass beds or algae meadows characterized the vegetated soft bottom fish assemblage, in particular two emperors (*Lethrinus variegatus* and *Lethrinus genivittatus*), one leather jacket (*Paramonacanthus japonicus*) and two wrasses (*Oxycheilinus bimaculatus* and *Suezichthys devisi*). The bare soft bottom fish assemblage was the least diverse. Its main characteristic was the absence of hard bottom species or vegetated soft bottom species. The only fish observed on these bottoms were specimens moving between the other habitats of the lagoon. However, this assemblage was characterized by the presence of spangled emperors (*Lethrinus nebulosus*) which frequent the large areas of the lagoon with a preference for sandy bottoms, where they find their food.

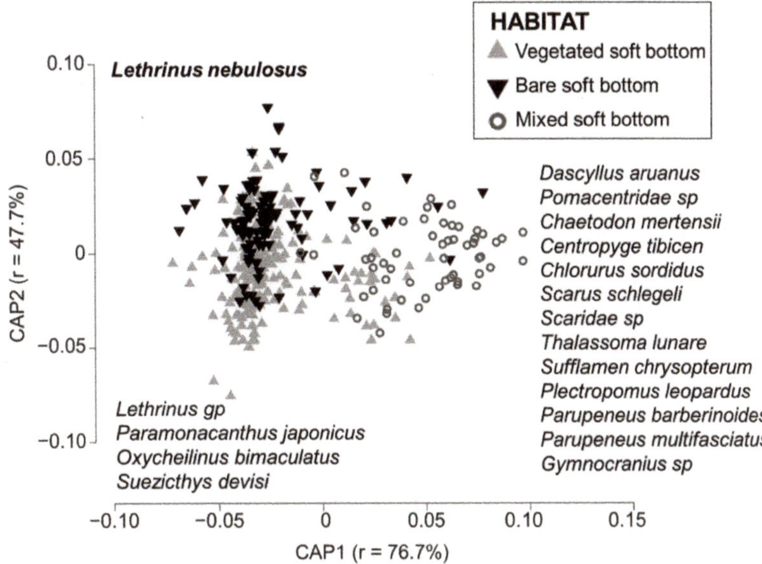

Figure 4. CAP of fish assemblage between stations, under constraint of habitat type. Species with a correlation ≥ 0.35 to the first factorial design are specified for each habitat.

3.4. Influence of Soak Time and Number of Stations Sampled on Fish Assemblages

The deployment duration had a significant effect on the species richness (SR) and abundance (MaxN) observed by station (Friedman test, $p < 0.001$). The average number of species observed per station increased from 1.2 ± 0.6 (SR $\pm$ SE) species with 30 sec of observation to 4.2 ± 1.5 species with 25 min (Figure 5A). The SR was stable after 7.5 min of observation (pairwise comparisons test, $p > 0.05$). The MaxN per station also increased significantly over time (MaxN $\pm$ SE = 8.0 ± 6.6 fish after 30 sec and 18.8 ± 10.5 fish after 25 min) (Friedman test, $p < 0.001$). The MaxN was stable after 1.5 min (pairwise comparisons test, $p > 0.05$). The SR increased very quickly over time at the beginning of the recording (Figure 5B), before dropping progressively to reach an asymptote corresponding to the total theoretical species richness according to the footage duration (Michaelis-Menten model, theoretical SR-time = 173 species) in the study area: 80% of the theoretical SR-time was observed after 5 min and 95% of the theoretical SR-time after 14 min (Table 4). The theoretical SR-time was not significantly different between habitats (Chi-squared test, $p > 0.05$). Within the vegetated and mixed soft bottoms, SR progressed very quickly at the beginning of the recordings: 80% of the theoretical SR-time was observed after 5 min on the vegetated soft bottoms and 4.5 min on the mixed soft bottoms (Table 4). In contrast, the SR on the bare soft bottoms increased more slowly at the beginning of the recordings: 11 min were necessary to observe 80% of the theoretical SR-time on this habitat. However, 95% of the theoretical SR-time on the bare soft bottoms was observed within 17 min, which was only 1.5 to 2.5 min more than for the other soft bottom habitats.

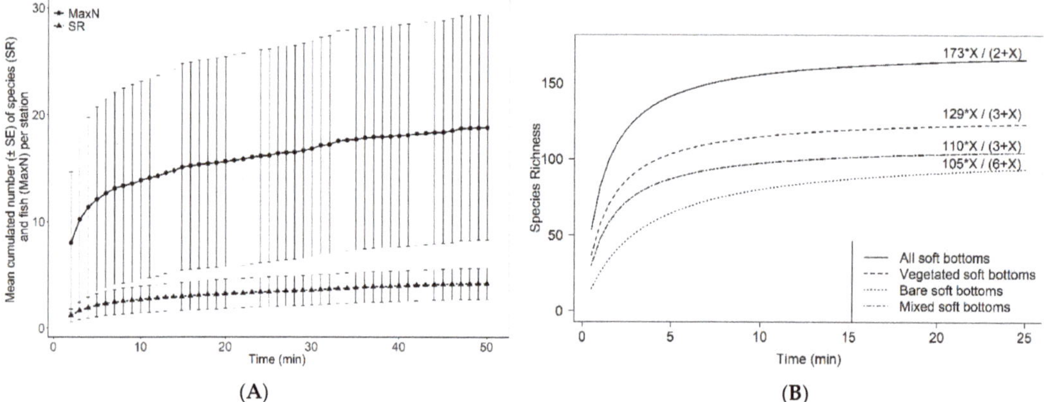

Figure 5. (**A**) Mean cumulated number (±SE) of species and fish per station in 30 sec time intervals up to 25 min. (**B**) Species accumulation curves in 30 s time intervals up to 25 min for all soft bottoms and by habitat (see legend). Equation of the curve for each habitat is given on the corresponding curves.

Table 4. Deployment duration necessary to observe 50, 80, 85, 90, 95% of the theoretical SR-time. Deployment durations were evaluated from the accumulation curves calculated as a function of time over all the stations and by habitat.

Proportion of the Theoretical SR-Time (%)	Deployment Duration			
	All Soft Bottoms	**Bare Soft Bottoms**	**Vegetated Soft Bottoms**	**Mixed Soft Bottoms**
50	1 min 06	3 min 15	1 min 15	1 min 18
80	5 min 00	11 min 00	5 min 00	4 min 30
85	7 min 00	14 min 00	9 min 00	7 min 30
90	10 min 00	15 min 30	11 min 30	10 min 30
95	14 min 00	17 min 00	14 min 30	15 min 30

There was no significant link between the number of stations and the estimates of SR or MaxN observed per station (Spearman correlation, $p > 0.05$). Indeed, the mean number of species observed per station was relatively stable, regardless of the number of stations sampled. It varied from 3.9 species on average per station with 2 stations to 4.1 species on average per station with 534 stations (Supplementary Materials Figure S1). On the other hand, the standard error (SE) decreased significantly as the number of stations increased (SE for 2 stations = 2.5 and SE for 534 stations = 0.2). The average abundance (MaxN) per station followed the same trend. It was relatively stable and ranged, on average, from 19.0 fish per station for 2 stations to 18.9 fish per station for 534 stations (Supplementary Materials Figure S1). The SE of relative abundance per station also decreased significantly as the number of stations increased (SE for 2 stations = 13.4 and SE for 534 stations = 1.4). The SR gradually increased depending on the number of stations sampled (Figure 6). The total theoretical SR according to the stations sampled (theoretical SR-station) estimated by the model within the study area was 195 species. Eighty percent of the theoretical SR-station was observed with 369 stations (6.2 stations/km^2 in the study area), while 88% was observed with all the stations sampled (534 stations or 9 stations/km^2) (Table 5). The theoretical SR-station was not significantly different between habitats (Chi-squared test, $p > 0.05$). Within the vegetated and mixed soft bottoms, SR progressed more quickly than for bare soft bottoms: 80% of the theoretical SR-station was observed with 265 stations (7.5 stations/km^2) on the vegetated soft bottoms and 70 stations (11.1 stations/km^2) on the mixed soft bottoms (Table 5). Again, in contrast, the SR on the bare soft bottoms increased more slowly depending of the number of stations sampled: 320 stations (17.9 stations/km^2) were necessary to observe 80% of the theoretical SR-station on this habitat.

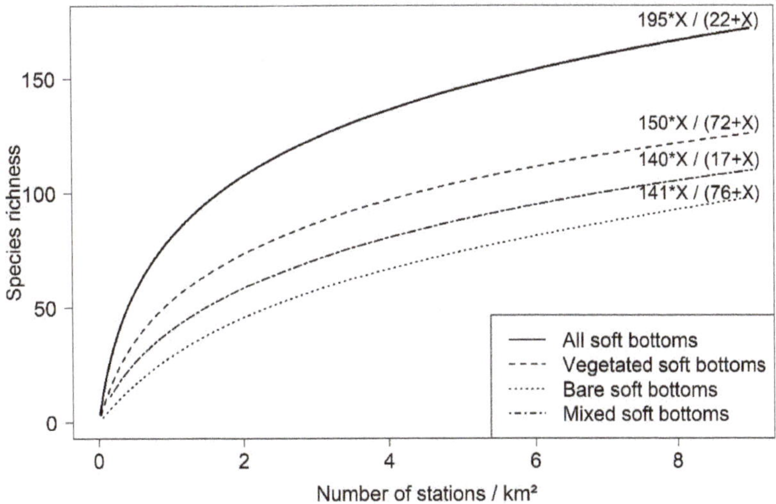

Figure 6. Species accumulation curves per number of stations/km² for all soft bottoms and by habitat (see legend). Equation of the curve for each habitat is given on the corresponding curves.

Table 5. Number of stations per km² required to observe 50, 80, 85, 90 and 95% of the theoretical SR-Scheme 25. min.

Proportion of the Theoretical SR-Station. (%)	Number of Stations per km²			
	All Soft Bottoms	**Bare Soft Bottoms**	**Vegetated Soft Bottoms**	**Mixed Soft Bottoms**
50	1.6	4.3	2	2.7
80	6.2	18	7.5	11.1
85	7.8	24.4	11.5	15.9
90	14.1	38.8	18.2	24.6
95	29.5	83.7	38.4	55.6

4. Discussion

An unbaited video technique was selected because it did not attract fish to the camera. Using bait to attract fish would modify the fish assemblage because fish species react differently to bait [21,33,43,44]. The objective was to get a less biased representation of the assemblage during daytime. A 360° video technique was selected to sample in all directions simultaneously and record all fish in the sampling area.

4.1. Fieldwork Implementation and Costs

The RUV360 was easy to use and particularly efficient, since 88% of the initially selected stations were successfully sampled. The approach appears to be more efficient than other unbaited, multidirectional video systems. For instance, the "STAVIRO" (rotating video system), described by Pelletier et al. [26], for use on hard- and soft-substrate habitats successfully sampled 70% of the stations during a pilot study and reached 81% validation in a subsequent studies. More recently, the "compact video lander" developed by Watson and Huntington [45] was used on rocky reefs, and successfully sampled 70% of the stations. When deploying video systems directly from a boat, one of the main causes of nonvalidation is an inappropriate orientation of the camera towards the seabed. Only 3.1% of nonvalidated stations of the present study were attributed to seafloor relief issues. Such problems were reduced with the RUV360 because (1) we targeted only soft bottoms which are less complex than hard substrate, and (2) the camera had a 235° vertical field of view (V-FOV), compared to cameras generally used in other video techniques (BRUV, RUV or

STAVIRO), that have a V-FOV of 60° for the wide angle lens of the latest Sony (specification of the model FDR-AX700 on www.sony.com) and 94.4° for the wide angle lens of the latest Gopro (specification of the model HERO8 Black on www.gopro.com). The RUV360 was also efficient in terms of other typical causes of nonvalidation. It was particularly stable (only 0.3% of nonvalidated stations were attributed to its instability) and could be used in channels where tide currents occurred. The impact of a low visibility was limited because the FOV was good (only 0.7% of nonvalidated stations were attributed to the visibility).

The cost associated with the use of the RUV360 method was evaluated as a combination of the time required for sampling and video analyses. Fieldwork was estimated for the simultaneous deployment of four RUV360s within a systematic sampling grid of 300 m width and a 25 min video recordings per station using one boat (<8 m, two persons minimum). The RUV360 appears to be an efficient alternative to other video systems, although comparisons are complicated, as very few studies provided cost information related to the use of their video system. From a literature review on video techniques, we found four studies providing details on the performance of their video systems: Pelletier et al. [26] for STAVIRO, Gladstone et al. [31], Santana-Garcon et al. [46] and Langlois et al. [47] for BRUVs. The size of the boat (small boat between 5 and 10 m), the number of persons required at sea (two persons minimum) and video analysis (one person assisted by experts as required) were common to all approaches. The number of stations sampled per day varied between studies (from 10 to 30 stations/day) depending on the number of systems used simultaneously, the duration of the footage and the distance between stations. Two to ten video systems were used per boat, with footage lasting from 9 min [26] to 180 min [46] and distance between stations varying from 200 to 500 m. The time required to analyze videos depends on the complexity of the habitat, as well as the diversity and abundance of fish. The analysis of RUV360 video was faster (49 min for a 25 min video on average, corresponding to 2 min per minute of video) than for STAVIRO video (43 min for a 9 min video on average, corresponding to 4 min 47 per minute of video) [26], because all fish present are visible within one frame, whereas six sectors of 60° are necessary for STAVIRO to get a 360° frame. The RUV360 takes longer to analyze than the BRUV (65 min for a 60 min video on average which correspond to 1 min per minute of video [31,46]) because fish are attracted to the camera with BRUV and are easier to identify, whereas greater zooming in is necessary with the RUV360 for species identification. The performance of the RUV360 was also linked to the nature of the videos analyzed, as soft bottom habitats are easier to analyze than complex habitats such as coral reefs.

4.2. Biodiversity Sampled

The RUV360 method was successful at sampling bare soft bottom habitats, seagrass beds, macroalgae meadows and mixed soft bottoms. The fish assemblages were significantly different according to the type of the soft bottom habitat. The differences were mainly driven by the presence of hard substrate, corroborating the observed relationship between the complexity of marine habitats and the composition of fish assemblages [2]. Structurally more diverse habitats are known to sustain fish communities which are more diverse and functionally complex in comparison with habitats with monotonous bare substrates [48]. The fish assemblages were first discriminated in the mixed soft bottoms (88% of correct classification), followed by vegetated or bare soft bottom habitats (59% of correct classification for each habitat). There were no clear boundaries between the vegetated soft bottom and the bare soft bottom assemblages, which form a continuum along a plant density gradient. Within the studied area, marine plants were common (only 3% of the stations had less than 10% plant cover). Therefore, even if bare soft bottom habitats were mainly composed of bare bottom, they also included marine plants to a lesser extent (<50%). The presence of marine plants on these habitats, and their associated species, can explain the difficulty of better discriminating fish assemblages between the vegetated and bare soft bottoms habitats. It appears that fish communities change along a gradient of marine plant abundance.

We recorded 10,007 fish belonging to 172 species (98 genera and 37 families), including 45 species (3365 individuals) targeted by fishing in New Caledonia and many emblematic species such as rays, sharks, turtles and dolphins; 104 sea snakes were also observed in the study area. For video analysis, several species were aggregated into groups, because they are similar in appearance and difficult to distinguish from each other. Grouping species which share specific traits in relation to their habitat, biology, behavior and ecology is common for studies using video techniques [31,46,49,50]. Another group of species seen in videos during this study could not be identified (26%), as they were too small or at the limit of the detectability (too far or too high in the water column from the camera). The observation of cryptic fish such as gobies (Gobiidae) and blennies (Blenniidae) is challenging using video, as they are too small and were often too far from the camera to be identified [21]. These two families represent a large number of species throughout New Caledonia (255 species on reefs and soft bottoms; [51] a number of the unidentified individuals in this study belonged to these two families. The difficulty of undertaking a census of cryptic fish species is not only related to the video analysis technique applied; it has also been reported in other, nonextractive sampling methods such as underwater visual censuses (e.g., [52]). Our results are consistent with previous knowledge of the biodiversity of lagoon soft bottoms in New Caledonia. Invertivores species dominate the assemblages, ahead of herbivores, piscivores and plankton feeders [1,11]. We observed 156 species out of the 542 species (28%) recorded on soft bottoms in New Caledonia using trawls or underwater visual census techniques (MK, pers. comm) [1,11,53]. The videos captured 16 additional species: 10 were hard bottom species observed on mixed soft bottoms, two were ubiquitous species, two were sharks and two were rays. The differences between videos and these other techniques are linked to the study area (location and size) and the techniques themselves. Bottom trawls census fewer hard bottom species because mixed soft bottoms cannot be trawled when the seafloor becomes too irregular, and most large species will avoid the trawl [53]. Video techniques are not adapted to census cryptic species [31,46,49,50]. Pelagic species are more frequently censused using UVCs than video techniques, and these species are seldom targeted by bottom trawls [54].

Consumer grade, spherical camera systems are significantly less expensive than high cost underwater cameras. However, the resolution may be sacrificed for the large field of view. Consequently, the range to which fish are identifiable will likely be reduced compared with high cost standard cameras, and this effect could be species-specific. Previous field tests using underwater benchmarks for distances indicated that we can identify fish at a typical maximum range of 8 m from the camera [47]. As species size also has an impact on detectability, we propose a list of species identifiable on the habitat sampled and have grouped similar species together.

4.3. Optimization of the Sampling Design Using RUV360

In order to optimize the sampling design (recording time per station and number of stations) using the RUV360, we had to collect representative, stable and reproducible data on soft bottom fish communities. During our study, 99% of the theoretical total species richness according to footage duration (="theoretical SR-time") was censused by the RUV360 in the area, using footage of 25 min. This demonstrates that it is not necessary to extend the duration of the footage, as 95% of the theoretical SR-time was observed within 14 min. The duration of footage varies greatly between studies, depending on the video technique used and the purpose of the study (from 8 min to several days) [21,33]. None of the studies referenced here specified the proportion of theoretical SR recorded, according to the duration of the footage taken. Therefore, subsequent results are strongly linked to the length of the selected footage. For example, according to the review on BRUVs by Whitmarsh et al. [33], 32% of BRUV studies used 60 min, 25% used 30 min and 17% used soak times greater than 90 min.

The RUV360 also recorded 88% of the theoretical species richness in the study area according to the station sampled, using nine stations/km^2. Very few studies using video investigated the optimal number of stations required to obtain stable observations of biodiversity, and none of them reported this number in relation to the surface of the area studied. To the best of our knowledge, no experiments have investigated the impact of replicate spacing on observed assemblages ([33] for BRUV). For example, Santana-Garcon et al. [46] gave an optimal sample size of at least eight replicates per treatment in sampling a pelagic fish assemblage with a BRUV technique, while Gladstone et al. [31] concluded that for BRUV, there is no optimal value related to sampling precision, with values needing to be set by researchers according to the specific objective.

When designing a sampling strategy for soft bottom fish communities using the RUV360, it is possible to adapt footage duration and sampling effort. Therefore, it is possible to favor a strategy of either "short videos on many stations", or "long videos on a limited number of stations". Based on the data obtained in this study, we propose two reference plots to help in this process (Figure 7 and Supplementary Materials Figure S2 for reference tables). The choice will be a compromise between achieving acceptable precision, the variables and/or species of interest, and the need to manage costs [31,55].

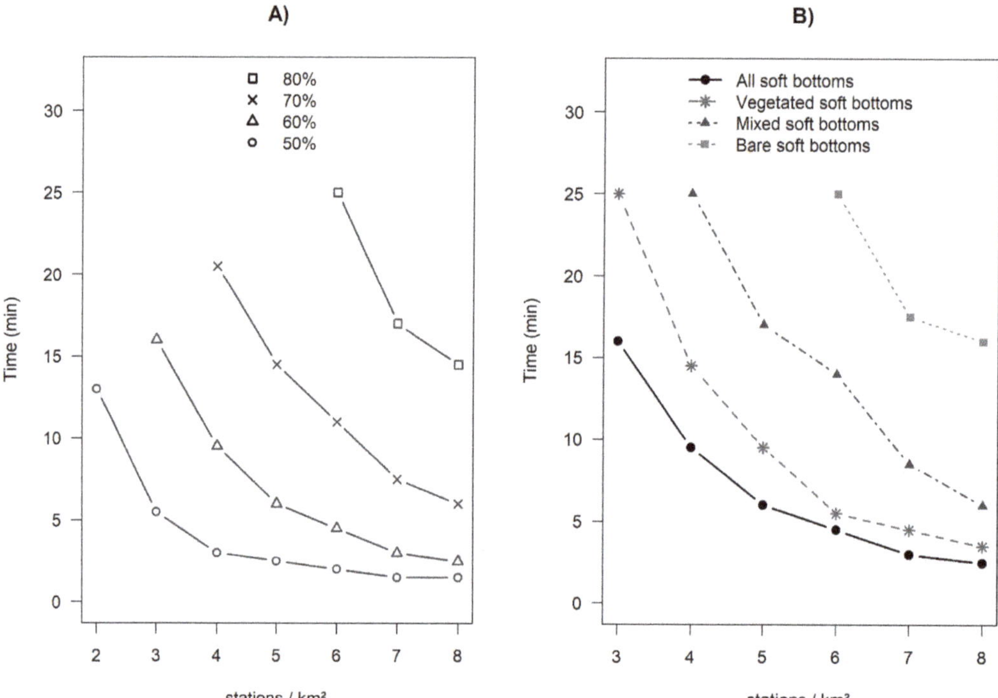

Figure 7. (**A**) Proportion of the theoretical SR-station depending on the duration and the number of video recorded within the studied area. (**B**) 60% of the theoretical SR-station per habitat depending on the duration and the number of video sampled.

5. Conclusions

The results of this study support the proposed sampling protocol to monitor fish communities in perireefal habitats during the daytime. To date, most attention in the scientific literature has focused on reefs, mangroves and seagrass habitats within the coral reef seascape. The sampling protocol described here offers the opportunity to obtain data on perireefal habitats that are comparable in space and time (specific richness, abundance) using a consumer grade 360° video camera. The results are consistent with the known characteristics of the lagoon soft bottoms fish assemblages, and the impacts of irregular seafloors, current and visibility were limited. We provide reference plots to estimate the proportion of the theoretical total species richness sampled, according to the number of stations or the duration of the footage are provided. Further development should include the refinement of the method to collect body-size data from stereo video or other means. Body-size and length data are valuable for a range of ecological studies, from those focused on the impact of fishing to those on ontogenetic shifts of fish assemblages.

Supplementary Materials: The following are available online at https://www.mdpi.com/article/10.3390/fishes6040050/s1, Table S1: List of families and species of fish sampled in the study area. freq: frequency of occurrence; H1: primary habitat; H2: secondary habitat; S: soft bottom; H: hard-bottom; S/H: soft bottom and/or hard-bottom; C: Commercial fish: o (Wantiez pers. comm.). Figure S1: Mean cumulated number (±SE) of species and fish per station depending of the number of stations sampled. "MaxN": Abundance of fish; "SR": species richness. Figure S2. Proportion of the theoretical SR-station depending on the duration and the number of video recorded within the studied area for all soft bottoms and by habitats. See legend for colors.

Author Contributions: Conceptualization, D.M. and L.W.; data collection at sea, D.M. and M.O.; video analysis, D.M., M.O., S.I and L.W; data analysis: D.M., S.I. and L.W.; validation: D.M., M.O., S.I., M.K., L.W.; writing–original Draft Preparation, D.M., M.O., S.I., M.K., L.W.; Project Administration, D.M. and L.W.; Funding Acquisition, D.M. and L.W. All authors have read and agreed to the published version of the manuscript.

Funding: This research was funded by VISIOON, the Foundation of the University of New Caledonia, the University of New Caledonia, the Research Institute for Development (IRD), the Southern Province of New Caledonia and ADECAL-TECHNOPOLE.

Institutional Review Board Statement: Not applicable.

Acknowledgments: We are grateful to Jo Scutt Phillips from the Pacific Community (SPC) for helpful comments and English checking. We thank the Aquarium des lagons of Noumea for its logistical support during the development of the video system. We also sincerely thank the entire team of boat pilots from IRD Noumea, Miguel CLARQUE, Samuel TEREUA and Philippe NAUDIN for their professionalism, their competence and their assistance at sea allowing us to optimize the sampling and quality of the data collected.

Conflicts of Interest: The authors declare no conflict of interest.

References

1. Kulbicki, M.; Vigliola, L.; Wantiez, L. La biodiversité des poissons côtiers. In *Bonvallot Jacques (coord.).Gay Jean-Christophe (coord.). Habert Elisabeth (coord.). Atlas de la Nouvelle Calédonie*; Congrès de la Nouvelle-Calédonie: Marseille, France, 2012; pp. 85–88.

2. Giakoumi, S.; Kokkoris, G.D. Effects of habitat and substrate complexity on shallow sublittoral fish assemblages in the Cyclades Archipelago. North-eastern Mediterranean Sea. *Mediterr. Mar. Sci.* **2013**, *14*, 58–68. [CrossRef]

3. Mumby, P.J. Connectivity of reef fish between mangroves and coral reefs: Algorithms for the design of marine reserves at seascape scales. *Biol. Conserv.* **2006**, *128*, 215–222. [CrossRef]

4. Alsaffar, Z.; Pearman, J.K.; Cúrdia, J.; Ellis, J.; Calleja, M.L.I.; Ruiz-Compean, P.; Roth, F.; Villalobos, R.; Jones, B.H.; Moran, X.A.G.; et al. The role of seagrass vegetation and local environmental conditions in shaping benthic bacterial and macroinvertebrate communities in a tropical coastal lagoon. *Sci. Rep.* **2020**, *10*, 13550. [CrossRef]

5. Fulton, C.J.; Abesamis, R.A.; Berkstom, C.; Depczynski, M.; Graham, N.A.J.; Holmes, T.H.; Kulbicki, M.; Noble, M.M.; Radford, B.T.; Tinkler, P.; et al. Form and function of tropical macroalgal reefs in the Anthropocene. *Funct. Ecol.* **2019**, *33*, 989–999. [CrossRef]

6. Fulton, C.J.; Berkstrom, C.; Wilson, S.K.; Abesamis, R.A.; Bradley, M.; Akerlund, C.; Barrett, L.T.; Bucol, A.A.; Chacin, D.H.; Chong-Seng, K.M.; et al. Macroalgal meadow habitats support fish and fisheries in diverse tropical seascapes. *Fish Fish.* **2020**, *21*, 700–717. [CrossRef]

7. Priester, C.R.; Martinez-Ramirez, L.; Erzini, K.; Abecasis, D. The impact of trammel nets as an MPA soft bottom monitoring method. *Ecol. Indic.* **2021**, *120*, 106877. [CrossRef]

8. Tripp-Valdez, A.; Arreguin-Sanchez, F.; Zetina-Rejon, M.J. The use of stable isotopes and mixing models to determine the feeding habits of soft bottom fishes in the southern Gulf of California. *Cah. Biol. Mar.* **2015**, *56*, 13–23.

9. Sousa, I.; Goncalves, J.M.S.; Claudet, J.; Coelho, R.; Goncalves, E.J.; Erzini, K. Soft-bottom fishes and spatial protection: Findings from a temperate marine protected area. *PeerJ* **2018**, *6*, e4653. [CrossRef] [PubMed]

10. Pérez-Castillo, J.; Barjau-González, E.; López-Vivas, J.M.; Armenta-Quintana, J.Á. Taxonomic diversity of the fish community associated with soft bottoms in a coastal lagoon of the west coast of Baja California Sur. México. *Int. J. Mar. Sci.* **2018**, *9*, 20–29.

11. Wantiez, L. Trophic networks of the soft bottom fish community in the lagoon of New Caledonia. *C. R. Acad. Sci. III* **1994**, *317*, 847–856.

12. Richer de Forges, B.; Bargibant, G. Le lagon nord de la Nouvelle Calédonie et les atolls de Huon et Surprise. *Rapp. Sci. Tech. Cent. Nouméa (Océanogr)* **1985**, *37*, 23.

13. Richer de Forges, B.; Bargibant, G.; Menou, J.L.; Garrigue, C. Le lagon sud-ouest de Nouvelle-Calédonie. Observations préalables à la cartographie bionomique des fonds meubles. *Rapp. Sci. Tech. Sci. Mer. Biol. Mar.* **1987**, *45*, 110.

14. Clavier, J.; Laboute, P. Connaissance et mise en valeur du lagon nord de la Nouvelle Calédonie: Premiers résultats concernant le bivalve pectinidé Amusium japonicum balloti (étude bibliographique, estimation de stock et données annexes). *Rapp. Sci. Tech. Sci. Mer. Biol. Mar.* **1987**, *48*, 73.

15. Kulbicki, M.; Bargibant, G.; Menou, J.L.; Mou-Tham, G.; Thollot, P.; Wantiez, L.; Williams, J. Evaluations des ressources en poissons du lagon d'Ouvéa: 3ème partie: Les poissons. *Rapp. Conv. Sci. Mer. Biol. Mar.* **1994**, *11*, 448.

16. Kulbicki, M.; Labrosse, P.; Letourneur, Y. Fish stock assessment of the northern New Caledonian lagoons: 2—Stocks of lagoon bottom and reef-associated fishes. *Aquat. Living Resour.* **2000**, *13*, 77–90. [CrossRef]

17. Kronen, M.; Sauni, S.; Magron, F.; Fay-Sauni, L. Status of reef and lagoon resources in the South Pacific—The influence of socioeconomic factors. In Proceedings of the 10th International Coral Reef Symposium, Okinawa, Japan, 28 June–2 July 2004; Secretariat of the Pacific Community: Noumea, New Caledonia, 2006; pp. 1185–1193.

18. Labrosse, P.; Ferraris, J.; Letourneur, Y. Assessing the sustainability of subsistence fisheries in the Pacific: The use of data on fish consumption. *Ocean. Coast. Manag.* **2006**, *49*, 203–221. [CrossRef]

19. Guillemot, N.; Léopold, M.; Cuif, M.; Chabanet, P. Characterization and management of informal fisheries confronted with socio-economic changes in New Caledonia (South Pacific). *Fish. Res.* **2009**, *98*, 51–61. [CrossRef]

20. Bicknell, A.W.J.; Godley, B.J.; Sheehan, E.V.; Votier, S.C.; Witt, M.J. Camera technology for monitoring marine biodiversity and human impact. *Front. Ecol. Environ.* **2016**, *14*, 424–432. [CrossRef]

21. Mallet, D.; Pelletier, D. Underwater video techniques for observing coastal marine biodiversity: A review of sixty years of publications (1952–2012). *Fish. Res.* **2014**, *154*, 44–62. [CrossRef]

22. Dalley, K.L.; Gregory, R.S.; Morris, C.J.; Cote, D. Seabed habitat determines fish and macroinvertebrate community associations in a subarctic marine coastal nursery. *Trans. Am. Fish. Soc.* **2017**, *146*, 1115–1125. [CrossRef]

23. Kiggins, R.S.; Knott, N.A.; Davis, A.R. Miniature baited remote underwater video (mini-BRUV) reveals the response of cryptic fishes to seagrass cover. *Environ. Biol. Fishes* **2018**, *101*, 1717–1722. [CrossRef]

24. Kiggins, R.S.; Knott, N.A.; New, T.; Davis, A.R. Fish assemblages in protected seagrass habitats: Assessing fish abundance and diversity in no-take marine reserves and fished areas. *Aquac. Fish.* **2019**, *5*, 213–223. [CrossRef]

25. Zarco-Perello, S.; Enríquez, S. Remote underwater video reveals higher fish diversity and abundance in seagrass meadows, and habitat differences in trophic interactions. *Sci. Rep.* **2019**, *9*, 1–11.

26. Pelletier, D.; Leleu, K.; Mallet, D.; Mou-Tham, G.; Hervé, G.; Boureau, M.; Guilpart, N. Remote High-Definition Rotating Video Enables Fast Spatial Survey of Marine Underwater Macrofauna and Habitats. *PLoS ONE* **2012**, *7*, e30536. [CrossRef] [PubMed]

27. Mallet, D. *Des systèmes vidéo rotatifs pour étudier l'ichtyofaune: Applications à l'analyse des variations spatiales et temporelles dans le lagon de Nouvelle-Calédonie. Ingénierie de l'environnement*; Université de la Nouvelle-Calédonie: Noumea, New Caledonia, 2014.

28. Letessier, T.B.; Juhel, J.B.; Vigliola, L.; Meewig, J.J. Low-cost small action cameras in stereo generates accurate underwater measurements of fish. *J. Exp. Mar. Biol. Ecol.* **2015**, *466*, 120–126. [CrossRef]

29. Juhel, J.B.; Vigliola, L.; Mouillot, D.; Kulbicki, M.; Letessier, T.B.; Meeuwig, J.J.; Wantiez, L. Reef accessibility impairs the protection of sharks. *J. Appl. Ecol.* **2018**, *55*, 673–683. [CrossRef]

30. Juhel, J.B.; Vigliola, L.; Wantiez, L.; Letessier, T.B.; Meewing, J.J.; Mouillot, D. Isolation and no-entry marine reserves mitigate anthropogenic impacts on grey reef shark behavior. *Sci. Rep.* **2019**, *9*, e2897. [CrossRef]

31. Gladstone, W.; Lindfield, S.; Coleman, M.; Kelaher, B. Optimisation of baited remote underwater video sampling designs for estuarine fish assemblages. *J. Exp. Mar. Biol. Ecol.* **2012**, *429*, 28–35. [CrossRef]

32. Schmid, K.; Reis-Filho, J.A.; Harvey, E.; Giarrizzo, T. Baited remote underwater video as a promising nondestructive tool to assess fish assemblages in Clearwater Amazonian rivers: Testing the effect of bait and habitat type. *Hydrobiologia* **2017**, *784*, 93–109. [CrossRef]

33. Whitmarsh, S.K.; Fairweather, P.G.; Huveneers, C. What is Big BRUVver up to? Methods and uses of baited underwater video. *Rev. Fish. Biol. Fish.* **2017**, *27*, 53–73. [CrossRef]

34. Andréfouët, S.; Cabioch, G.; Flamand, B.; Pelletier, B. A reap-praisal of the diversity of geomorphological and genetic processes of New Caledonian coral reefs: A synthesis from optical remote sensing, coring and acoustic multibeam observations. *Coral Reefs* **2009**, *28*, 691–707. [CrossRef]

35. Mallet, D.; Vigliola, L.; Wantiez, L.; Pelletier, D. Diurnal temporal patterns of the diversity and the abundance of reef fishes in a branching coral patch in New Caledonia. *Austral Ecol.* **2016**, *41*, 733–744. [CrossRef]

36. Clua, E.; Legendre, P.; Vigliola, L.; Magron, F.; Kulbicki, M.; Sarramegna, S.; Labrosse, P.; Galzin, R. Medium scale approach (MSA) for improved assessment of coral reef fish habitat. *J. Exp. Mar. Biol. Ecol.* **2006**, *333*, 219–230. [CrossRef]

37. Cappo, M.; Harvey, E.; Malcolm, H.; Speare, P. Potential of video techniques to monitor diversity, abundance and size of fish in studies of Marine Protected Areas. Aquatic Protected Areas—What Works Best and How Do We Know? In Proceedings of the World Congress on Aquatic Protected Areas, Cairns, Australia, 17 August 2002; pp. 455–464.

38. Lebart, L.; Morineau, A.; Piron, M. *Statistique Exploratoire Multidimensionnelle*, 2nd ed.; Dunod: Paris, France, 1997; p. 456.

39. Clarke, K.R.; Gorley, R.N.; Somerfield, P.J.; Warwick, R.M. *Change in Marine Communities: An Approach to Statistical Analysis and Interpretation*, 3rd ed.; Plymouth Primer-E Ltd.: Plymouth, UK, 2014; p. 262.

40. Colwell, R.K.; Chao, A.; Gotelli, N.J.; Lin, S.Y.; Mao, C.X.; Chazdon, R.L.; Longino, J.T. Models and estimators linking individual-based and sample-based rarefaction, extrapolation and comparison of assemblages. *J. Plant Ecol.* **2012**, *5*, 3–21. [CrossRef]

41. Chiarucci, A.; Bacaro, G.; Rocchini, D. Discovering and rediscovering the sample-based rarefaction formula in the ecological literature. *Community Ecol.* **2008**, *9*, 121–123. [CrossRef]

42. Colwell, R.K.; Coddington, J.A. Estimating terrestrial biodiversity through extrapolation. *Philos. Trans. R. Soc. Lond. B.* **1994**, *345*, 101–118.

43. Martinez, I.; Jonez, E.G.; Davie, S.L.; Neat, F.C.; Wigham, B.D.; Priede, I.G. Variability in behaviour of four fish species attracted to baited underwater cameras in the North Sea. *Hydrobiologia* **2011**, *670*, 23–34. [CrossRef]

44. Dunlop, K.M.; Marian Scott, E.; Parsons, D.; Bailey, D.M. Do agonistic behaviours bias baited remote underwater video surveys of fish? *Mar. Ecol.* **2015**, *36*, 810–818. [CrossRef]

45. Watson, J.L.; Huntington, B.E. Assessing the performance of a cost-effective video lander for estimating relative abundance and diversity of nearshore fish assemblages. *J. Exp. Mar. Biol. Ecol.* **2016**, *483*, 104–111. [CrossRef]

46. Santana-Garcon, J.; Newman, S.J.; Harvey, E.S. Development and validation of a mid-water baited stereo-video technique for investigating pelagic fish assemblages. *J. Exp. Mar. Biol. Ecol.* **2014**, *452*, 82–90. [CrossRef]

47. Langlois, T.; Goetze, J.; Bond, T.; Monk, J.; Abesamis, R.A.; Asher, J.; Barrett, N.; Bernard, A.T.F.; Bouchet, P.J.; Birt, M.J.; et al. A field video annotation guide for baited remote underwater stereo-video surveys of demersal fish assemblages. *Methods Ecol. Evol.* **2020**, *11*, 1401–1409. [CrossRef]

48. Gratwicke, B.; Speight, M.R. Effects of habitat complexity on Caribbean marine fish assemblages. *Mar. Ecol. Prog. Ser.* **2005**, *292*, 301–310. [CrossRef]

49. Fréon, P.; Cury, P.; Shannon, L.; Roy, C. Sustainable exploitation of small pelagic fish stocks challenged by environmental and ecosystem changes: A review. *Bull. Mar. Sci.* **2005**, *76*, 385–462.

50. Mallet, D.; Wantiez, L.; Lemouellic, S.; Vigliola, L.; Pelletier, D. Complementarity of rotating video and underwater visual census for assessing species richness, frequency and density of reef fish on coral reef slopes. *PLoS ONE* **2014**, *9*, e84344. [CrossRef]

51. Fricke, R.; Kulbicki, M.; Wantiez, L. Cheklist of the fishes of New Caledonia, and their distribution in the Southwest Pacific Ocean (Pisces). *Stuttg. Beitr. Nat. Kd. A* **2011**, *4*, 341–463.

52. Willis, T.J. Visual census methods underestimate density and diversity of cryptic reef fishes. *J. Fish Biol.* **2001**, *59*, 1408–1411. [CrossRef]

53. Wantiez, L. Comparison of the fish assemblages as determined by a shrimp trawl net and a fish trawl net, in St. Vincent Bay, New Caledonia. *J. Mar. Biolog. Assoc. UK* **1996**, *76*, 759–775. [CrossRef]

54. Kulbicki, M.; Wantiez, L. Comparison Between Fish by catch from Shrimp Trawl net and Visual Censuses in St. Vincent Bay, New Caledonia. *Fish. Bull.* **1990**, *88*, 667–675.

55. Andrew, N.L.; Mapstone, B.D. Sampling and the description of spatial pattern in marine ecology. *Oceanogr. Mar. Biol.* **1987**, *25*, 39–90.

Article

Conservation-Status Gaps for Marine Top-Fished Commercial Species

Imanol Miqueleiz [1], Rafael Miranda [1,*], Arturo Hugo Ariño [1] and Elena Ojea [2]

[1] Biodiversity Data Analytics and Environmental Quality Group, Department of Environmental Biology, School of Sciences, University of Navarra, 31008 Pamplona, Spain; imiqueleiz@alumni.unav.es (I.M.); artarip@unav.es (A.H.A.)

[2] Future Oceans Lab, CIM—UVigo, Campus Lagoas—Marcosende, Universidade de Vigo, 36310 Vigo, Spain; elenaojea@uvigo.es

* Correspondence: rmiranda@unav.es; Tel.: +34-948-425-600 (ext. 806449)

Abstract: Biodiversity loss is a global problem, accelerated by human-induced pressures. In the marine realm, one of the major threats to species conservation, together with climate change, is overfishing. In this context, having information on the conservation status of target commercial marine fish species becomes crucial for assuring safe standards. We put together fisheries statistics from the FAO, the IUCN Red List, FishBase, and RAM Legacy databases to understand to what extent top commercial species' conservation status has been assessed. Levels of assessment for top-fished species were higher than those for general commercial or highly commercial species, but almost half of the species have outdated assessments. We found no relation between IUCN Red List traits and FishBase Vulnerability Index, depreciating the latter value as a guidance for extinction threat. The RAM database suggests good management of more-threatened species in recent decades, but more data are required to assess whether the trend has reverted in recent years. Outdated IUCN Red List assessments can benefit from reputed stock assessments for new reassessments. The future of IUCN Red List evaluations for commercial fish species relies on integrating new parameters from fisheries sources and improved collaboration with fisheries stakeholders and managers.

Keywords: fishing importance; FAO; IUCN Red List; RAM Legacy; overfishing; sustainability

Citation: Miqueleiz, I.; Miranda, R.; Ariño, A.H.; Ojea, E. Conservation-Status Gaps for Marine Top-Fished Commercial Species. *Fishes* **2022**, *7*, 2. https://doi.org/10.3390/fishes7010002

Academic Editors: Maria Angeles Esteban, Bernardo Baldisserotto and Eric Hallerman

Received: 25 October 2021
Accepted: 21 December 2021
Published: 23 December 2021

Publisher's Note: MDPI stays neutral with regard to jurisdictional claims in published maps and institutional affiliations.

1. Introduction

For millennia, mankind has had an especially close bond with the sea. Oceans contribute significantly to the support of many of the Sustainable Development Goals (SDGs) [1], as they provide people with food, as well as other ecosystem services that contribute to health, well-being, cultural identity, and to the economy of societies [2].

Marine fisheries are the main contributors of seafood (referred to as finfish and marine invertebrates) for human consumption [3,4], with almost six-billion tons of fish and invertebrates taken from the oceans since the 1950s [5], contributing to 17% of global human protein intake and sustaining millions of jobs [6]. In this context, their importance is closely linked to their long-term sustainability. The Convention on Biological Diversity (CBD), through the Aichi targets, aimed to achieve by 2020 both sustainable management of existent fish stocks (target 6) and prevention of the extinction and improvement of the conservation status of threatened species (target 12) [7]. In the same line, SDG14 on "Life below water" had the same goal of effectively regulating overfishing and rebuilding stocks to levels that produce maximum sustainable yield by 2020 (sub-target 14-4).

Unfortunately, overfishing is yet one of the main threats to marine biodiversity [8]. Increasing human pressures in response to rising demands for food [9] have led to marine fish populations' declines over the last decades [10]. Historically, humanity has failed in preventing fish-population collapses and has not taken conservation biology of marine fishes seriously enough, resulting in declines in species diversity and abundance [11,12].

Globally, a few fish species dominate catches, owing to several factors such as natural abundances, consumer preference, geography, history, and ease to catch [13]. Population declines of as much as 90% have been reported for pelagic fish species, which can cause a range of ecological impacts, restructuring communities with cascade top-down effects on other species and population assemblages [14]. For example, the selective extraction of species and individuals of higher commercial value leads to the disappearance of higher trophic levels of the marine food webs, implying an increased fishery reliance on organisms at the low levels of the food webs [15].

Assessing the extent to which stocks are being overexploited is a technically difficult and controversial issue. According Food and Agriculture Organization (FAO) assessments, the number of sustainably harvested stocks is decreasing [6]. However, there are disagreements about whether the FAO's data are reflecting the reality of stock assessments [16] or, conversely, whether using catch data overestimates the number of overexploited stocks [17]. Furthermore, other authors suggest that the FAO's catches are underestimating the amount of fish extracted from the sea as they may be misled by the omission of small-scale and recreational fisheries [3] and discards, as well as manufactured or altered data that locally would increase catches [18]. Acknowledging the limitations of its data but also offering detailed insight about how capture data are treated to produce fishing statistics [19], the FAO welcomes studies comparing their data and outputs with the those available in other databases [20].

Stock assessments, such as the RAM Legacy database [21], provide biomass estimates and management reference points for exploited aquatic populations by combining catch data with indices of stock status. However, they are usually only available for industrially exploited fisheries, leaving aside species captured in recreational or artisanal fisheries, which account for an important part of fishing effort [22].

From a biodiversity conservation point of view, the IUCN (International Union for Conservation of Nature) is the institution responsible for assessing the global conservation status of plants and animals through their periodically reviewed Red List of Threatened Species [23]. Overfishing is considered by the IUCN Red List as a threat to many marine fish species under the threat 5.4 "fishing & harvesting aquatic resources", acknowledging that it can lead to population declines, which is one of the criteria to classify a species under the IUCN Red List categories [24]. The IUCN Red List is recognized as the most authoritative institution in addressing species conservation status [25], but also valuable for informing natural resource policy and management [26]. Despite this, the status of only a small fraction of marine animal species has been evaluated by the IUCN Red List. FishBase [27], the most comprehensive database compiling fish species information, has more than 40% of "commercial" and "highly commercial" species unassessed in the IUCN Red List [28]. Other previous studies have also suggested that the IUCN Red List may not be adequately covering fished species [20]. Conservationists and fisheries scientists generally agree on the statuses of exploited marine fishes [29], but maintaining complete and up-to-date assessments (both in IUCN Red List and stock-management agencies) is essential to appropriately manage responses for species and populations of mutual concern.

Considering the existing level of overfishing in marine commercial species, together with its direct extinction risk, and the indirect impacts in communities, monitoring of the conservation status of top-fished marine commercial species is a priority. To identify the gaps in the conservation status knowledge of these species, we compared the information contained about the FAO's top-fished fishes in the IUCN Red List, FishBase, and RAM Legacy databases. We hypothesized that species of higher fishing importance would be better assessed in IUCN Red List than those of general fishing importance. Furthermore, for the subset of top-fished species, we hypothesized that more-threatened species in the IUCN Red List would have been more effectively managed in recent decades. The objective of the present study is, therefore, to understand how top-fished marine commercial species are categorized in the IUCN Red List and evaluate the degree of knowledge we have on their conservation status.

2. Materials and Methods

In this study, we explored several databases comprising information about fisheries trends, species conservation status, and other traits. We first identified the top-fished species globally from FAO statistics. Then, we collected information on IUCN Red List status for these highly fished species (with assessments between 1996 and 2021) and from FishBase about their commercial importance and vulnerability to fishing pressure. Finally, we explored the stock-assessment data available for these species in the RAM Legacy database.

2.1. FAO Data

In the late 1940s, the FAO began collecting global fishing statistics. In recent years, the FAO has produced several Yearbooks of Fishery Statistics and reports about the State of World Fisheries and Aquaculture (SOFIA). Countries submit their "best scientific estimates" of their annual landings, but the FAO acknowledges the presence of uncertainties in the reports that they receive [19]. Based on the FAO's work, we used the 2018 FAO Yearbook of Fishery and Aquaculture Statistics [30] that identifies the 70 top-fished species (based in landings data) at the global level. The subset of data analyzed represents almost half (48.3%) of the global landings in 2018 according to FAO statistics, considering not only fish sensu stricto but also squids and crustaceans (Table S1 in Supplementary Materials). Most of these species corresponded to marine species, with only three of them being freshwater-restricted fish. Henceforth, we will refer to this subset of most-important commercial species as top-fished species.

2.2. IUCN Red List and FishBase

To represent the conservation status of the top-fished species, we used the IUCN Red List [23]. The IUCN Red List establishes extinction risk of species, assigning them to a category according to their conservation status, assessing them against a series of criteria based on the size and decline rate of the population and home range [24]. From lesser to greater risk, these categories are Least Concern (LC), Near Threatened (NT), Vulnerable (VU), Endangered (EN), Critically Endangered (CR), Extinct in the Wild (EW), and Extinct (EX). Moreover, there is a category for those species with insufficient information to assess their conservation status, Data Deficient (DD). These assessments are regularly updated (IUCN Red List recommends revaluating species every 10 years [31]) and contain information relevant for species conservation (population trends, biogeography, threats, or conservation actions). From the IUCN Red List database, we obtained five parameters: (1) Current conservation category; (2) Assessment date; (3) Population trends: Increasing, stable, decreasing, or unknown; (4) Current threats, focusing on Threat 5.4 "fishing & harvesting aquatic resources" (henceforth "fishing pressure threat"); and (5) Number of IUCN Red List assessments conducted on the species. IUCN Red List assessments are generated mainly at the global scale, and less frequently at the regional scale [32]. In our study, we considered only IUCN Red List global evaluations for top-fished species and not regional evaluations (e.g., European or Mediterranean evaluations).

We searched the top-fished species in FishBase [27], the reference tool for fish studies. From this database, we obtained the Vulnerability Index of top-fished species, which estimates intrinsic extinction vulnerabilities of marine fishes to fishing [33] based on species life-history traits, including maximum length, age at first maturity, longevity, natural mortality rate, fecundity, strength of spatial behavior, and geographic range. This index assigns each species a value between 0 (low vulnerability) and 100 (high vulnerability) and a vulnerability category following a fuzzy logic approach [33]. We also retrieved from FishBase a list of fish classified as of "commercial" or "highly commercial" interest, for direct comparison with the top-fished species list. For these subsets of species, we obtained their Vulnerability Index values and their IUCN Red List status.

2.3. RAM Legacy Database

The last database we included was the RAM Legacy database. This database develops assessments on specific geographic and/or genetically distinct populations of a species, the so called "stocks". These assessments are used to set management reference points to work towards sustainable fisheries management [20].

This dataset contains information about 1433 stocks belonging to 387 unique species [34]. From the 70 FAO top-fished species, only 40 were included in the dataset. Among the several metrics provided by the database for each stock, we decided to use for comparison among different IUCN Red List categories the ratio between the catches and the maximum sustainable yield (MSY), henceforth catches/MSY ratio. This dimensionless metric is the coefficient between the catches of one stock and its MSY, the number of catches that ensure the maximum yield sustained over time [21]. We selected this value as we found it the most suitable way of addressing the questions of whether an individual stock was being fished above safe biological limits and allowed comparison among different stocks. In some stocks, different catches/MSY ratio values were available owing to different assessment methodologies. Thus, in these cases, the average of all the assessments conducted for the given stock was calculated to obtain the stock's catches/MSY ratio value. We selected stock-assessment data between 1996, when the current IUCN Red List classification scheme was created, and 2020. All data from the RAM Legacy database were retrieved using the associated R package *ramlegacy* [34] in R software version 4.1.1 [35] and can be found in the Table S2 in Supplementary Materials.

2.4. Taxonomic Checking

We joined the databases at the species level. Owing to the different sources of the data (four different databases), we corrected possible discrepancies in the taxonomy used. Discrepancies in the scientific name for a given species may result from differences in data entry across databases, revisions of species classifications owing to new taxonomies, or misspelling. In order to best match species names among the datasets, we checked each species in the Eschmeyer Catalog of Fishes [36] for fish species, and SeaLife (https://www.sealifebase.ca/, accessed on 15 September 2021) database for the remaining species, to update taxonomy to the most recent valid name and allow the correct alignment of the information among the databases.

2.5. Analysis

We compared the proportion of assessed species from the subset of top-fished species with the assessment rates of FishBase commercial categories (commercial and highly commercial) through chi-square tests for analyses of frequencies. We also analyzed differences in the Vulnerability Index for fish species between the three species subsets (top-fished, highly commercial, and commercial species) trough Kruskal–Wallis tests, owing to the non-normality of the data.

Focusing on the subset of top-fished species, we examined the number of top-fished species within each IUCN Red List category. Furthermore, for each IUCN Red List category, we calculated the number of species under fishing pressure (IUCN Red List threat 5.4) and their IUCN Red List population trend. We performed Kruskal–Wallis rank-sum tests to assess differences in the FishBase Vulnerability Index for top-fished species according to the different explanatory variables: IUCN Red List conservation status, population trends, and threat 5.4. All analyses were performed using the *stats* package in R software version 4.1.1 [35].

We analyzed changes in the catches/MSY ratio for different IUCN Red List categories and population trends between 1996 and the most recent data for each stock, as not all of them were available until 2020. We performed linear (regression) models with the *stats* package in R software, to analyze the influence of the IUCN Red List category and population trend with the observed changes in the catches/MSY ratio.

3. Results

From the 70 top-fished species, twenty (28.6%) were not assessed by the IUCN Red List [23, including two of the ten most fished species (*Gadus chalcogrammus* and *Micromesistius poutassou*). We compared the IUCN Red List assessment rates and Vulnerability Index values (FishBase VI see Table S1) of top-fished species with the larger commercial species groups (commercial and highly commercial species). IUCN Red List assessment rates are higher in top-fished species than in commercial categories (χ^2 test $p < 0.05$) (Table 1). Vulnerability Index values were not significantly different among top-fished species and those considered to have a commercial or highly commercial interest ($p = 0.052$) according to FishBase's classification. Nevertheless, Vulnerability Index values for top-fished species tended to be lower than in the other categories (Table 1).

Table 1. Number and proportion of IUCN Red List-assessed and -unassessed species for each species sub-set, and their average vulnerability indices.

	Top-Fished Species	**Commercial Species**	**Highly Commercial Species**
IUCN Red List-assessed species	50 (71.4%)	1217 (59.2%)	126 (56%)
Unassessed species	20 (28.6%)	839 (40.8%)	99 (44%)
Vulnerability Index (Median, Q1, and Q3)	34.47 (24.7–50.67)	40 (30–56)	55.5 (33.3–63.75)

IUCN Red List assessment dates ranged between 1996 and 2021, with twenty species (40% of assessed species) assessed in 2011 and previous years, and thus having outdated assessments. Within the assessed species, we also found cases of deficient evaluation, with six of them classified as Data Deficient (DD), five fish and one cephalopod (Table 2). We also found two species with old assessments (*Gadus morhua* and *Melanogrammus aeglefinus*, dating from 1996) that do not have complete information. Most assessed species, 33 out of 50 (66.6%), were under fishing pressure (IUCN threat 5.4). Almost all assessed species (48 out of 50) had information on population trends, but many of them (23, 48%) had unknown trends. Most IUCN Red List-assessed species have only been assessed once (38 out of 50), whilst multiple assessment has been conducted twice for nine species and three times for three species, all of them from the genus Thunnus (Table S1 in Supplementary Materials).

Table 2. Number and percentage of top-fished species assessed in the IUCN Red List, and their population trends and vulnerability indices. IUCN Red List categories: DD (Data Deficient), LC (Least Concern), NT (Near Threatened), VU (Vulnerable), NE (Not Evaluated).

	IUCN Red List Conservation Status				
	DD	**LC**	**NT**	**VU**	**NE**
Top-fished species	6 (8.6%)	37 (48.6%)	4 (7.1%)	3 (5.7%)	20 (30.0%)
IUCN Red List fishing pressure threat	4 (66.7%)	31 (83.8%)	4 (100%)	1 (33.3%)	
Vulnerability index (VI) median (Q1, Q3)	47.9 (21.5–69.7)	36 (20.2–51.6)	37.1 (30–68.5)	45.5 (37.5–57.9)	41.4 (25.5–49.6)

No significant differences were found in the Vulnerability Index for different IUCN categories ($\chi^2 = 4.8$, d.f. $= 3$, $p = 0.19$), populations trends ($\chi^2 = 2.52$, d.f. $= 3$, $p = 0.47$), or IUCN Red List fishing-pressure threat ($\chi^2 = 1.35$, d.f. $= 1$, $p = 0.25$).

Only 62 stocks, comprising a total of 92 assessments and belonging to 23 individual fish species, had catches/MSY ratio values in the RAM Legacy database. The number of assessments with catches/MSY ratio value has constantly declined since a maximum of 91 assessments in the 1998–2008 decade, and only 10 and 3 stocks had this value in 2019 and 2020, respectively.

All IUCN Red List categories have reduced their stocks' catches/MSY ratio in the 1996–2020 period, but the reduction has only been significant for species assessed as Near Threatened and Vulnerable (Table 3). Regarding the IUCN Red List population trends, significant reductions in the catches/MSY ratio were only found in species with decreasing and increasing population trends in the period. However, data from the most recent years suggest an increase in the ratio, especially for Least Concern species' stocks, with catches/MSY ratio values over 1. Nevertheless, the low number of stocks assessed in recent years prevented us from considering these data as completely reliable.

Table 3. Linear models for the trends in catches/MSY ratio between 1996 and 2020 for IUCN Red List categories and population trends. LC: Least Concern, NT: Near Threatened, VU: Vulnerable, NE: Not Evaluated.

IUCN Red List Trait		Slope (Catches/MSY Year-1)	t	p
	NE	−0.001	−0.367	0.717
Category	LC	−0.004	−1.892	0.071
	NT	−0.033	−8.560	* <0.001
	VU	−0.009	−2.391	* 0.027
	Decreasing	−0.012	−5.557	* <0.001
	Increasing	−0.048	−5.221	* <0.001
Population trend	NA	−0.002	−0.398	0.695
	Stable	0.003	0.389	0.701
	Unknown	−0.001	−0.329	0.745

* $p < 0.05$.

We examined the catches/MSY ratio trend in NT and VU species and observed that one species, *Melanogrammus aeglefinus*, had increased this ratio since 2013 and currently was over 1 (Figure 1). The remaining species have stocks managed more sustainably, according to the RAM database.

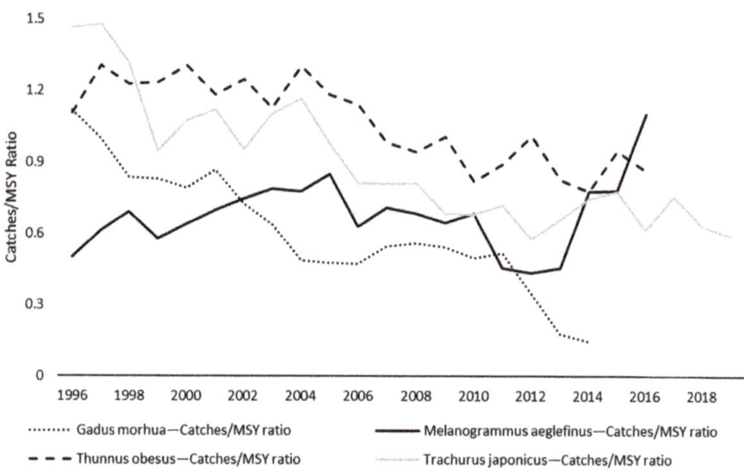

Figure 1. Trends in catches/MSY ratio between 1996 and 2020 for species in IUCN Red List categories NT (Near Threatened, grey line) and VU (Vulnerable, black lines).

4. Discussion

The revision of the conservation status and population trends of the main fish species of commercial interest is urgent and mandatory. Solutions for restoring marine ecosystems and the fish species that live in them are still under debate, but scholars and international organizations agree that sustainable management is becoming more and more urgent in several fish stocks [37]. With a horizon of human population increase in the coming years, assessing species conservation status is more important than ever, and only possible if all players do their part to manage fisheries sustainably and sustain the oceans and their biodiversity. Instead of using catch data to assess the state of marine ecosystems, which has been previously criticized [38], we compared the information of different databases to envision which knowledge gaps are more urgent to fill to ensure species conservation and fisheries sustainability.

Since fishes sensu stricto represent most of the top-fished species, and some values such as the FishBase Vulnerability Index or commercial category were only available for them, we have focused the discussion on this taxonomic group. However, we acknowledge that many of the problems detected for fishes can be present in other groups. For instance, cephalopods have increased in commercial importance in the last decades [39] and yet face several threats from unregulated fisheries, bycatch, and poor life-history knowledge [40].

We found progress in the higher assessment rates for top-fished species, as they were significantly better assessed compared with the commercial or highly commercial counterparts (Table 2), acknowledging recent IUCN Red List efforts in evaluating some commercially exploited groups [23]. The absence of significant differences in the Vulnerability Index among IUCN Red List categories had already been noticed [41], and our results reinforce those findings. We have also demonstrated that the categorization of a species under fishing threat, or a declining population trend, are not related to a higher Vulnerability Index. In this sense, we consider that the Vulnerability Index may not be accurately measuring the extinction risk of a species, as fishing pressure is not only driven by biological traits and other factors are present too [13]. We consider that the IUCN Red List categorization provides us with more accurate information about species extinction risk.

Among assessed species, *Gadus morhua* (Atlantic cod) and *Melanogrammus aeglefinus* (haddock) have extremely outdated IUCN Red List assessments, classified as Vulnerable in 1996 (25 years ago). After the stocks' collapse in the late 1980s and early 1990s, some studies have stated that some cod stocks have not yet recovered from the collapse in northwest Atlantic [42], whereas IUCN Red List European assessment classifies *G. morhua* as LC, with increasing populations in some stocks [43]. In such cases, we acknowledge the big difference between assessing a whole species, as IUCN Red List does, and quantitative stock or population assessments [44], but we support the labor of regional IUCN Red List assessments integrating stocks statistics into this conservation tool, promoting both stock management and species conservation.

Apart from these two species assessed in 1996, remaining assessments dated from 2007 and onwards. IUCN Red List estimates assessments to be outdated after 10 years [31], so priority reassessments should be conducted for 40% of the top-fished species (most of them classified as LC or DD). In this sense, data deficiency not only implies "inadequate information to make a direct, or indirect, assessment of its risk of extinction based on its distribution and/or population status" [24] but also affects conservation priorities, which rely upon threatened-species lists [45]. The uncertainty associated with data deficiency affects extinction-risk patterns and should be solved through a reassessment of DD species. The opposite situation is found in three *Thunnus* species: *T. alabacares*, *T. obesus*, and *T. alalunga*. These species have been assessed three times, with the last assessment in 2021. This is the result of the work carried by the IUCN Tuna and Billfishes Specialist (https://www.iucn.org/commissions/ssc-groups/fishes/tuna-and-billfishes), highlighting the importance of having organized resources to ensure high-quality and updated assessments.

Almost all species assessed (96%) had information about population trends in the IUCN Red List, with haddock and Atlantic cod lacking these data because of their outdated assessments. Almost half of IUCN Red List-assessed species had unknown populations trends, and as subjects of intensive fishing, we consider that they should be reassessed in the short term to examine if current data allow us to establish their population trends, at least at regional level. Recent regional IUCN Red List evaluations, such as those conducted for *G. morhua* and *M. aeglefinus* in Europe, show detailed information about populations trends in the different stocks [43,46] and are a good example to follow. Not all top-fished assessed species were under fishing pressure according to the IUCN Red List (Table 2). Certainly, species may indeed have abundant populations or be subjects of sustainable fishing or fishing quotas. However, FAO data indicate increasing overexploited stocks in recent years [6], which could lead to regional threats to extensively harvested stocks.

Previous studies have stated the importance of analyzing together IUCN Red List assessments with stock assessments, such as the RAM Legacy Stock Assessment Database, resulting in high agreement in the conservation status of exploited marine fishes [29]. In this sense, data from stock assessments have several advantages when compared with other data based on reported catches or reconstructions [47], whose utility to assess the collapse or overexploitation of fish stocks is also dubious [17]. When combining RAM database data with IUCN Red List evaluations, we observed a general improvement in the stock for all the categories in recent decades, having reached catches/MSY ratio values under 1 in the 2010s. However, this declining trend is only significant for Vulnerable and Near Threatened species, since those classified as Least Concern or not evaluated seem to have increased in their catches/MSY ratio in recent years. We found that stock assessments were substantially scarcer in recent years, and thus the reverted trend observed for LC and NE species can be the result of this bias. Thus, we support the labor conducted by stock-assessment agencies such as RAM Legacy and the value that these data can have in evaluating the conservation status of top-fished species. Specifically, we consider that the haddock *Melanogrammus aeglefinus* should be promptly assessed by the IUCN Red List, as some stocks have been increasingly exploited in recent years, and its outdated assessments may not be adequately reflecting the current conservation status of the species and its different populations. Similar to the regional assessment conducted for its European population [46], we consider that other populations or the global conservation status of the species should be evaluated using current information.

IUCN Red List assessments and criteria, despite having proven not to be biased towards exaggerating marine fishes' threat status [29], can pose problems when evaluating them, especially commercial fishes [48,49]. In particular, Criterion A, related to the reduction in population size, discards information that would be included under Criterion E and could help estimate recent trends and risk to extinction (such as age or length structure, recruitment of juveniles, and exploitation rate [48]). Considering these potential flaws, we suggest that further investigation is required on the adaptation of IUCN Red List categories and criteria for commercial fish species. The IUCN Red List is widely recognized among the general public as an authoritative source about species conservation status [25], but we support the inclusion of other methods such as management-strategy evaluation (MSE) to make more precise evaluations [50]. As previously stated, we consider that future IUCN Red List assessments for fished species should be developed at both global and regional level. Regional assessments can better reflect the situation and conservation status of regional stocks, in contrast to a global vision of the species conservation status, which may lack relevance in terms of stocks management.

Having failed in meeting most CDB Aichi targets [51–53], the sustainability of one of our main food sources is at stake. Climate change and global warming affect fish-populations' viability and compromise the subsistence of human communities linked to them [54]. Moreover, the growing proportions of unidentified catches in key regions such as Asia calls for better fisheries monitoring and management [20]. In the context of a global threat to marine species, where human and climate pressures combine to jeopardize them, we consider that the voice of conservation initiatives, such as the IUCN Red List, should be extensively heeded by fishing authorities. Governments, conservation agencies, and fishing authorities must work together to achieve useful conservation objectives and policies. The latest SOFIA report states that collaboration efforts between the FAO, CITES, and IUCN are taking place [6], but unless urgent measures are taken, the overexploitation of our seas can lead again to stock depletion and compromise not only species conservation but also food security and the way of life in many regions.

5. Conclusions

In this paper, we have demonstrated how the knowledge on the conservation status of marine top-fished species is far from being adequate. FishBase Vulnerability Index has proved to be a poor proxy for predicting species extinction risk. Several species have never been assessed by IUCN Red List (28.6% of top-fished species) or have outdated assessments that should be redone promptly (40 % of the assessed ones).

To this end, IUCN Red List evaluations can benefit from data stock assessments' data like the RAM Legacy database, especially in some cases with deficient evaluations and increasing fishing pressure as the haddock. In a context of increasing food demand by human population, fisheries sustainability is essential to ensure both human food security and species conservation.

Supplementary Materials: The following are available online at https://www.mdpi.com/article/10.3390/fishes7010002/s1. Table S1: FAO top-fished species and related traits used in the study. Detailed information about columns can be found in the metadata sheet. Table S2: RAM Legacy catches/MSY ratio data used in the study, corresponding to FAO top-fished species. Detailed information about columns can be found in the metadata sheet.

Author Contributions: Conceptualization, I.M. and R.M.; methodology, I.M. and E.O.; software, I.M.; formal analysis, I.M.; investigation, I.M.; resources, R.M., A.H.A.; data curation, I.M.; writing—original draft preparation, I.M.; writing—review and editing, I.M., R.M., A.H.A. and E.O.; visualization, I.M.; supervision R.M., A.H.A. and E.O.; project administration, I.M.; funding acquisition, I.M. All authors have read and agreed to the published version of the manuscript.

Funding: I.M. was funded by the Association of Friends of the University of Navarra. E.O. acknowledges financial support from the European Research Council through the project CLOCK ("Climate Adaptation to Shifting Stocks"; ERC Starting Grant Agreement n8679812; EU Horizon 2020), and the GAIN Oportunius program, Xunta de Galicia.

Data Availability Statement: The data that support the findings of this study are available in the Supplementary Materials provided in the manuscript.

Conflicts of Interest: The authors declare no conflict of interest. The funders had no role in the design of the study; in the collection, analyses, or interpretation of data; in the writing of the manuscript, or in the decision to publish the results.

References

1. FAO. *The State of World Fisheries and Aquaculture 2018. Meeting the Sustainable Development Goals*; FAO: Rome, Italy, 2018.
2. Hughes, R.M. Recreational fisheries in the USA: Economics, management strategies, and ecological threats. *Fish. Sci.* **2014**, *81*, 1–9. [CrossRef]
3. Pauly, D.; Zeller, D. Catch reconstructions reveal that global marine fisheries catches are higher than reported and declining. *Nat. Commun.* **2016**, *7*, 1–9. [CrossRef]
4. Hicks, C.C.; Cohen, P.J.; Graham, N.A.J.; Nash, K.L.; Allison, E.H.; D'Lima, C.; Mills, D.J.; Roscher, M.; Thilsted, S.H.; Thorne-Lyman, A.L.; et al. Harnessing global fisheries to tackle micronutrient deficiencies. *Nature* **2019**, *574*, 95–98. [CrossRef]
5. Sea Around Us Concepts, Design and Data. Available online: Seaaroundus.org (accessed on 25 June 2021).
6. FAO. *The State of World Fisheries and Aquaculture 2020: Sustainability in Action*; FAO: Rome, Italy, 2020; ISBN 9789251326923.
7. SCBD. Decision adopted by the Conference of the Parties to the Convention on Biological Diversity at its tenth meeting. In *X/2. The Strategic Plan for Biodiversity 2011–2020 and the Aichi Biodiversity Targets*; SCBD: Nagoya, Japan, 2010.
8. Morato, T.; Watson, R.; Pitcher, T.; Pauly, D. Fishing down the deep. *Fish Fish.* **2006**, *7*, 24–34. [CrossRef]
9. Halpern, B.S.; Frazier, M.; Afflerbach, J.; Lowndes, J.S.; Micheli, F.; Scarborough, C.; Selkoe, K.A. Recent pace of change in human impact on the world's ocean. *Sci. Rep.* **2019**, *9*, 11609. [CrossRef] [PubMed]
10. WWF. *Living Blue Planet Report: Species, Habitats and Human Well-Being*; Tanzer, J., Phua, C., Lawrence, A., Gonzales, A., Roxburgh, T., Gamblin, P., Eds.; WWF: Gland, Switzerland, 2015.
11. Butchart, S.H.M.; Walpole, M.; Collen, B.; van Strien, A.; Scharlemann, J.P.W.; Almond, R.E.A.; Baillie, J.E.M.; Bomhard, B.; Brown, C.; Bruno, J.; et al. Global biodiversity: Indicators of recent declines. *Science* **2010**, *328*, 1164–1169. [CrossRef]
12. Roberson, L.A.; Watson, R.A.; Klein, C.J. Over 90 endangered fish and invertebrates are caught in industrial fisheries. *Nat. Commun.* **2020**, *11*, 4764. [CrossRef]

13. Sadovy de Mitcheson, Y.; Craig, M.T.; Bertoncini, A.A.; Carpenter, K.E.; Cheung, W.W.L.; Choat, J.H.; Cornish, A.S.; Fennessy, S.T.; Ferreira, B.P.; Heemstra, P.C.; et al. Fishing groupers towards extinction: A global assessment of threats and extinction risks in a billion dollar fishery. *Fish Fish.* **2013**, *14*, 119–136. [CrossRef]
14. Smith, A.D.M.; Brown, C.J.; Bulman, C.M.; Fulton, E.A.; Johnson, P.; Kaplan, I.C.; Lozano-Montes, H.; Mackinson, S.; Marzloff, M.; Shannon, L.J.; et al. Impacts of fishing low-trophic level species on marine ecosystems. *Science* **2011**, *333*, 1147–1150. [CrossRef]
15. Szuwalski, C.S.; Burgess, M.G.; Costello, C.; Gaines, S.D. High fishery catches through trophic cascades in China. *Proc. Natl. Acad. Sci. USA* **2016**, *114*, 717–721. [CrossRef]
16. Froese, R.; Zeller, D.; Kleisner, K.; Pauly, D. What catch data can tell us about the status of global fisheries. *Mar. Biol.* **2012**, *159*, 1283–1292. [CrossRef]
17. Branch, T.A.; Jensen, O.P.; Ricard, D.; Ye, Y.; Hilborn, R. Contrasting Global Trends in Marine Fishery Status Obtained from Catches and from Stock Assessments. *Conserv. Biol.* **2011**, *25*, 777–786. [CrossRef]
18. Pauly, D.; Zeller, D. Comments on FAOs State of World Fisheries and Aquaculture (SOFIA 2016). *Mar. Policy* **2017**, *77*, 176–181. [CrossRef]
19. Ye, Y.; Barange, M.; Beveridge, M.; Garibaldi, L.; Gutierrez, N.; Anganuzzi, A.; Taconet, M. FAO's statistic data and sustainability of fisheries and aquaculture: Comments on Pauly and Zeller (2017). *Mar. Policy* **2017**, *81*, 401–405. [CrossRef]
20. Blasco, G.D.; Ferraro, D.M.; Cottrell, R.S.; Halpern, B.S.; Froehlich, H.E. Substantial Gaps in the Current Fisheries Data Landscape. *Front. Mar. Sci.* **2020**, *7*, 612831. [CrossRef]
21. Ricard, D.; Minto, C.; Jensen, O.P.; Baum, J.K. Examining the knowledge base and status of commercially exploited marine species with the RAM Legacy Stock Assessment Database. *Fish Fish.* **2012**, *13*, 380–398. [CrossRef]
22. Rousseau, Y.; Watson, R.A.; Blanchard, J.L.; Fulton, E.A. Evolution of global marine fishing fleets and the response of fished resources. *Proc. Natl. Acad. Sci. USA* **2019**, *16*, 12238–12243. [CrossRef] [PubMed]
23. IUCN. The IUCN Red List of Threatened Species. *Version 2021-2*. Available online: https://www.iucnredlist.org/ (accessed on 13 September 2021).
24. IUCN. *IUCN Red List Categories and Criteria, Version 3.1*, 2nd ed.; IUCN: Gland, Switzerland; Cambridge, UK, 2012; ISBN 2831707862.
25. Rodrigues, A.S.L.; Pilgrim, J.D.; Lamoreux, J.F.; Hoffmann, M.; Brooks, T.M. The value of the IUCN Red List for conservation. *Trends Ecol. Evol.* **2006**, *21*, 71–76. [CrossRef] [PubMed]
26. Bennun, L.; Regan, E.C.; Bird, J.; van Bochove, J.W.; Katariya, V.; Livingstone, S.; Mitchell, R.; Savy, C.; Starkey, M.; Temple, H.; et al. The Value of the IUCN Red List for Business Decision-Making. *Conserv. Lett.* **2018**, *11*, e12353. [CrossRef]
27. Fishbase. Available online: www.fishbase.org (accessed on 9 April 2021).
28. Miqueleiz, I.; Bohm, M.; Ariño, A.H.; Miranda, R. Assessment gaps and biases in knowledge of conservation status of fishes. *Aquat. Conserv. Mar. Freshw. Ecosyst.* **2020**, *30*, 225–236. [CrossRef]
29. Davies, T.D.; Baum, J.K. Extinction risk and overfishing: Reconciling conservation and fisheries perspectives on the status of marine fishes. *Sci. Rep.* **2012**, *2*, 1–9. [CrossRef] [PubMed]
30. FAO. FAO Yearbook. In *Fishery and Aquaculture Statistics 2018/FAO Annuaire. Statistiques Des pêches et de L'aquaculture 2018/FAO Anuario. Estadísticas de Pesca y Acuicultura 2018*; FAO: Rome, Italy, 2020.
31. Rondinini, C.; Di Marco, M.; Visconti, P.; Butchart, S.H.M.; Boitani, L. Update or outdate: Long-term viability of the IUCN Red List. *Conserv. Lett.* **2014**, *7*, 126–130. [CrossRef]
32. IUCN. *Guidelines for Application of IUCN Red List Criteria at Regional and National Levels: Version 4.0*; IUCN: Gland, Switzerland; Cambridge, UK, 2012.
33. Cheung, W.W.L.; Pitcher, T.J.; Pauly, D. A fuzzy logic expert system to estimate intrinsic extinction vulnerabilities of marine fishes to fishing. *Biol. Conserv.* **2005**, *124*, 97–111. [CrossRef]
34. RAM Legacy Stock Assessment Database RAM Legacy Stock Assessment Database v4.495. Available online: https://zenodo.org/record/4824192 (accessed on 28 September 2021).
35. R Development Core Team. *R: A Language and Environment for Statistical Computing*; Version 4.1.1; R Foundation for Statistical Computing: Vienna, Austria, 2020.
36. Fricke, R.; Eschmeyer, W.N.; Van der Laan, R. Eschmeyer's Catalog of Fishes: Genera, Species, References. San Francisco, CA, USA. 2020. Available online: https://researcharchive.calacademy.org/research/ichthyology/catalog/fishcatmain.asp (accessed on 15 September 2021).
37. Froese, R.; Winker, H.; Coro, G.; Demirel, N.; Tsikliras, A.; Dimarchopoulou, D.; Scarcella, G.; Palomares, M.; Dureuil, M.; Pauly, D. Estimating stock status from relative abundance and resilience. *ICES J. Mar. Sci.* **2019**, *77*, 527–538. [CrossRef]
38. Branch, T.A.; Watson, R.; Fulton, E.A.; Jennings, S.; McGilliard, C.R.; Pablico, G.T.; Ricard, D.; Tracey, S.R. The trophic fingerprint of marine fisheries. *Nature* **2010**, *468*, 431–435. [CrossRef] [PubMed]
39. Pita, C.; Roumbedakis, K.; Fonseca, T.; Matos, F.L.; Pereira, J.; Villasante, S.; Pita, P.; Bellido, J.M.; Gonzalez, A.F.; García-Tasende, M.; et al. Fisheries for common octopus in Europe: Socioeconomic importance and management. *Fish. Res.* **2021**, *235*, 105820. [CrossRef]
40. Lishchenko, F.; Perales-Raya, C.; Barrett, C.; Oesterwind, D.; Power, A.M.; Larivain, A.; Laptikhovsky, V.; Karatza, A.; Badouvas, N.; Lishchenko, A.; et al. A review of recent studies on the life history and ecology of European cephalopods with emphasis on species with the greatest commercial fishery and culture potential. *Fish. Res.* **2021**, *236*, 105847. [CrossRef]

41. Miranda, R. The misguided comparison of vulnerability and conservation status. *Aquat. Conserv. Mar. Freshw. Ecosyst.* **2017**, *27*, 898–899. [CrossRef]
42. Neuenhoff, R.D.; Swain, D.P.; Cox, S.P.; McAllister, M.K.; Trites, A.W.; Walters, C.J.; Hammill, M.O. Continued decline of a collapsed population of Atlantic cod (Gadus morhua) due to predation-driven allee effects. *Can. J. Fish. Aquat. Sci.* **2019**, *76*, 168–184. [CrossRef]
43. Cook, R.; Fernandes, P.; Florin, A.-B.; Lorance, P.; Nedreaas, K. Gadus morhua. The IUCN Red List of Threatened Species 2015. Available online: https://www.iucnredlist.org/species/8784/45097319 (accessed on 3 August 2021).
44. Fernandes, P.G.; Ralph, G.M.; Nieto, A.; García Criado, M.; Vasilakopoulos, P.; Maravelias, C.D.; Cook, R.M.; Pollom, R.A.; Kovačić, M.; Pollard, D.; et al. Coherent assessments of Europe's marine fishes show regional divergence and megafauna loss. *Nat. Ecol. Evol.* **2017**, *1*, 0170. [CrossRef]
45. Bland, L.M.; Böhm, M. Overcoming data deficiency in reptiles. *Biol. Conserv.* **2016**, *204A*, 16–22. [CrossRef]
46. Cook, R.; Fernandes, P.; Florin, A.-B.; Lorance, P.; Nedreaas, K. Melanogrammus aeglefinus. The IUCN Red List of Threatened Species 2015. Available online: https://www.iucnredlist.org/species/13045/3406968 (accessed on 3 August 2021).
47. Pauly, D.; Zeller, D. *Catch Reconstruction: Concepts, Methods, and Data Sources*; Online Publication, 2015.
48. Millar, S.; Dickey-Collas, M. *Report on IUCN Assessments and Fisheries Management Approaches*; ICES CM 2018/ACOM: Copenhagen, Denmark, 2018.
49. Collen, B.; Dulvy, N.K.; Gaston, K.J.; Gärdenfors, U.; Keith, D.A.; Punt, A.E.; Regan, H.M.; Böhm, M.; Hedges, S.; Seddon, M.; et al. Clarifying misconceptions of extinction risk assessment with the IUCN Red List. *Biol. Lett.* **2016**, *12*, 1424–1442. [CrossRef]
50. Punt, A.E.; Butterworth, D.S.; de Moor, C.L.; De Oliveira, J.A.A.; Haddon, M. Management strategy evaluation: Best practices. *Fish Fish.* **2016**, *17*, 303–334. [CrossRef]
51. UNEP-WCMC; IUCN; NGS. *Protected Planet Report 2018*; UNEP-WCMC: Cambridge, UK; IUCN: Gland, Switzerland; NGS: Washington, DC, USA, 2018; ISBN 9789280737219.
52. Tittensor, D.P.; Walpole, M.; Hill, S.L.L.; Boyce, D.G.; Britten, G.L.; Burgess, N.D.; Butchart, S.H.M.; Leadley, P.W.; Regan, E.C.; Alkemade, R.; et al. A mid-term analysis of progress toward international biodiversity targets. *Science* **2014**, *346*, 241–243. [CrossRef] [PubMed]
53. Nature. New biodiversity targets cannot afford to fail. *Nature* **2020**, *578*, 337–338. [CrossRef]
54. Cheung, W.W.L. The future of fishes and fisheries in the changing oceans. *J. Fish Biol.* **2018**, *92*, 790–803. [CrossRef] [PubMed]

Article

Feed Supplementation with the GHRP-6 Peptide, a Ghrelin Analog, Improves Feed Intake, Growth Performance and Aerobic Metabolism in the Gilthead Sea Bream *Sparus aurata*

Leandro Rodríguez-Viera [1], Ignacio Martí [2], Rebeca Martínez [3], Erick Perera [4], Mario Pablo Estrada [3], Juan Miguel Mancera [2] and Juan Antonio Martos-Sitcha [2,*]

[1] Center for Marine Research, University of Havana, Havana 10300, Cuba; leokarma@gmail.com
[2] Department of Biology, Faculty of Marine and Environmental Sciences, Instituto Universitario de Investigación Marina (INMAR), University of Cadiz, Campus de Excelencia Internacional del Mar (CEIMAR), 11519 Puerto Real, Cadiz, Spain; martignacio16@icloud.com (I.M.); juanmiguel.mancera@uca.es (J.M.M.)
[3] Aquatic Biotechnology Project, Center for Genetic Engineering and Biotechnology, Havana 10600, Cuba; rebeca.martinez@cigb.edu.cu (R.M.); mario.pablo@cigb.edu.cu (M.P.E.)
[4] Andalusian Institute of Marine Sciences (ICMAN), Spanish National Research Council (CSIC), 11519 Puerto Real, Cadiz, Spain; erick.perera@csic.es
* Correspondence: juanantonio.sitcha@uca.es

Citation: Rodríguez-Viera, L.; Martí, I.; Martínez, R.; Perera, E.; Estrada, M.P.; Mancera, J.M.; Martos-Sitcha, J.A. Feed Supplementation with the GHRP-6 Peptide, a Ghrelin Analog, Improves Feed Intake, Growth Performance and Aerobic Metabolism in the Gilthead Sea Bream *Sparus aurata*. Fishes **2022**, 7, 31. https://doi.org/10.3390/fishes7010031

Academic Editor: Geneviève Corraze

Received: 5 December 2021
Accepted: 26 January 2022
Published: 28 January 2022

Publisher's Note: MDPI stays neutral with regard to jurisdictional claims in published maps and institutional affiliations.

Abstract: The aquaculture sector has experienced rapid and important growth with the subsequent increase of feeding and nutritional issues for sustaining this activity, mainly related to the use of high quality, safe and environmentally friendly feed ingredients. The use of additives in aquafeeds has proven to be a suitable option to improve different productive indicators in farmed fish. In the present study, the effect of adding the GHRP-6 peptide, a ghrelin analog, to a commercial diet of gilthead sea bream (*Sparus aurata*) was studied at two proportions (100 or 500 µg/kg of feed). Both experimental diets show an increase in growth performance, as well as in feed efficiency after 97 days of experiment. The lower inclusion of GHRP-6 (100 µg/kg) results in a better aerobic metabolism, while the higher inclusion significantly increased plasma GH levels in agreement with the GH secretagogue effects of ghrelin. Similar growth outcome and differences between GHRP-6 levels in aerobic metabolism and GH stimulation suggest that improvements in culture performance by this peptide may occur through different mechanisms. Taken together, this compound can be considered as a viable dietary supplement for increasing production efficiency of sea bream aquaculture, although a better understanding of its dose-specific effects is still required.

Keywords: aquaculture; ghrelin; gilthead sea bream; growth hormone secretagogue; feed intake; metabolism; welfare

1. Introduction

Aquaculture is the fastest growing animal food-producing sector in the world [1]. While there are several phenotypic traits that are currently being improved in farmed fish through genetic selection, feed composition, management and farming practices, the improvement of growth rates and feed efficiency remain as the main goal for most species. Accordingly, intense research effort has been made to understand the internal and external factors regulating feed intake [2] and growth [3] in fish. The use of additives such as short- or medium-chain fatty acids [4,5] or nutraceutical compounds from algae [6], among others, have been proven to be suitable options to improve different productive indicators in farmed fish. However, it has been suggested that the use of endogenous feeding and growth regulatory factors as feed additives should be also explored to modulate growth rate and yield in cultured species [2]. In addition, the use of synthetic compounds that stimulate feed intake, feeding efficiency, and physiological pathways related with growth, metabolism or welfare, may open new avenues to increase the competence of this productive activity. Few studies are available in farmed fish in this regard, especially those targeting the regulation of somatic growth.

The growth hormone (GH), mainly produced by the pituitary, is a key regulator of growth, although it is also involved in regulating nutrition, reproduction, physical activity, neuroprotection, immunity, and osmotic pressure [3]. The main action of GH is the stimulation of hepatic insulin-like growth factors (IGFs), which promote growth, protein synthesis, cell proliferation and metabolism [3]. GH releasing hormone (GHRH) is the principal stimulator of GH synthesis and secretion, somatostatin is a potent noncompetitive inhibitor of the release of GH, and ghrelin has a marked growth hormone-stimulating activity, the last linking gastrointestinal-pituitary axis [7]. Ghrelin is a 28 amino acid peptide that was discovered in rat stomach [8] as the endogenous ligand to the previous orphan GHS-receptor (GHS-R), and thus a potent stimulator of pituitary GH release in vertebrates [9], including fish [10]. Ghrelin, often called a "hunger hormone", plays key roles not only in the control of GH release but also in the regulation of feed intake, energy metabolism, and immune responses in vertebrates [11–14], even from early stages [15,16]. In gilthead sea bream, ghrelin is highly expressed in the stomach and pyloric caeca [14] as described in other fish [17].

GH secretagogues (GHSs) are a family of synthetic, non-natural peptides, initially termed GH-releasing peptides (GHRPs) [18], which are recognized by the GHS-R. GHSs have no structural homology with GHRH and act via specific receptors present in the pituitary and maybe also at the hypothalamic level [19]. The action of several synthetic GHS on GH secretion has been studied in different animals [20,21], including fish [22]. Among this family, the Growth Hormone-Releasing Peptide 6 (GHRP-6) is a six amino acid synthetic peptide (His-(D-Trp)-Ala-Trp-(D-Phe)-Lys-NH2, MW = 872.44 Da) first described by [23], and it is considered a strong GHS. Few studies have been performed in fish regarding the effects of this synthetic peptide. Moreover, most available information comes from studies in freshwater species. GHRP-6 mimics the orexigenic action of ghrelin in goldfish [24] and increases pituitary Gh secretion both in vitro and in vivo in juvenile tilapia [25–27], whereas intraperitoneal administration of GHRP-6 induces Igf-i expression in the liver, and stimulates growth rate when administered by a plastic tube to the pharyngeal cavity of juvenile tilapia [27,28]. In addition, GHRP-6 increased body weight when it was administered by immersion baths to tilapia larvae [26], enhancing their non-specific immunity [27]. In addition, in freshwater fishes (i.e., tilapia and rainbow trout), a related synthetic peptide secretagogue (GHRP-2) increased the levels of Gh after intraperitoneal injections [29,30]. However, to the best of our knowledge, no study has reported similar effects in a marine fish. In the present study, the effect of adding the GHRP-6 peptide to a commercial diet for a marine fish was studied for the first time. The gilthead sea bream (*Sparus aurata*) was used as the model species, which is one of the most important farmed fish species in Europe [31]. After a feeding trial, different biometric, somatic and feed efficiency indexes were concomitantly analyzed with plasma levels of Gh and Igf-i hormones, as well as several parameters related to the metabolic and welfare status of the animals.

2. Materials and Methods

2.1. Animal Maintenance

Gilthead sea bream (*S. aurata*) juveniles were provided by a commercial source (PRE-DOMAR, Carboneras, Almeria) and acclimated to the indoor experimental facilities at the *Servicios Centrales de Investigación en Cultivos Marinos* (SC-ICM, CASEM, University of Cadiz, Puerto Real, Cadiz, Spain) with seawater in controlled conditions of salinity (36 ppt), temperature (19 °C), and under natural photoperiod at our latitude (36°31′45″ N, 6°11′31″ W, from October 2019 until January 2020). All assay procedures were conducted in their experimental facilities (Spanish Operational Code REGA ES11028000312).

2.2. Diets

Based on a standard commercial feed for gilthead sea bream (BioMar, Palencia; INICIO Plus 805–Crude protein: 50.0%, Crude fat: 18.0%, Digestible carbohydrates: 16.1%, Crude cellulose: 2.4%, Ash: 8.0%, phosphorous: 1.1%), experimental diets were prepared by its supplementation with the GHRP-6 peptide at the rate of 100 µg GHRP-6/kg of feed (D100) and 500 µg GHRP-6/kg of feed (D500), and a control diet in which only the excipients used for the addition of the peptide were added (see below). GHRP-6 (His-(D-Trp)-Ala-Trp-(D-Phe)-Lys-NH2, MW = 872.44 Da) with a purity >99% was provided by Sigma-Aldrich, USA. Aquafeeds were prepared as described by Adelmann et al. [32] with some modifications. Briefly, peptide reconstituted in PBS was added to a mixture of aluminum hydroxide (Al(OH)$_3$, 10%, Sigma-Aldrich) and polyethylene glycol 1000 (PEG, Sigma-Aldrich) previously melted at 37 °C in a water bath. The pulverized commercial diet was enriched with this suspension containing a final dose of 0, 100 and 500 µg GHRP-6/kg of feed, and then pellets were prepared by cold extrusion using a manual extruder. Subsequently, the feed was dried at 22 °C.

2.3. Experimental Design and Sampling Procedure

Fish with an average initial body mass of 20.6 ± 0.5 g and body length of 10.55 ± 0.08 cm (*n* = 270) were randomly distributed in nine 400 L tanks (*n* = 30 fish per tank, 90 fish per experimental diet, three repetitions per treatment) and maintained under constant conditions as described above and fed for 97 days. Experimental diets were offered to visual satiety (ad libitum) two times per day, ensuring that the amount offered in each experimental unit was completely ingested. The feeding test was carried out "blindly", in such a way that the three feeds were labelled with different colors but with no reference to its composition, eliminating any source of subjectivity when feeding the animals to obtain final results regarding the acceptance and growth performance. Fish were group-weighed and measured after 27 and 57 days of the beginning of the feeding trial. The feed intake was recorded every week, allowing to calculate the feeding efficiency = 100 x (wet weight gain/dry feed intake) for each experimental replicate. No mortalities were registered during the trial.

At the end of the trial (day 97), overnight fasted fish (4 fish per tank, 12 per experimental diet) were randomly selected and deeply anesthetized with 1 mL of 2-phenoxyethanol/L of seawater [6]. After being weighed and measured individually for body mass and length, blood and tissue samples were obtained. Blood was drawn from caudal vein with heparinized syringes, centrifuged at 3000× *g* for 20 min at 4 °C, and plasma samples were snap-frozen in liquid nitrogen and stored at −80 °C until biochemical and hormone analysis. Before tissue collection, fish were euthanized by cervical section, and livers and perivisceral fat were removed and weighed. Samples of liver and white skeletal muscle were rapidly taken, snap-frozen in liquid nitrogen, and stored at −80 °C until biochemical analyses. Intestine was taken for length measurements.

2.4. Growth Performance and Biometric Parameters

Growth parameters were also evaluated according to the following equations: (i) Condition Factor (K) = 100 × (body mass/fork length); (ii) weight gain (WG, %) = 100 × (body weigh increase/initial body weight); (iii) specific growth rate (SGR, %·day^{-1}) = [100 × (ln final body mass− ln initial body mass)]/days; and (iv) feed efficiency (FE) = weight gain/total feed intake. Survival was calculated by estimating the number of fish at the end of the experiment with respect to the number of fish at the beginning of the experiment.

Biometric indices were estimated in accordance with the following equations: (i) Hepatosomatic index (HSI) = 100 × (liver weight/fish body mass); (ii) Mesenteric fat index (MSI) = 100 × (mesenteric fat/fish body mass); and (iii) Intestine length index (ILI) = 100 × (Li/Lb), where Li and Lb are the intestine and fork body length, respectively.

2.5. Metabolites in Plasma and Tissues

Plasma total protein concentration was determined with a BCA Protein Assay Kit (PIERCE, Thermo Fisher Scientific, Rockford, IL, USA, #23225) using BSA as the standard, whereas plasma glucose (Ref. 1001200), lactate (Ref. 1001330), triglycerides (Ref. 1001311) and cholesterol (Ref. 41021) levels were measured using commercial kits from SPINREACT (Girona, Spain) adapted to 96-well microplates.

Frozen tissues used for the assays of metabolites were homogenized by ultrasonic disruption in 7.5 volumes of ice-cold 0.6 N perchloric acid, neutralized using 1 M KCO_3, and centrifuged (30 min, $3220\times g$ at 4 °C). Supernatants were used to measure tissue metabolites. Prior to the centrifugation, an aliquot was removed and frozen at -80 °C for triglyceride determination. Tissue triglycerides and lactate levels were determined spectrophotometrically with commercial kits (SPINREACT, see above). Tissue glycogen concentration was quantified using the method described by [33], where glucose obtained after glycogen breakdown with amyloglucosidase (Sigma-Aldrich A7420) was determined with a commercial kit (SPINREACT, Girona, Spain). All assays were performed using a Bio-Tek Power Wave 340 Microplate spectrophotometer using KCjunior Data Analysis Software (Bio-Tek Instruments, Winooski, VT, USA).

2.6. Hormones

Plasma cortisol levels were measured with a commercial Cortisol Enzyme Immunoassay Kit from ARBORASSAYS (Ref. #K003), whereas plasma Gh and Igf-i were measured through competitive inhibition ELISA using commercial kits (CSB-E12121Fh for GH and CSB-E12122Fh for IGF-I, CUSABIO). All assays were performed following the manufacturer's protocols.

2.7. Statistical Analyses

All data were checked for normality and homogeneity of variance using Kolmogorov–Smirnov and Levene's tests, respectively, with $p \leq 0.05$. Differences among treatments (CTRL, D100, D500) were evaluated using the one-way ANOVA ($p \leq 0.05$), except for the growth evolution (Figure 1), where significant differences were analyzed using two-way ANOVA followed by Tukey's test taking (i) diet and (ii) experimental time as the mean factors. In all cases, the Tukey's test ($p \leq 0.05$) was used to determine differences among means. For the values of the somatic index (MSI), the premises of a parametric test were not met, and a non-parametric Kruskal–Wallis test was performed. All results are expressed as the mean $\pm$ SEM (standard error of the mean). The software package GraphPad Prism 8.0 (GraphPad Software, Inc., San Diego, CA, US) was used for all tests performed and generated figures.

3. Results

3.1. Growth Performance and Biometric Parameters

In general, all fish groups increased their length from 10.5 cm to 14–15 cm and their weight from 20 g to 50–59 g, with an overall weight gain of 144−190% and specific growth rates of 0.92–1.10%·day^{-1} (Table 1). All the experimental groups presented similar growth trajectory during the first feeding periods (eight weeks), but reached a higher biomass at the end of the experiment in those fish groups that ingested the diets supplemented with the peptide (D100, 56.53 $\pm$ 1.56 g and D500, 58.50 $\pm$ 0.46 g), compared to the control diet (49.57 $\pm$ 1.45 g) (Figure 1). Both the WG and SGR were significantly higher with the diets supplemented with the peptide (Table 1). However, no significant differences were found in final length among the three treatments (Table 1). The analysis of the Condition Factor (K) showed significant differences in the weight–furcal length relationship, with an increase in this index in the experimental groups D100 and D500 with respect to the control group (Table 1).

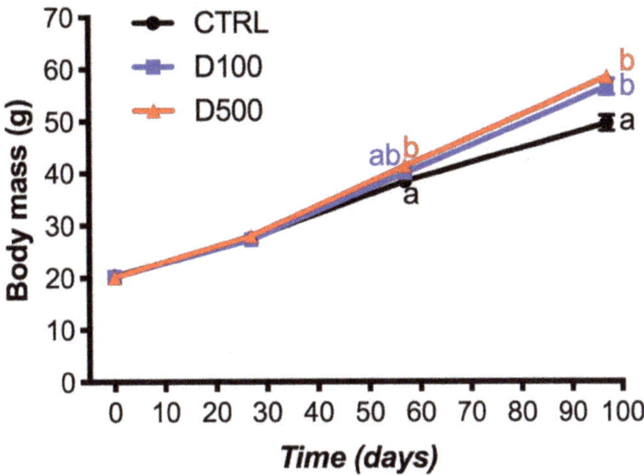

Figure 1. Weight increase as a function of days per experimental group. The results are expressed as the mean ± SEM of the triplicate tanks for each experimental group. Different letters in each group indicate significant differences among treatments (two-way ANOVA, followed by Tukey's test; $p <$ 0.05). CTRL: control; D100: 100 µg GHRP-6/kg of feed; D500: 500 µg GHRP-6/kg of feed.

Table 1. Growth performance and somatic indexes of juvenile gilthead sea breams fed to visual satiety from October 2019 to January 2020 (14 weeks). Data on body weight, feed intake and growth indexes are the mean ± SEM of triplicate tanks. Data on somatic indexes are the mean ± SEM of 12 fish. Different superscript letters in each row indicate significant differences among dietary treatments based on one-way ANOVA and Tukey's test ($p <$ 0.05). CTRL: control; D100: 100 µg GHRP-6/kg aquafeed; D500: 500 µg GHRP-6/kg aquafeed.

	CTRL	**D100**	**D500**	p [1]
Initial body weight (g)	20.32 ± 0.08	20.21 ± 0.05	20.13 ± 0.03	0.129
Final body weight (g)	49.57 ± 1.45 [a]	56.53 ± 1.56 [b]	58.50 ± 0.46 [b]	0.006
Final fork length (cm)	14.58 ± 0.14	14.50 ± 0.20	14.85 ± 0.20	0.374
K [2]	1.85 ± 0.03 [a]	1.99 ± 0.04 [b]	2.00 ± 0.05 [b]	0.015
Weight gain (%) [3]	144.0 ± 7.1 [a]	179.8 ± 7.6 [b]	190.6 ± 2.32 [b]	0.004
SGR (%) [4]	0.92 ± 0.03 [a]	1.07 ± 0.03 [b]	1.10 ± 0.01 [b]	0.004
Feed intake (g DM/fish)	44.67 ± 0.71 [a]	47.01 ± 0.69 [b]	48.36 ± 0.58 [b]	0.017
FE (%) [5]	63 ± 2 [a]	77 ± 2 [b]	79 ± 1 [b]	<0.001
HSI (%) [6]	1.59 ± 0.07	1.62 ± 0.08	1.72 ± 0.10	0.491
MSI (%) [7]	0.69 ± 0.05	0.59 ± 0.12	0.56 ± 0.05	0.120
ILI (%) [8]	107.4 ± 5.8	105.0 ± 7.9	109.3 ± 7.5	0.914

[1] Values resulting from one-way analysis of variance, [2] Condition Factor = (100 × body weight)/fork length, [3] weight gain (%) = (100 × (body weigh increase)/initial body weight, [4] specific growth rate (%) = 100 × (ln final body weight − ln initial body weight)/days, [5] feed efficiency (%) = 100 × (wet weight gain/dry feed intake), [6] Hepatosomatic index = (100 × liver weight)/fish weight, [7] Mesenteric fat index = (100 × mesenteric fat weight)/fish weight, [8] Intestine length index = (100 × intestine length)/standard length.

No differences were found in feed intake among the two supplemented diets (D100 and D500). However, fish fed with these two diets exhibited a significantly higher fed intake with respect to the CTRL group. In addition, feed efficiency increased significantly with GHRP-6 peptide supplementation, from 0.63 in fish fed the CTRL diet to 0.77 and 0.79 for D100 and D500 groups, respectively (Table 1). No differences were observed among treatments in the calculated organosomatic indexes HIS, MSI, and ILI (Table 1).

3.2. Metabolites in Plasma and Tissues

No differences were found among the three experimental diets in relation to circulating levels of glucose, triglycerides and proteins (Table 2). However, lactate values showed a significant decrease in fish fed with the D100 diet (2.39 ± 0.22 mM) compared to the CTRL group (3.14 ± 0.24 mM) and to the experimental diet D500 (3.25 ± 0.19 mM) (Table 2). Plasma cholesterol was significantly higher in fish fed both supplemented diets with respect to the CTRL group (Table 2). In the liver, no differences were found on the content of triglycerides and glucose among treatments (Table 2). Nevertheless, a significant improvement in glycogen reserves was detected in fish ingesting the diet with the highest dose of the peptide (D500) (Table 2). No changes were found in the level of glucose, glycogen, triglycerides and lactate in the white skeletal muscle among fish fed the two different experimental diets and the control (Table 2).

Table 2. Blood and tissue biochemistry of juvenile gilthead sea breams fed to visual satiety from October 2019 to January 2020 (14 weeks). Data are the mean $\pm$ SEM of 12 fish. Different superscript letters in each row indicate significant differences among dietary treatments based on one-way ANOVA and Tukey's test ($p < 0.05$). CTRL: control; D100: 100 µg GHRP-6/kg aquafeed; D500: 500 µg GHRP-6/kg aquafeed. [1] Values resulting from one-way analysis of variance.

	CTRL	**D100**	**D500**	p [1]
Plasma glucose (mM)	1.67 ± 0.05	1.59 ± 0.02	1.69 ± 0.03	0.089
Plasma lactate (mM)	3.14 ± 0.24 [a]	2.39 ± 0.22 [b]	3.25 ± 0.19 [a]	0.013
Plasma triglycerides (mM)	2.61 ± 0.24	2.41 ± 0.21	2.29 ± 0.14	0.529
Plasma proteins (mg·mL^{-1})	38.85 ± 2.79	40.25 ± 1.65	41.34 ± 1.58	0.699
Plasma cholesterol (mg·dL^{-1})	258.7 ± 8.9 [a]	326.7 ± 12.4 [b]	318.1 ± 10.9 [b]	<0.001
Plasma cortisol (ng·mL^{-1})	15.19 ± 1.14	16.04 ± 0.95	15.84 ± 0.73	0.808
Hepatic glucose (µmol·gww^{-1})	2.87 ± 0.20	2.56 ± 0.24	3.31 ± 0.34	0.150
Hepatic glycogen (µmol·gww^{-1})	19.77 ± 0.50 [a]	20.52 ± 0.73 [a]	24.33 ± 1.49 [b]	0.007
Hepatic triglycerides (µmol·gww^{-1})	137.6 ± 13.1	148.6 ± 11.0	147.0 ± 14.0	0.807
Muscular glucose (µmol·gww^{-1})	1.94 ± 0.19	1.97 ± 0.22	2.05 ± 0.20	0.939
Muscular glycogen (µmol·gww^{-1})	1.02 ± 0.19	1.13 ± 0.26	1.38 ± 0.27	0.556
Muscular triglycerides (µmol·gww^{-1})	80.7 ± 7.9	78.1 ± 5.7	81.4 ± 7.2	0.941
Muscular lactate (µmol·gww^{-1})	51.66 ± 3.54	56.36 ± 2.71	62.80 ± 3.12	0.057

3.3. GH, IGF-I, and Cortisol in Plasma

In general, low amplitude in the response of plasma GH and IGF-I was observed after the supplementation of feed with the peptide, although a positive and significant effect was observed in GH plasma levels in a dose-dependent manner (Figure 2A). IGF-I values on fish fed with D100 diet increased slightly over those observed in controls, while the level of IGF-I was significantly lower in fish ingesting the D500 diet with respect to those who ingested the D100 diet (Figure 2B). This resulted in a statistically significant reduction in IGF-I/GH ratio in animals ingesting the D500 diet with respect to both CTRL and D100 diets (Figure 2C). No differences were found in the circulant levels of cortisol among experimental diets (Table 2).

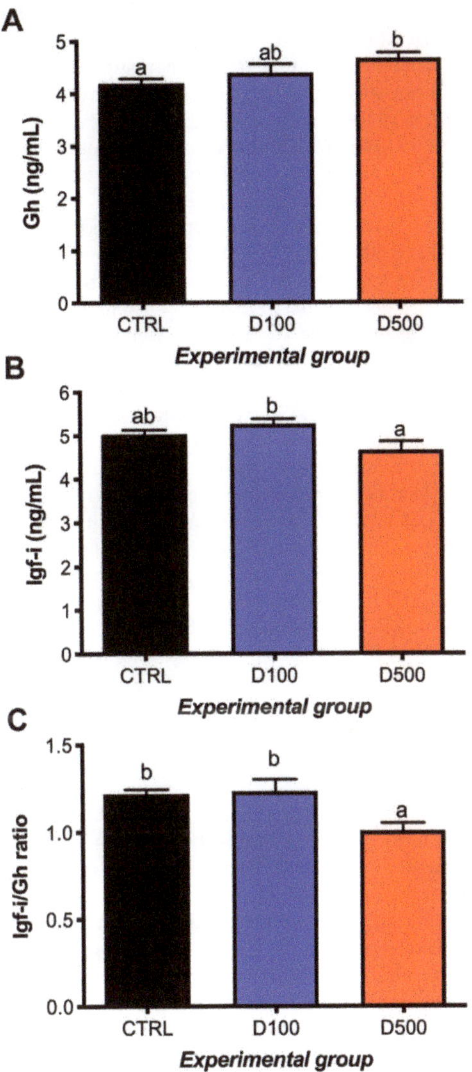

Figure 2. Plasma Gh (**A**) and Igf-ilevels (**B**) in gilthead seabream juveniles fed with GHRP-6 peptide supplementation. Igf-i/Gh ratio (**C**). The results are expressed as the mean ± SEM of 12 fish for each experimental group. Different letters in each group indicate significant differences among treatments (one-way ANOVA, followed by Tukey's test; $p < 0.05$). CTRL: control; D100: 100 µg GHRP-6/kg of feed; D500: 500 µg GHRP-6/kg of feed.

4. Discussion

There are currently numerous studies on the effects of different additives, both natural and synthetic, incorporated into diets to increase growth performance in fish species of interest for aquaculture production. The addition of the peptide GHRP-6 or other ghrelin homologues in commercial feed may represent a suited change in the composition of future diets aimed at maximizing fish growth performance, at least for some developmental stages, which would translate into a greater aquaculture production and efficiency. However, this approach has not been assessed before in a marine fish. In this study, we analyzed for the first time the effect of this peptide on growth performance for a marine fish, the gilthead sea bream (*S. aurata*). In general, all the growth indices showed an increase with the addition of the peptide to the diets, compared to the non-supplemented control diet (Table 1), whereas survival was not affected during the trial with any of the doses tested (no mortality was observed in any of the experimental tanks).

Even at the lowest inclusion level, this peptide had a positive effect on the final body mass, as well as on the specific growth rate when included in diets and offered in a medium-term feeding trial (Figure 1, Table 1). Similar results were found in tilapia juveniles (*Oreochromis* sp.), where the groups supplemented with concentrations of 100 µg GHRP-6/kg of feed and 500 µg GHRP-6/kg of feed showed a significant increase in their body weight [27]. Furthermore, in previous studies carried out with other ghrelin homologues, such as the A233 peptide in a proportion of 600 µg A233/kg of feed, groups supplemented with the peptide also increased their body weight compared to the control diet [34]. In this regard, it is important to relate the data obtained on the increase in body weight (WG) with the increase in the size of the animals to know if isometric growth is occurring, evidenced by an increase in biomass proportional to the size, or not [35]. To confirm this relationship, the Condition Factor (K) was used, which represents the relationship between the body mass of each individual versus their length (furcal) [36]. This index is commonly used to compare the condition or wellbeing of fish, whereas its optimum dependents on the species and age/size. Optimum values in the case of the gilthead sea bream vary from 1.5 to 2.5 [37]. Our results reveal that all groups presented a Condition Factor index within the optimal range for this species, although diets supplemented with the peptide GHRP-6 had a higher value (Table 1). Thus, there may be an added benefit of these supplemented feeds if they promote fast muscle growth instead of fat deposition. Indeed, a strong positive correlation is known to occur between K and total lipid content in fish [38]. Other main contributing factors such as gonad weight (i.e., maturation) and gut fill are not issues in our experiment, as fish were immature juveniles, and they were fasted overnight before samplings. It is worthy to note that HSI and MSI indices did not show significant variations, indicating that none of the diets tested caused higher hepatic or mesenteric accumulation of fat than that produced by the control diet. This is also supported by the absence of differences in triglyceride content among dietary treatment in the tissues examined. Thus, the observed increase in K factor can be considered be to results from somatic growth, and indicate that supplemented diets orchestrated a dietary energy partitioning that favors metabolism and growth rather than accumulation, thus not implying undesirable fatter fish. Indeed, this issue is strongly suggested since triglycerides did not show variations in muscle. Whether this effect could be produced by hyperplasia and/or hypertrophy deserves further studies.

Growth enhancement was accompanied in our study by an increase in plasma Gh levels in a dose-dependent manner with respect to the peptide inclusion. This effect was expected, as the GHRP-6 is considered a strong GHS. Previous studies showed that the GHRP-6 is able to stimulate Gh secretion in tilapia primary cultures of pituitary cells [25,26]. The GHS dependent increase in serum Gh in tilapia juveniles also increased transcription of Igf-i in the liver, when the peptide was administered intraperitoneally and orally [27]. However, although we did not measure gene expression in liver, we found no evidence of this effect. Indeed, in our study plasma Igf-i levels decreased with the D500 diet (Figure 2B). This agrees with the previously reported inverse correlation between circulating Gh and Igf-i in sea bream, likely due to an Igf-i feedback inhibition on pituitary Gh synthesis and secretion [39]. In spite of this, the improved growth response was somewhat unexpected given that plasma Igf-i reflects differences in growth potentiality between *S. aurata* strains [40] and genetic families [41], as occurs in other fish because of its widely known stimulating effect on myogenic cell proliferation, differentiation, and protein synthesis. Yet, it has been suggested that the correlation between Igf-i and growth can reflect scarce variations in fish [14], and it can be variable across fish species and physiological contexts due to the actions of a wide range of endogenous and exogenous factors [14,42]. For instance, a certain seasonal lag has been found among circulating Igf levels and growth rates in the production cycle of gilthead sea bream [43]. In our study, the inclusion of the peptide induced the secretion of Gh but also increased feed intake (see below), and this may promote different physiological responses depending on the doses of the peptide. For instance, higher ingestion may lead to higher energy availability and storage, but concomitantly an increase in Gh levels may favor a flux of lipids from adipose tissue toward the skeletal muscle to fuel growth, because of its lipolytic action [39]. Even so, more studies are required on the mode of action of GHRP-6 in marine fish, which may differ to some extent from freshwater fish.

On the other hand, both experimental diets (D100 and D500) showed higher feeding efficiency (FE), which translates into better feed conversion. The control group showed 62% conversion, while the experimental groups presented higher values with 79% and 80% conversion for groups D100 and D500, respectively, which represents an improvement of 17–18%. These results are interesting from the aquaculture industry perspective, given that the efficiencies shown here at a constant temperature of 19 °C are close to those reported for this species of fish grown in summer conditions, with temperature between 23 °C and 27 °C, and with higher metabolic rates and feed intake [5,40,41]. However, the peptide produced an increment in feed intake at 19 °C with respect to that of control fish, which can be understood as a likely promoted metabolism and growth at low temperature. Indeed, plasma cholesterol and hepatic glycogen reflected the variations observed in feed intake. In line with this, ghrelin is involved in the regulation of feed intake and is known as the hunger hormone [14]. Previous studies in other species such as golden carp (*Carassius auratus*) or tilapia using ghrelin analog peptides through different procedures (e.g., oral administration or intraperitoneal injection) produce increases in the feed intake [27,28,34,44,45]. These previous results are in accordance with the observations of this work. The positive effect of GHRP-6 on feed intake of the gilthead sea bream can be considered as a good feature in terms of production together with the higher FE registered.

Finally, the ILI did not show differences among the three treatment groups. It is known that there is a strong correlation between diet and ILI, where animals that are naturally fed diets based mainly on plant compounds (herbivores) have longer intestines than animals naturally fed higher levels of animal protein (piscivores) [46]. However, some carnivore fish exhibit certain flexibility, and when exposed to changes in their usual diet toward more vegetable proteins, their intestines undergo morphological adaptations for a better use of the new diet. This feature has been previously shown in the gilthead sea bream [6,41]. In our study, we used a specific diet for this species with a high content of animal protein, which was only supplemented with the studied peptide and without introducing major changes in its formulation. Therefore, it was expected that the animals did not have to adapt their absorption surface (i.e., intestinal length). However, more studies are required to know if the improvement on FE index may be influenced by other changes at the intestine, such as changes in the conformation of the enterocytes (e.g., increased surface area or length of the microvilli), changes in paracellular routes that improve the absorption of nutrients, or even changes in the trans-epithelial selectivity of the intestine [4].

In addition, no significant differences were found in glucose values in the tissues analyzed (Table 2). It has been proposed that glycolysis in fish is more important as a supplier of biosynthetic products than as a way of producing pyruvate for its subsequent oxidation [47]. The glucose values obtained, being similar for the three diets, show that the experimental peptide GHRP-6 incorporated by ingestion does not alter the metabolic pathways related to glucose utilization. This result matches with previous observations in studies focused on the incorporation of other additives in the teleost diet, such as tryptophan, heptanoate or compounds extracted from medicinal herbs and microalgae [5,48,49], where the animals under non-stressful conditions presented a balance in glucose values similar to those observed in this experiment. Lactate, on the other hand, presents lower values for the D100 diet than in the CTRL and D500 diets. This result suggests that the D100 diet may favor oxidative over anaerobic metabolism in white skeletal muscle, or that lactate uptake and elimination by the liver or other tissues is encouraged. Thus, lactate production or accumulation after high metabolic demand caused by stress, physical exercise or oxygen concentrations below that needed to sustain mitochondrial aerobic activity has been previously demonstrated in this fish species [50,51]. Although, our results were obtained in free-swimming and resting fish, and low plasma cortisol levels demonstrated that all groups maintain a homeostatic state. The positive effect of the D100 diet on aerobic metabolism may be the result of more metabolites being metabolized in the mitochondria through aerobic processes, such as fatty acids or glucose. This fact could suggest a greater energy production without the need to require anaerobic routes more than its production and withdrawal for extra energy supply, thus presenting a homeostatic load with similar levels in the other energy substrates analyzed (glucose, TAG and proteins) with the control.

5. Conclusions and Open Issues

In the present study, we demonstrated for the first time that GHRP-6 stimulates growth performance in juveniles of a marine fish, the gilthead sea bream, when administered as a feed additive. In general, growth is favored by diets supplemented with both concentrations of the peptide, even under a temperature condition (19 °C) far from the optimal for this species growth. On the other hand, no physiological or metabolic alterations were detected in studied individuals, finding even better aerobic food management by the specimens fed the D100 diet. The higher inclusion of GHRP-6 (500 µg/kg) significantly increased plasma Gh levels with respect to controls, in agreement with the GH secretagogue effects of ghrelin. However, similar growth outcome and differences between GHRP-6 inclusion levels in aerobic metabolism (D100), and Gh and Igf-i plasma levels, suggest that improvements in culture performance may have occurred through different mechanisms. Thus, this peptide can be considered as a viable supplement for increasing the production efficiency of gilthead sea bream although more studies are required to better understand its mode of action at different inclusion levels, including to delve into the biological effects of GHS on marine fish growth and feed efficiency. On the other hand, it must be considered that the use of this peptide as an additive during the processing of commercial foods can be developed to benefit animal health and nutrition [52], and though steam cooking and extrusion at high temperatures can lead to its partial denaturation, results reported in this work strongly suggest the stability of the compound during manipulation. Future studies that address the appropriate way to incorporate this peptide in commercial diets are necessary. For example, a recent study in rainbow trout tested the inclusion of the neuropeptide PACAP (pituitary adenylate cyclase activating peptide) in the diet through an oil-based preformulation of the peptide using water-in-oil (W/O) emulsions from a dispersed aqueous phase and a continuous oily phase, with good results [53]. This work opens a new path for deeper investigations of the effects and possible applications of the GHRP-6 peptide as an additive in diets for farmed marine fish.

Author Contributions: Conceived and designed the experiment, L.R.-V., R.M., J.A.M.-S., J.M.M. and M.P.E.; performed the experiments, L.R.-V., I.M., J.A.M.-S. and E.P.; analyzed the data, L.R.-V., J.A.M.-S., I.M., E.P. and R.M.; interpreted the results, L.R.-V., J.A.M.-S., R.M., E.P. and J.M.M.; wrote the manuscript, L.R.-V., J.A.M.-S., R.M. and E.P.; writing—review and editing, L.R.-V., J.A.M.-S., R.M., E.P., J.M.M. and M.P.E.; project administration and funding acquisition, J.A.M.-S. and J.M.M. All authors have read and agreed to the published version of the manuscript.

Funding: This research was supported under the Fellowship for short stays to prestigious researchers awarded by the University of Cadiz (reference EST2019-021) to L.R.-V. Additional funding was obtained by the projects AGL2016-76069-C2-1-R of MICINN (Spain) to J.M.M, and co-financed by the European Union under the 2014–2020 ERDF Operational Programme and by the Department of Economic Transformation, Industry, Knowledge, and Universities of the Regional Government of Andalusia (Project reference: FEDER-UCA18-107182) to J.A.M.-S. J.M.M. and J.A.M.-S. belong to the Fish Welfare and Stress Network (AGL2016-81808-REDT) supported by the Spanish Ministry of Economy, Industry and Competitiveness.

Institutional Review Board Statement: The study was conducted following the guidelines for experimental procedures in animal research from the Ethics and Animal Welfare Committee of the University of Cadiz, according to the Spanish (RD53/2013) and European Union (2010/63/UE) legislation. The Ethical Committee from the Autonomous Andalusian Government approved the experiments (Junta de Andalucía reference number 04/04/2019/056).

Data Availability Statement: The data that support the findings of this study are all presented in the figures and tables, as well as available from the authors upon reasonable request.

Acknowledgments: The authors wish to thank Servicios Centrales de Investigación en Cultivos Marinos (SCI-CM, CASEM, University of Cádiz, Puerto Real, Cádiz, Spain) for providing experimental fish and for their excellent technical assistance. In addition, we acknowledge the support of the University of Almería (Experimental Feeds Service, https://www.ual.es/universidad/serviciosgenerales/stecnicos/perifericos-convenio/piensos-experimentales; accessed on 10 January 2022) on aquafeed elaboration.

Conflicts of Interest: The authors declare no conflict of interest.

References

1. FAO. Meeting the sustainable development goals. In *The State of World Fisheries and Aquaculture 2018*; FAO: Rome, Italy, 2018; Available online: https://www.fao.org/documents/card/es/c/I9540EN (accessed on 26 January 2022).
2. Bertucci, J.I.; Blanco, A.M.; Sundarrajan, L.; Rajeswari, J.J.; Velasco, C.; Unniappan, S. Nutrient Regulation of Endocrine Factors Influencing Feeding and Growth in Fish. *Front. Endocrinol.* **2019**, *10*, 83. [CrossRef] [PubMed]
3. Vélez, E.J.; Unniappan, S. A Comparative Update on the Neuroendocrine Regulation of Growth Hormone in Vertebrates. *Front. Endocrinol.* **2021**, *11*, 1174. [CrossRef] [PubMed]
4. Estensoro, I.; Ballester-Lozano, G.; Benedito-Palos, L.; Grammes, F.; Martos-Sitcha, J.A.; Mydland, L.T.; Calduch-Giner, J.A.; Fuentes, J.; Karalazos, V.; Ortiz, A.; et al. Dietary butyrate helps to restore the intestinal status of a marine teleost (*Sparus aurata*) fed extreme diets low in fish meal and fish oil. *PLoS ONE* **2016**, *11*, e0166564. [CrossRef]
5. Martos-Sitcha, J.A.; Simó-Mirabet, P.; Piazzon, M.C.; de las Heras, V.; Calduch-Giner, J.A.; Puyalto, M.; Tinsley, J.; Makol, A.; Sitjà-Bobadilla, A.; Pérez-Sánchez, J. Dietary sodium heptanoate helps to improve feed efficiency, growth hormone status and swimming performance in gilthead sea bream (*Sparus aurata*). *Aquac. Nutr.* **2018**, *24*, 1638–1651. [CrossRef]
6. Perera, E.; Sánchez-Ruiz, D.; Sáez, M.I.; Galafat, A.; Barany, A.; Fernández-Castro, M.; Antonio Jesús Vizcaíno, A.J.; Fuentes, J.; Martínez, T.F.; Mancera, J.M.; et al. Low dietary inclusion of nutraceuticals from microalgae improves feed efficiency and modifies intermediary metabolisms in gilthead sea bream (*Sparus aurata*). *Sci. Rep.* **2020**, *10*, 1–14. [CrossRef]
7. Khatib, N.; Gaidhane, S.; Gaidhane, A.M.; Khatib, M.; Simkhada, P.; Gode, D.; Zahiruddin, Q.S. Ghrelin: Ghrelin as a regulatory Peptide in growth hormone secretion. *J. Clin. Diagn Res.* **2014**, *8*, MC13–MC17. [CrossRef]
8. Kojima, M.; Hosoda, H.; Date, Y.; Nakazato, M.; Matsuo, H.; Kangawa, K. Ghrelin is a growth-hormone-releasing acylated peptide from stomach. *Nature* **1999**, *402*, 656–660. [CrossRef]
9. Kaiya, H.; Kangawa, K.; Miyazato, M. What is the general action of ghrelin for vertebrates? Comparisons of ghrelin's effects across vertebrates. *Gen. Comp. Endocrinol.* **2013**, *181*, 187–191. [CrossRef]
10. Riley, L.G.; Hirano, T.; Grau, E.G. Rat ghrelin stimulates growth hormone and prolactin release in the tilapia, *O. mossambicus*. *Zool. Sci.* **2002**, *19*, 797–800. [CrossRef]
11. Kaiya, H.; Miyazato, M.; Kangawa, K.; Peter, R.E.; Unniappan, S. Ghrelin: Amultifunctional hormone in non-mammalian vertebrates. *Comp. Biochem. Physiol. A* **2008**, *149*, 109–128. [CrossRef]
12. Kaiya, H.; Miyazato, M.; Kangawa, K. Recent advances in the phylogenetic study of ghrelin. *Peptides* **2011**, *32*, 2155–2174. [CrossRef]
13. Jönsson, E. The role of ghrelin in energy balance regulation in fish. *Gen. Comp. Endocrinol* **2013**, *187*, 79–85. [CrossRef] [PubMed]
14. Perelló-Amorós, M.; Vélez, E.J.; Vela-Albesa, J.; Sánchez-Moya, A.; Riera-Heredia, N.; Hedén, I.; Fernández-Borràs, J.; Blasco, J.; Calduch-Giner, J.A.; Navarro, I.; et al. Ghrelin and Its Receptors in Gilthead Sea Bream: Nutritional Regulation. *Front. Endocrinol.* **2018**, *9*, 399. [CrossRef] [PubMed]
15. Navarro-Guillén, C.; Dias, J.; Rocha, F.; Castanheira, M.F.; Martins, C.I.; Laizé, V.; Gavaiaa, P.J.; Engrola, S. Does a ghrelin stimulus during zebrafish embryonic stage modulate its performance on the long-term? *Comp. Biochem. Physiol. A* **2019**, *228*, 1–8. [CrossRef] [PubMed]
16. Navarro-Guillén, C.; Yufera, M.; Engrola, S. Ghrelin in Senegalese sole (*Solea senegalensis*) post-larvae: Paracrine effects on food intake. *Comp. Biochem. Physiol. A* **2017**, *204*, 85–92. [CrossRef]
17. Terova, G.; Rimoldi, S.; Bernardini, G.; Gornati, R.; Saroglia, M. Sea bass ghrelin: Molecular cloning and mRNA quantification during fasting and refeeding. *Gen. Comp. Endocrinol.* **2008**, *155*, 341–351. [CrossRef]
18. Momany, F.A.; Bowers, C.Y.; Reynolds, G.A.; Chang, D.; Hong, A.; Newlander, K. Design, synthesis and biological activity of peptides which release growth hormone, in vitro. *Endocrinology* **1981**, *108*, 31–39. [CrossRef]

19. Camanni, F.; Ghigo, E.; Arvat, E. Growth Hormone-Releasing Peptides and Their Analogs. *Front. Neuroendocrinol.* **1998**, *19*, 47–72. [CrossRef]

20. Hashizume, T.; Sasaki, K.; Sakai, M.; Tauchi, S.; Masuda, H. The effect of new growth hormone-releasing peptide (KP 102) on the release of growth hormone in goats. *Anim. Sci. Technol.* **1997**, *68*, 247–256.

21. Roh, S.G.; Lee, H.G.; Phung, L.T.; Hidari, H. Characterization of Growth Hormone Secretion to Growth Hormone releasing Peptide-2 in Domestic Animals-A Review. *Asian-Australas J. Anim. Sci.* **2002**, *15*, 757–766. [CrossRef]

22. Chan, C.B.; Fung, C.K.; Fung, W.; Margaret, C.L.; Cheng, C.H. Stimulation of growth hormone secretion from seabream pituitary cells in primary culture by growth hormone secretagogues is independent of growth hormone transcription. *Comp. Biochem. Physiol. C* **2004**, *139*, 77–85. [CrossRef] [PubMed]

23. Bowers, C.Y.; Momany, F.A.; Reynolds, G.A.; Hong, A. On the in vitro and in vivo activity of a new synthetic hexapeptide that acts on the pituitary to specifically release growth hormone. *Endocrinology* **1984**, *114*, 1537–1545. [CrossRef] [PubMed]

24. Yahashia, S.; Kang, K.S.; Kaiya, H.; Matsudaa, K. GHRP-6 mimics ghrelin-induced stimulation of food intake and suppression of locomotor activity in goldfish. *Peptides* **2012**, *34*, 324–328. [CrossRef] [PubMed]

25. Lugo, J.M.; Rodríguez, A.; Helguera, Y.; Morales, R.; González, O.; Acosta, J.; Besada, V.; Sánchez, A.; Estrada, M.P. Recombinant novel pituitary adenylate cyclase activating polypeptide (PACAP) from African catfish (*Clarias gariepinus*) authenticates its biological function as a growth promoting factor in lower vertebrates. *J. Endocrinol.* **2008**, *197*, 583–597. [CrossRef]

26. Martinez, R.; Ubieta, K.; Herrera, F.; Forellat, A.; Morales, R.; de la Nuez, A.; Rodriguez, R.; Reyes, O.; Oliva, A.; Estrada, M.P. A novel GH secretagogue, A233, exhibits enhanced growth activity and innate immune system stimulation in teleosts fish. *J. Endocrinol.* **2012**, *214*, 409–419. [CrossRef]

27. Martinez, R.; Carpio, Y.; Morales, A.; Lugo, J.M.; Herrera, F.; Zaldívar, C.; Carrillo, O.; Arenal, A.; Pimentel, E.; Estrada, M.P. Oral administration of the growth hormone secretagogue-6 (GHRP-6) enhances growth and non-specific immune responses in tilapia (*Oreochromis* sp.). *Aquaculture* **2016**, *452*, 304–310. [CrossRef]

28. Lugo, J.M.; Oliva, A.; Morales, A.; Reyes, O.; Garay, H.E.; Herrera, F.; Cabrales, R.; Perez, E.; Estrada, M.P. The biological role of pituitary adenylate cyclase-activating polypeptide (PACAP) in growth and feeding behavior in juvenile fish. *J. Pept. Sci.* **2010**, *16*, 633–643. [CrossRef]

29. Shepherd, B.S.; Eckert, S.M.; Parhar, I.S.; Vijayan, M.M.; Wakabayashi, I.; Hirano, T. The hexapeptide KP-102 (D-ala-D-beta-Nal-ala-trp-D-phe-lys-NH(2) stimulates growth hormone release in a cichlid fish (*Oreochromis mossambicus*). *J. Endocrinol.* **2000**, *167*, 7–10. [CrossRef]

30. Shepherd, B.S.; Johnson, J.K.; Silverstein, J.T.; Parhar, I.S.; Vijayan, M.M.; McGuire, A.; Weber, G.M. Endocrine and orexigenic actions of growth hormone secretagogues in rainbow trout (*Oncorhynchus mykiss*). *Comp. Biochem. Physiol. A* **2007**, *146*, 390–399. [CrossRef]

31. Rigos, G.; Kogiannou, D.; Padrós, F.; Cristofol, C.; Florio, D.; Fioravanti, M.; Zarza, C. Best therapeutic practices for the use of antibacterial agents in finfish aquaculture: A particular view on European seabass (*Dicentrarchus labrax*) and gilthead seabream (*Sparus aurata*) in Mediterranean aquaculture. *Rev. Aquac.* **2021**, *13*, 1285–1323. [CrossRef]

32. Adelmann, M.; Kollner, B.; Bergmann, S.M.; Fischer, U.; Lange, B.; Weitschies, W.; Enzmann, P.J.; Fichtner, D. Development of an oral vaccine for immunisation of rainbow trout (*Oncorhynchus mykiss*) against viral haemorrhagic septicaemia. *Vaccine* **2008**, *26*, 837–844. [CrossRef] [PubMed]

33. Keppler, D.; Decker, K. Glycogen. Determination with amyloglucosidase. In *Methods of Enzymatic Analysis*; Bergmeyer, H.U., Ed.; Academic Press: New York, NY, USA, 1974; pp. 1127–1131.

34. Martínez, R.; Morales, C.; Arenal, A.; Morales, A.; Herrera, F.; González, V.; Estrada, M.P. Growth Hormone Secretagogue (A233) Improves Growth and Changes the Tissue Fatty Acid Profile in Juvenile Tilapia (*Oreochromis niloticus*). *Lipids* **2018**, *53*, 429–436. [CrossRef] [PubMed]

35. Cifuentes, R.; González, J.; Montoya, G.; Jara, A.; Ortíz, N.; Piedra, P.; Habit, E. Relación longitud-peso y factor de condición de los peces nativos del río San Pedro (cuenca del río Valdivia, Chile). *Gayana (Concepción)* **2012**, *76*, 86–100. [CrossRef]

36. Froese, R. Cube law, condition factor and weight–length relationships: History, meta-analysis and recommendations. *J. Appl. Ichthyol.* **2006**, *22*, 241–253. [CrossRef]

37. Ortega, A. *Cultivo de Dorada (Sparus aurata)*; Fundación Observatorio Español de Acuicultura, Consejo Superior de Investigaciones Científicas, Ministerio de Medio Ambiente y Medio Rural y Marino: Madrid, Spain, 2008; pp. 1–44.

38. Herbinger, C.M.; Friars, G.W. Correlation between condition factor and total lipid content in Atlantic salmon, *Salmo salar* L., parr. *Aqua Res.* **1991**, *22*, 527–529. [CrossRef]

39. Pérez-Sánchez, J. The involvement of growth hormone in growth regulation, energy homeostasis and immune function in the gilthead sea bream (*Sparus aurata*): A short review. *Fish Physiol Biochem.* **2000**, *22*, 135–144. [CrossRef]

40. Simó-Mirabet, P.; Felip, A.; Estensoro, I.; Martos-Sitcha, J.A.; de las Heras, V.; Calduch-Giner, J.; Puyalto, M.; Karalazos, V.; Sitjà-Bobadilla, A.; Pérez-Sánchez, J. Impact of low fish meal and fish oil diets on the performance, sex steroid profile and male-female sex reversal of gilthead sea bream (*Sparus aurata*) over a three-year production cycle. *Aquaculture* **2018**, *490*, 64–74. [CrossRef]

41. Perera, E.; Simó-Mirabet, P.; Shin, H.S.; Rosell-Moll, E.; Naya-Catalá, F.; de las Heras, V.; Martos-Sitcha, J.A.; Karalazos, V.; Armero, E.; Pérez-Sánchez, J. Selection for growth is associated in gilthead sea bream (*Sparus aurata*) with diet flexibility, changes in growth patterns and higher intestine plasticity. *Aquaculture* **2019**, *507*, 349–360. [CrossRef]

42. Triantaphyllopoulos, K.A.; Cartas, D.; Miliou, H. Factors influencing GH and IGF-I gene expression on growth in teleost fish: How can aquaculture industry benefit? *Rev. Aquac.* **2020**, *12*, 1637–1662. [CrossRef]
43. Mingarro, M.; de Celis, S.V.R.; Astola, A.; Pendón, C.; Valdivia, M.M.; Pérez-Sánchez, J. Endocrine mediators of seasonal growth in gilthead sea bream (*Sparus aurata*): The growth hormone and somatolactin paradigm. *Gen. Comp. Endocrinol.* **2002**, *128*, 102–111. [CrossRef]
44. Unniappan, S.; Lin, X.; Cervini, L.; Rivier, J.; Kaiya, H.; Kangawa, K.; Peter, R.E. Goldfish ghrelin: Molecular characterization of the complementary deoxyribonucleic acid, partial gene structure and evidence for its stimulatory role in food intake. *Endocrinology* **2002**, *143*, 4143–4146. [CrossRef] [PubMed]
45. Unniappan, S.; Peter, R.E. In vitro and in vivo effects of ghrelin on luteinizing hormone and growth hormone release in goldfish. *Am. J. Physiol.—Regul. Integr. Comp. Physiol.* **2004**, *286*, 1093–1101. [CrossRef] [PubMed]
46. Wagner, C.E.; McIntyre, P.B.; Buels, K.S.; Gilbert, D.M.; Michel, E. Diet predicts intestine length in Lake Tanganyika's cichlid fishes. *Funct. Ecol.* **2009**, *23*, 1122–1131. [CrossRef]
47. Polakof, S.; Panserat, S.; Soengas, J.L.; Moon, T.W. Glucose metabolism in fish: A review. *J. Comp. Physiol. B* **2012**, *182*, 1015–1045. [CrossRef]
48. Saccol, E.M.H.; Parrado-Sanabria, Y.A.; Gagliardi, L.; Jerez-Cepa, I.; Mourão, R.H.V.; Heinzmann, B.M.; Baldisserotto, B.; Pavanato, M.A.; Mancera, J.M.; Martos-Sitcha, J.A. *Myrcia sylvatica* essential oil in the diet of gilthead sea bream (*Sparus aurata* L.) attenuates the stress response induced by high stocking density. *Aquac. Nutr.* **2018**, *24*, 1381–1392. [CrossRef]
49. Azeredo, R.; Machado, M.; Martos-Sitcha, J.A.; Martínez-Rodríguez, G.; Moura, J.; Peres, H.; Oliva-Teles, A.; Afonso, A.; Mancera, J.M.; Costas, B. Dietary tryptophan induces opposite health-related responses in the Senegalese sole (*Solea senegalensis*) reared at low or high stocking densities with implications in disease resistance. *Front. Physiol.* **2019**, *10*, 508. [CrossRef]
50. Magnoni, L.J.; Martos-Sitcha, J.A.; Queiroz, A.; Calduch-Giner, J.A.; Gonçalves, J.F.M.; Rocha, C.M.; Abreu, H.T.; Scharma, J.W.; Ozorio, R.O.A.; Pérez-Sánchez, J. Dietary supplementation of heat-treated *Gracilaria* and *Ulva* seaweeds enhanced acute hypoxia tolerance in gilthead sea bream (*Sparus aurata*). *Biol. Open* **2017**, *6*, 897–908. [CrossRef]
51. Naya-Català, F.; Martos-Sitcha, J.A.; de las Heras, V.; Simó-Mirabet, P.; Calduch-Giner, J.À.; Pérez-Sánchez, J. Targeting the Mild-Hypoxia Driving Force for Metabolic and Muscle Transcriptional Reprogramming of Gilthead Sea Bream (*Sparus aurata*) Juveniles. *Biology* **2021**, *10*, 416. [CrossRef]
52. Poudel, P.; Levesque, C.L.; Samuel, R.; St-Pierre, B. Dietary inclusion of Peptiva, a peptide-based feed additive, can accelerate the maturation of the fecal bacterial microbiome in weaned pigs. *BMC Vet. Res.* **2020**, *16*, 1–13. [CrossRef]
53. Herrera, F.; Velazquez, J.; Lugo, J.M.; Orellana, P.; Ruiz, J.; Vega, M.; Romero, A.; Santos, N.; Ramses, G.; Rodríguez-Ramos, T.; et al. Oral Pituitary Adenylate Cyclase Activating Polypeptide (PACAP) formulation modified muscle fatty acid profile and cytokines transcription in head kidney in rainbow trout (*Oncorhynchus mykiss*) fingerlings. *Aquac. Rep.* **2021**, *20*, 100772. [CrossRef]

Article

Effect of Dietary Plant Feedstuffs and Protein/Carbohydrate Ratio on Gilthead Seabream (*Sparus aurata*) Gut Health and Functionality

Catarina Basto-Silva [1,2,*,†], Irene García-Meilán [3,†], Ana Couto [1,2], Cláudia R. Serra [1,2], Paula Enes [1,2], Aires Oliva-Teles [1,2], Encarnación Capilla [3] and Inês Guerreiro [1]

[1] CIIMAR—Interdisciplinary Centre of Marine and Environmental Research, University of Porto, 4050-208 Matosinhos, Portugal
[2] FCUP—Department of Biology, Faculty of Sciences, University of Porto, 4169-007 Porto, Portugal
[3] Department of Cell Biology, Physiology and Immunology, Faculty of Biology, University of Barcelona, 08028 Barcelona, Spain
* Correspondence: bastosilva.c@gmail.com
† These authors contributed equally to this work.

Citation: Basto-Silva, C.; García-Meilán, I.; Couto, A.; Serra, C.R.; Enes, P.; Oliva-Teles, A.; Capilla, E.; Guerreiro, I. Effect of Dietary Plant Feedstuffs and Protein/Carbohydrate Ratio on Gilthead Seabream (*Sparus aurata*) Gut Health and Functionality. *Fishes* **2022**, *7*, 59. https://doi.org/10.3390/fishes7020059

Academic Editors: Maria Angeles Esteban, Bernardo Baldisserotto and Eric Hallerman

Received: 14 January 2022
Accepted: 4 March 2022
Published: 7 March 2022

Publisher's Note: MDPI stays neutral with regard to jurisdictional claims in published maps and institutional affiliations.

Abstract: This study investigated, for the first time, the integrated effects of dietary protein source and protein/carbohydrate (P/CH) ratio on gilthead seabream gut histomorphology, microbiota composition, digestive enzymes activity, and immunological and oxidative stress-related gene expressions. Four isolipidic diets: two fishmeal-based (FM) and two plant feedstuff (PF)-based diets, with P/CH ratios of 50/10 or 40/20 each (FM-P50/CH10; FM-P40/CH20; PF-P50/CH10; PF-P40/CH20), were tested. PF-based diets lead to more histomorphological alterations than FM-based diets. P/CH ratio had no relevant effect on gut histomorphology. Gut mucosa of fish fed PF-based diets presented a higher number of operational taxonomic units, and richness and diversity indices, while the P/CH ratio did not affect those parameters. The α-amylase activity was lower in fish fed with PF-based diets and in fish fed the P40/CH20 diets. Regarding the immune-related genes, only *cyclooxygenase-2* was affected, being higher in fish fed the P50/CH10 diets than the P40/CH20 diets. Fish fed the FM-based diets presented higher expression of *glutathione reductase* and *glutathione peroxidase*, while fish fed the P50/CH10 diet had higher expression of *superoxide dismutase*. In conclusion, PF-based diets can compromise gut absorptive and digestive metabolism, but decreasing the dietary P/CH ratio had little effect on the parameters measured.

Keywords: alternative ingredients; digestive enzymes; gut digesta; gut histomorphology; gut mucosa

1. Introduction

Fishmeal (FM) was traditionally used as the main and most adequate protein source for carnivorous fish due to its high quality, high digestibility, and good palatability [1–3]. Presently, its use is in a clear downward trend [4]. This reduction is largely due to supply and price variation, coupled with the continuously increasing demand from the aquafeed industry [4]. Hence, the use of plant feedstuffs (PF) and the inclusion of carbohydrate (CH) sources in fish feeds have been good alternatives to, respectively, decreasing dietary FM inclusion as a protein source and spare protein use for growth [5–10]. Gilthead seabream (*Sparus aurata*), one of the species with higher production in Europe, seems able to cope with a total replacement of dietary FM by PF [9]. This species requires about 45% of dietary protein [11]. However, if digestible CHs are provided in a suitable quantity, dietary protein might be spared for growth instead of being used as an energy source and, therefore, reduce nitrogen wastes and dietary costs [6,7,12]. Nonetheless, the maximum dietary CH inclusion that does not cause negative effects in gilthead seabream is limited to 20% [7]. Higher dietary CH inclusion may compromise growth and the digestive and absorptive capacities [6,12]. Several studies with gilthead seabream were already conducted to separately

Fishes **2022**, *7*, 59. https://doi.org/10.3390/fishes7020059   https://www.mdpi.com/journal/fishes

evaluate the effects of dietary inclusion of PF and the protein-sparing by CHs. Overall, results showed that PF-based diets often promoted gut morphological changes, modifications on microbiota composition, decreases in gut enzymatic activity, and increases in oxidative stress of fish [9,13–21]. The inclusion of 20% or more of dietary CHs also affected fish growth performance, digestive enzyme activities, and antioxidant status [7,22–24]. However, the interactive effects on gut functionality and the health of gilthead seabream fed diets with lower P/CH ratios and the replacement of FM by PF as a major dietary protein source has not received much attention, and the available information is somehow dispersed. For instance, Castro et al. [25] did not observe major changes in gut histomorphology, microbiota, α-amylase, and lipase activities of gilthead seabream fed diets with highly different P/CH ratios (50/17 and 66/0). Similarly, in the same species, Couto et al. [26] and Fountoulaki et al. [27] also did not find an effect of the dietary P/CH ratio on the proteolytic and amylolytic activities, nor did Castro et al. [23] on the gut oxidative status, or antioxidant enzymes activities. All these studies evaluating different dietary P/CH ratios were made with FM as the main dietary protein source. To our knowledge, only one study is available that evaluated dietary P/CH ratios using PF as the main protein source [12]. In this study, the authors reported that fish fed a P40/CH39 diet had higher lipase and trypsin activities and lower α-amylase activity than those fed a P46/CH19 diet.

Recently, we assessed the effects of FM- or PF-based diets with different P/CH ratios (50/10 and 40/20) in gilthead seabream growth, feed utilization, appetite regulation, and intermediary metabolism [28]. Results showed that diets only slightly modified fish appetite and metabolic parameters, although growth was higher in fish fed the FM-P50/CH10 diet than those fed the FM-P40/CH20 diet. Further, reducing the dietary P/CH ratio led to a decrease in the feed efficiency and an increase in the protein efficiency ratio.

The present study is a follow-up to our previous study [28]. While the previous study aimed to evaluate the effect of dietary protein sources (FM vs. PF) and P/CH ratio on gilthead seabream appetite regulation and intermediary metabolism, the present study aims to evaluate, for the first time, the effects of these factors (dietary protein source and P/CH ratio) on gilthead seabream gut function and health, by assessing gut histomorphology, gut microbiota composition, digestive enzymes activity, and gut immunological and oxidative stress genes expression.

2. Materials and Methods

2.1. Diets

Four isolipidic diets (18% crude lipids) were formulated to contain 100% FM or 20% FM + 80% PF as protein sources, and protein to carbohydrate (P/CH) ratios of P50/CH10 or P40/CH20 (diets FM-P50/CH10, FM-P40/CH20, PF-P50/CH10, and PF-P40/CH20). Details of diets, ingredient composition, and a proximate analysis are presented in the supplementary material (Table S1).

2.2. Experimental Conditions and Sampling

Fish-rearing conditions are described in detail in Basto-Silva et al. [28]. Briefly, 180 gilthead seabream (140 ± 0.1 g, initial body weight) were randomly distributed to twelve 300-L water capacity tanks in a temperature-controlled recirculation life-support system. The diets were randomly distributed to triplicate groups, and fish were fed with the corresponding diet by hand until apparent visual satiation—two meals per day, for 41 days, 6 days a week. The length of the trial was chosen based on previous studies conducted on fish, also including gilthead seabream, which show that this duration was enough to induce dietary effects at intestinal level [17,29].

At the end of the 41 days, 6 fish per tank were sampled 5 h after the first meal of the day and euthanized with a sharp blow to the head (Figure 1). Three fish were sampled for midgut, pyloric caeca (PC), and stomach, all with digestive content, for digestive enzymes evaluation. From the same fish, midgut and PC were also collected for histomorphology evaluation. The remaining 3 fish were sampled to collect midgut to perform gene expression analysis. Two of these three fish were also sampled for allochthonous (digesta) and autochthonous (mucosa) microbiota characterization. Digesta samples were collected by squeezing the entire gut, and mucosa samples were obtained by scrapping the internal surface of gut. Midgut was considered as the portion which began after the PC and finished before the hindgut, which is the final section of the gut [30], and the portions collected were the ones from the beginning of the midgut. Samples for enzymes activity and microbiota characterization were immediately frozen in liquid nitrogen and stored at −80 °C until analyses. Histology and gene expression samples were freed from the adjacent adipose and connective tissue, rinsed in phosphate-buffered saline (PBS), and the excess PBS was removed using a paper towel before being stored. Histology samples were fixed in phosphate-buffered formalin (4%, pH 7.4) for 24 h and then transferred to ethanol (70%) until further processing. Samples for gene expression were stored in RNA later, left at 4 °C overnight, and afterwards stored at −80 °C until analysis.

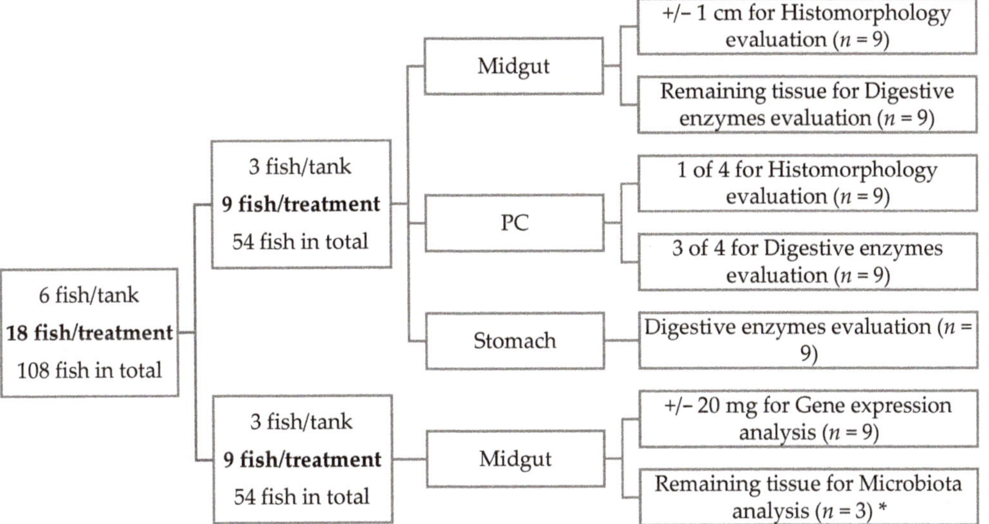

Figure 1. Schematic representation of sampling methodology applied in the present work. * In microbiota, only 2 of 3 fish per tank were used, and the samples were pooled to reduce individual variation, accounting for *n* = 3 per treatment.

2.3. Histological Evaluation

PC and midgut samples were processed and sectioned using standard histological techniques, stained with hematoxylin and eosin, and evaluated through a blinded semi-quantitative method, as described in Castro et al. [25], with slight modifications, namely, considering the nucleus position and hyper-vacuolization within the enterocytes. A score of 1 was given to the tissue with the least changes, and subsequent scores (up to 5) accounted for increasing histomorphological alterations, as described by Penn et al. [31]. Digital images were acquired with Zen software (Blue Edition; Zeiss, Jena, Germany), and using a light microscope Axio Imager.A2 (Zeiss, Oberkochen, Germany).

2.4. Microbial Diversity Analysis

Digesta and mucosa samples of the 2 fish per tank were pooled to reduce individual variation, accounting for *n* = 3 per treatment, each representing the microbial community of 6 fish. DNA extractions, polymorphism analyses of 16S rRNA genes by denaturing gradient gel electrophoresis (DGGE), band excisions, and re-amplifications were performed as described by Castro et al. [25], with each PCR product being loaded on a polyacrylamide gel at 8%, made of a denaturing gradient of 30 to 60% 7 M urea/40% formamide. Amplicons were sequenced to identify microbiota operational taxonomic units (OTUs), and a phylogenetic analysis was performed to identify the closest known species as described in Castro et al. [25].

2.5. Digestive Enzyme Activities and Zymograms

All samples were individually homogenized with a Ystral homogenizer—Laboratory Series X10 (Ballrechten-Dottingen, Germany) in 4 parts of ice-cold 50-mM Tris-HCl buffer pH 7.5, containing 0.1 mM EDTA (reference code E5134, Sigma-Aldrich, Sintra, Portugal), and 0.1% (v/v) Triton X-100 (reference code T8787, Sigma-Aldrich, Sintra, Portugal). Homogenates were centrifuged (30,000× g, 30 min, 4 °C) and supernatants were recovered and stored at −80 °C until use.

Pepsin activity was measured in the stomach, as described in Alarcón et al. [32], total protease activity was measured in PC and midgut, as described in Moyano et al. [33], and lipase (EC 3.1.1.3) and α-amylase (EC 3.2.1.1) activities were measured in PC and midgut using commercial kits from Spinreact (Girona, Spain), with code #1001275 and #41201, respectively.

Pepsin and proteolytic activities were expressed as units (U) per mg of soluble protein, and α-amylase and lipase as mU per mg of soluble protein, with one U of enzyme activity defined as the amount of enzyme that catalyzes the hydrolysis of 1 µmol/min of the substrate at the assay temperature.

Protein concentration of the samples was measured according to Bradford [34], using a Sigma-Aldrich (Sintra, Portugal) protein assay kit (reference code B6916) and albumin bovine serum (BSA; reference code A4503, Sigma-Aldrich, Sintra, Portugal) as standard.

All enzyme activities were measured in a Multiskan GO microplate reader (model 51119200; Thermo Scientific, Nanjing, China).

Alkaline protease zymograms were obtained after resolving, by SDS-PAGE, the homogenates, as described in Castro et al. [35]. The commercial Precision Plus Protein™ All Blue Prestained Standard (reference code 1610373, Bio-Rad Laboratories Lda., Amadora, Portugal) was used to estimate the proteins' molecular weight. The specific trypsin-like and chymotrypsin-like activities were identified based on García-Meilán et al. [24], where 6 bands with protease activity were identified in gilthead seabream. Coomassie-stained gels were imaged with a ChemiDoc XRS+ (Bio-Rad Laboratories Lda., Amadora, Portugal), and qualitatively evaluated by the presence or absence of bands.

2.6. RNA Extraction, cDNA Synthesis and Quantitative Real-Time PCR (qPCR)

The total RNA extraction from intestinal samples, the RNA concentration, the purity and integrity evaluation, the cDNA synthesis, and the quantitative real-time PCR (qPCR) were performed as described in Basto-Silva et al. [28]. The forward and reverse primers used (Table 1) were searched in the GenBank database [36], and their efficiency curves were evaluated according to the assay conditions. Most of the primers' amplification efficiencies were between 90% and 110%, which are the recommended efficiency values [37]. However, as not all used primers conform to this criteria, we used the Pfaffl method [38] to ensure the robustness of the data. The Bio-Rad CFX Manager 3.1 (California, CA, USA) was the software used to measure the expression levels. *Elongation factor 1α* (*ef1α*) and *ribosomal protein S18* (*rps18*) were used as reference genes.

Table 1. Genes and primers used for qPCR.

ID Primer	Sequence (5′–3′)	[1] Accession n°	Tm (°C)	Efficiency (%)
ef1α	F: CTTCAACGCTCAGGTCATCAT R: GCACAGCGAAACGACCAAGGGGA	AF184170	60	87.2
rps18	F: GGGTGTTGGCAGACGTTAC R: CTTCTGCCTGTTGAGGAACCA	AM490061.1	60	88.0
hsp70	F: AATGTTCTGCGCATCATCAA R: GCCTCCACCAAGATCAAAGA	EU805481	57	90.1
cat	F: TTCCCGTCCTTCATTCACTC R: CTCCAGAAGTCCCACACCAT	JQ308823	60	98.5
cox2	F: GAGTACTGGAAGCCGAGCAC R: GATATCACTGCCGCCTGAGT	AM296029	60	94.6
gpx1	F: GAAGGTGGATGTGAATGGAAAAGATG R: CTGACGGGACTCCAAATGATGG	DQ524992	60	91.2
gr	F: TGTTCAGCCACCCACCCATCGG R: GCGTGATACATCGGAGTGAATGAAGTCTTG	AJ937873	60	97.0
igM	F: CAGCCTCGAGAAGTGGAAAC R: GAGGTTGACCAGGTTGGTGT	AM493677	60	87.0
Il1β	F: GGGCTGAACAACAGCACTCTC R: TTAACACTCTCCACCCTCCA	AJ277166	60	99.0
sod	F: CCTGACCTGACCTACGACTATGG R: AGTGCCTCCTGATATTTCTCCTCTG	JQ308833	60	91.6
tnf α	F: TCGTTCAGAGTCTCCTGCAG R: CATGGACTCTGAGTAGCGCGA	AJ413189	60	96.0

cat: catalase; cox2: cyclooxygenase 2; ef1α: translation elongation factor 1α; F: forward; gpx1: glutathione peroxidase; gr: glutathione reductase; hsp70: 70 kilodalton heat shock proteins; igM: immunoglobulin M heavy chain; il1β: interleukin 1β; R: reverse; rps18: ribosomal protein S18; sod: superoxide dismutase; Tm: melting temperature; tnf α: tumor necrosis factor α. [1] from the GenBank database [36].

2.7. Statistical Analysis

Statistical analyses were completed using SPSS 25 software package for Windows (IBM® SPSS® Statistics, New York, NY, USA). Homogeneity of variances and data normality were tested by the Levene and Shapiro–Wilk tests, respectively. When normality was not verified, data were transformed before ANOVA. However, all data are presented as the mean and standard error of the mean (SEM), without any transformation. Differences were considered statistically significant at $p < 0.05$.

Since histological data was not normal nor homogenous even after transformation, statistical analysis of the histomorphology evaluation was completed by the non-parametric Kruskal–Wallis test, followed by all-pairwise comparisons. Furthermore, the significance values were adjusted by the Bonferroni correction for multiple tests.

The remaining data were evaluated by two-way ANOVA tests, with the protein source and P/CH ratios as factors. In the case of interaction between factors, one-way ANOVA was performed for the P/CH ratio within each protein source, and for the protein source within each P/CH ratio.

Statistical analysis related to the DGGE was performed as described in Castro et al. [25].

3. Results

During the trial, all experimental diets were well-accepted by the fish, and the fish survival rate was 100%. Results of the rearing trial were not the aim of this study and are presented elsewhere [28].

Regarding the PC histomorphology, fish fed the PF-P50/CH10 diet presented a higher total mean score (2.23) than those in the remaining experimental conditions, where the total mean score ranged between 1.78 and 1.88 (Table 2). Lamina propria width was higher in fish fed the PF-P50/CH10 diet than in those fed the FM-based diets (Figure 2I). Fish fed the PF-P50/CH10 diet also presented higher submucosa widths than those remaining in the experimental conditions (Figure 2II). Lamina propria cellularity was higher in fish fed the FM-P50/CH10 and PF-P50/CH10 diets than the FM-P40/CH20 diet. The enterocytes vacuolization was higher in fish fed the PF-based diets.

I. Mucosa villi analysis

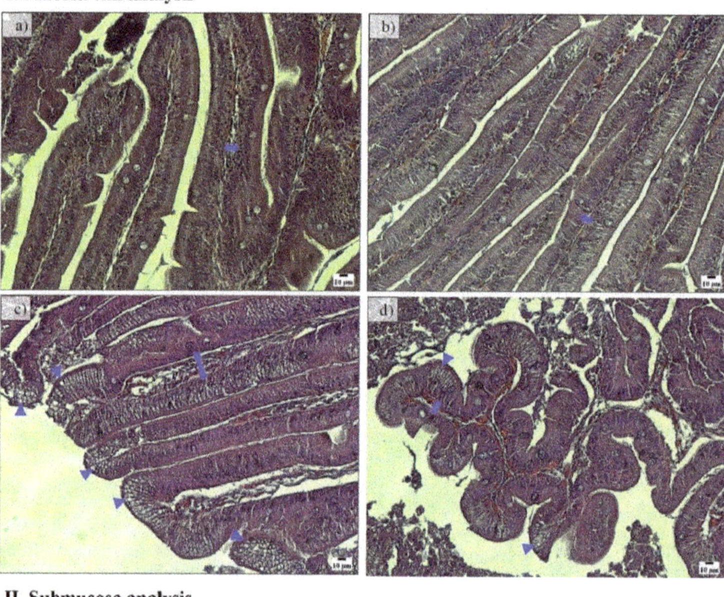

II. Submucosa analysis

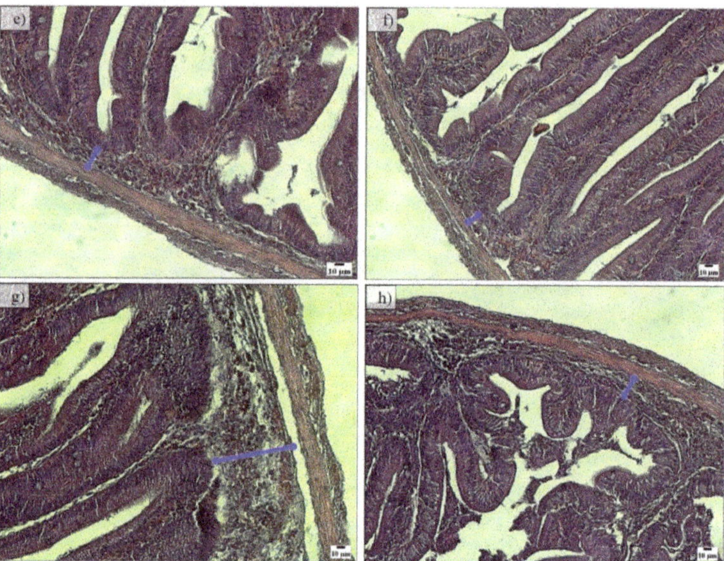

Figure 2. Representative histological sections of pyloric caeca mucosa villi (**I**) and submucosa (**II**) stained with hematoxylin and eosin of fish fed FM-P50/CH10 (**a,e**), FM-P40/CH20 (**b,f**), PF-P50/CH10 (**c,g**), and PF-P40/CH20 (**d,h**). ▶enterocytes vacuolization. (**I**): Lamina propria width was higher in fish fed the PF-P50/CH10 diet (**c**) than fish fed the FM-based diets (**a,b**). Enterocytes vacuolization was higher in fish fed the PF-based diets (**c,d**) than those in the remaining conditions (**a,b**). (**II**): Submucosa width was higher in fish fed diet PF-P50/CH10 (**g**), than those in the remaining conditions (**e–h**).

Table 2. Details of the score-based evaluation of the pyloric caeca histology of gilthead seabream fed the experimental diets.

Protein Source	FM		PF		SEM	*p*-Value
P/CH Ratio	**50/10**	**40/20**	**50/10**	**40/20**		
Gut fold height	1.44	1.33	1.72	1.50	0.14	0.97
Lamina propria—width	1.61 [a]	1.61 [a]	2.22 [b]	1.94 [a,b]	0.09	0.04
Lamina propria—cellularity	2.22 [b]	1.56 [a]	2.61 [b]	2.00 [a,b]	0.12	0.03
Submucosa—width	1.44 [a]	1.39 [a]	2.00 [b]	1.50 [a]	0.08	0.04
Submucosa—cellularity	1.94	2.00	2.11	1.61	0.10	0.28
Intraepithelial leucocytes infiltration	2.78	2.83	2.67	2.06	0.13	0.11
Eosinophilic granulocytes presence	2.11	1.94	2.44	1.89	0.13	0.33
Enterocytes nucleus alignment	2.33	2.28	2.44	2.61	0.09	0.66
Enterocytes vacuolization	1.00 [a]	1.11 [a]	1.83 [b]	1.72 [b]	0.10	0.00
Mean score	1.88 [a]	1.78 [a]	2.23 [b]	1.87 [a]	0.06	0.03

Values presented as means (*n* = 9) and standard error of the mean (SEM). Different lower-case letters stand for statistical differences across dietary groups as determined by the Kruskal–Wallis all-pairwise comparisons. Furthermore, the significance values have been adjusted by the Bonferroni correction for multiple tests. CH: carbohydrate; FM: fishmeal; PF: plant feedstuffs; P: protein.

Regarding midgut histomorphology, fish fed the PF-based diets presented a higher total mean score (2.77) and gut fold height than fish fed FM-based diets, which have a total mean score of 2.28 (Figure 3 and Table 3). No further differences between groups were detected.

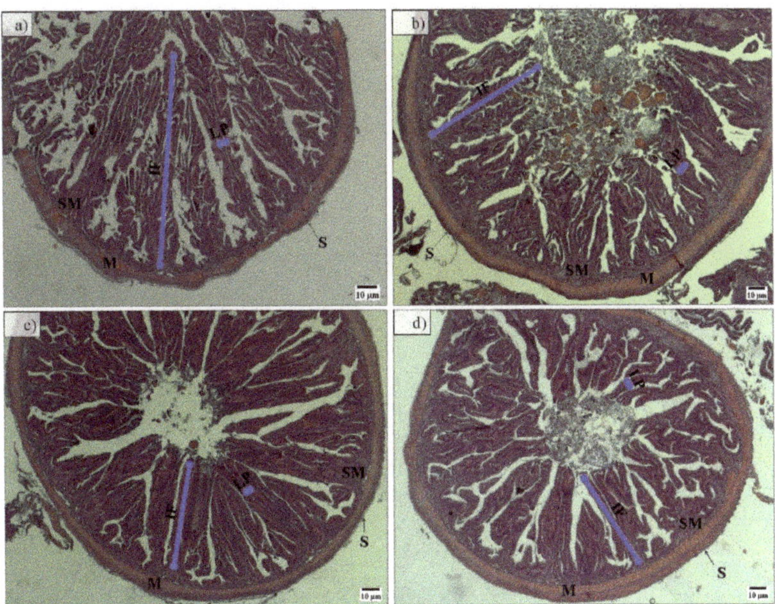

Figure 3. Representative hematoxylin and eosin-stained histological sections of midgut from fish fed FM-P50/CH10 (**a**), FM-P40/CH20 (**b**), PF-P50/CH10 (**c**), and PF-P40/CH20 (**d**). IF, intestine fold; LP, lamina propria; M, muscularis layer; S, serosa layer; SM, submucosa layer. Intestine fold height showed higher histomorphology deformations in fish fed the PF-based diets (**c,d**) than in fish fed the FM-based diets (**b**), except for fish fed the FM-P50/CH10 diet (**a**), which was not significantly different from fish fed the PF-P40/CH20 (**d**).

Table 3. Details of the score-based evaluation of the midgut histology of gilthead seabream fed the experimental diets.

Protein Source	FM		PF		SEM	*p*-Value
P/CH Ratio	**50/10**	**40/20**	**50/10**	**40/20**		
Gut fold height	1.50 [a,b]	1.22 [a]	2.33 [c]	2.00 [b,c]	0.14	0.02
Lamina propria—width	2.67	2.00	2.89	2.44	0.14	0.15
Lamina propria—cellularity	3.00	2.67	3.11	2.78	0.14	0.62
Submucosa—width	2.88	2.11	3.13	3.29	0.18	0.16
Submucosa—cellularity	2.75	2.44	3.25	3.29	0.15	0.08
Intraepithelial leucocytes infiltration	2.78	2.72	3.56	3.22	0.23	0.44
Eosinophilic granulocytes presence	3.11	2.78	3.13	3.56	0.14	0.39
Enterocytes nucleus alignment	2.44	2.11	2.56	2.89	0.14	0.29
Enterocytes vacuolization	1.00	1.00	1.22	1.22	0.05	0.22
Mean score	2.44 [a,b]	2.12 [a]	2.79 [b]	2.74 [b]	0.09	0.01

Values presented as means ($n = 9$) and standard error of the mean (SEM). Different lower-case letters stand for statistical differences across dietary groups as determined by the Kruskal–Wallis all-pairwise comparisons. Furthermore, the significance values have been adjusted by the Bonferroni correction for multiple tests. CH: carbohydrate; FM: fishmeal; PF: plant feedstuffs; P: protein.

DGGE fingerprints of the hypervariable V3 region of the 16S rRNA genes present in digesta and mucosa gut samples revealed that, independently of the dietary treatment, gut bacterial communities maintained a similarity, near 40% within both gut samples (Figure 4). Moreover, two clusters were observed in both gut microbiota regions, corresponding to samples recovered from fish fed the FM- and the PF-based diets, except for the FM-P50/CH10 diet in the digesta, which did not cluster with the remaining FM-based diets, and the PF-P50/CH10 diet in the mucosa, which did not cluster with the remaining PF-based diets. Despite this clear cluster separation, in digesta samples, the dietary composition did not affect the average number of OTUs, richness, and diversity indices (Table 4). Only the similarity index was higher in fish fed PF-P50/CH10 than in fish fed the FM-P50/CH10 diet. In mucosa samples, PF-based diets led to a higher number of gut OTUs, richness, and diversity indices than FM-based diets, while the similarity index was not different between groups. Sequence analysis from DGGE-selected bands showed that the dominant allochthonous and autochthonous bacteria detected were either corresponding to uncultured bacteria not yet assigned to a specific taxon or were closely related to genera belonging to the phylum Firmicutes and Proteobacteria, namely, Lactobacillus, Pseudomonas, Klebsiella, and Vibrio (Table 5 and Figure 4). Except for band 15, which was only found in digesta, all other bands were detected in digesta and mucosa samples.

Concerning digestive enzymes, α-amylase activity was lower in fish fed the PF-based diets, for both PC and midgut, and in fish fed the P40/CH20 diet only in the PC (Table 6). Proteolytic activity was higher in the PC of fish fed the P50/CH10 diet, but only within the PF-based diet-fed fish. Pepsin and lipase activities were not affected by dietary composition.

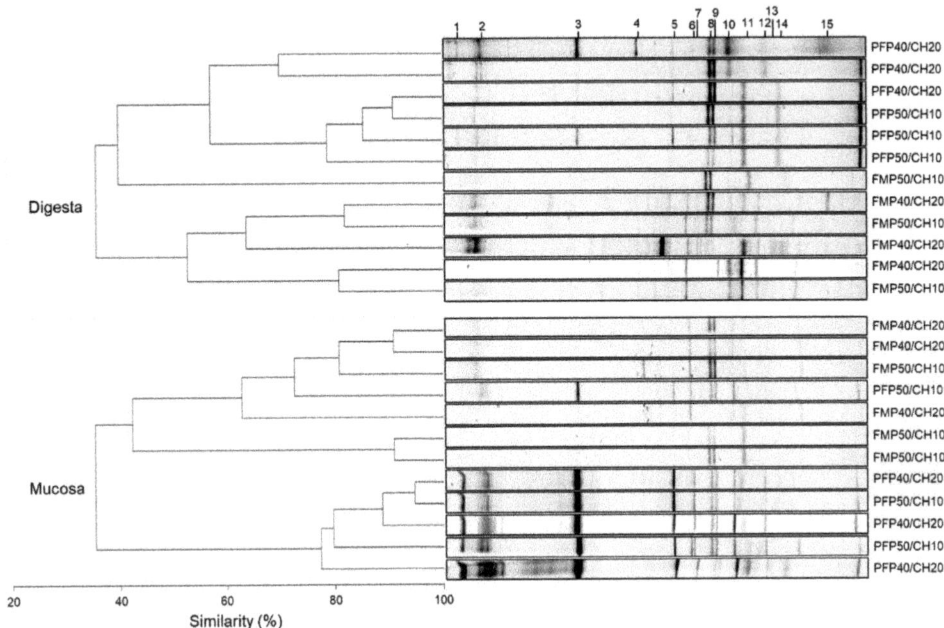

Figure 4. Dendrogram and PCR-DGGE fingerprints of the microbiota found in digesta and mucosa samples recovered from the gut of gilthead seabream fed the experimental diets. Numbers (1–15) on top of the figure correspond to the gel bands sequenced to identify the corresponding bacterial species, described on Table 5.

Table 4. Ecological parameters obtained from PCR- DGGE fingerprints of gut microbiota of gilthead seabream fed the experimental diets.

PS	FM		PF		SEM	Two-Way ANOVA		
P/CH Ratio	50/10	40/20	50/10	40/20		PS	P/CH Ratio	I
Digesta								
OTUs	8.7	13.7	10.0	11.3	0.9	0.76	0.08	0.29
Richness [1]	0.88	1.38	1.02	1.14	0.09	0.75	0.10	0.28
Diversity [2]	2.08	2.56	2.24	2.37	0.09	0.94	0.11	0.33
SIMPER Similarity (%) [3]	34.1 [A]	57.0	80.4 [B]	65.9	6.0	0.01	0.59	0.04
Mucosa								
OTUs	6.0	8.3	14.0	11.7	1.1	0.00	1.00	0.11
Richness [1]	0.60	0.87	1.41	1.15	0.11	0.00	0.97	0.08
Diversity [2]	1.67	2.11	2.59	2.39	0.12	0.01	0.48	0.09
SIMPER Similarity (%) [3]	65.3	71.2	72.8	83.8	4.3	0.29	0.37	0.78

Values presented as means (n = 3 per treatment pooled from 6 fish), and standard error of the mean (SEM). Different upper-case letters denote significant differences between dietary protein sources. In the case of interaction between factors, one-way ANOVA was performed for the P/CH ratio within each protein source, and for the protein source within each P/CH ratio. The significant interactions between the factors are presented in the upper part of the table. CH: carbohydrate; FM: fishmeal; I: interaction; OTUs: average number of operational taxonomic units; PF: plant feedstuffs; P: protein; PS: protein source. [1] Margalef species richness: d = (S − 1)/log(N). [2] Shannon's diversity index: H′ = −∑(pi(lnpi)). [3] SIMPER: similarity percentage within group replicates.

Table 5. Identified bacterial species from the DNA sequencing of the allochthonous and autochthonous gut bacteria communities of gilthead seabream fed the experimental diets.

Band	Closest Known Species (BLAST)	Phylum	Similarity (%)	Accession Number of Nearest Neighbor
1	Uncultured bacterium from Turkey fecal microbial community	-	99	EU873831.1
2	Uncultured *Pseudomonas* sp.	Proteobacteria	100	LC032367.1
3	*Lactobacillus aviarius* subsp. *aviarius*	Firmicutes	96	LC071825.1
4	Uncultured marine bacterium	-	96	HM437606.1
5	Uncultured *Lactobacillus* sp.	Firmicutes	97	LT571746.1
6	Uncultured *Pseudomonas* sp.	Proteobacteria	99	GU250534.1
7	Uncultured bacterium from gut microbiota of Atlantic salmon (*Salmo salar* L.)	-	100	EU009390.1
8	*Klebsiella pneumoniae*	Proteobacteria	100	CP031798.1
9	Uncultured *Klebsiella* sp.	Proteobacteria	97	MH767054.1
10	Uncultured bacterium from gut bacterial communities of *Mythimna separata*	-	80	JQ013040.1
11	Uncultured *Vibrio* sp.	Proteobacteria	97	HM214586.1
12	Uncultured bacterium from environmental samples	-	95	FJ785825.1
13	Uncultured bacterium from environmental samples	-	100	LT720113.1
14	Uncultured bacterium from intestine of Atlantic cod (*Gadus morhua*)	-	98	HM115943.1
15	Uncultured bacterium from environmental samples	-	100	KC527347.1

Table 6. Specific activity of pepsin (U mg protein^{-1}) in the stomach, and α-amylase, lipase (mU mg protein^{-1}), and proteolytic activity (U mg protein^{-1}) in the pyloric caeca, and midgut of gilthead seabream fed the experimental diet.

PS	FM		PF		SEM	Two-Way ANOVA		
P/CH Ratio	50/10	40/20	50/10	40/20		PS	P/CH Ratio	I
				Stomach				
Pepsin	34.7	23.7	22.4	18.6	3.7	0.26	0.34	0.64
				Pyloric caeca				
α-Amylase	45.2	27.1	19.0	6.3	4.0	0.00	0.01	0.42
Lipase	0.56	0.45	0.61	0.42	0.05	0.91	0.17	0.71
Proteolytic activity	17.4	16.5	45.4 [b]	11.4 [a]	4.7	0.67	0.09	0.03
				Midgut				
α-Amylase	207.3	191.0	57.6	52.2	24.6	0.00	0.69	0.39
Lipase	3.58	4.19	2.86	3.52	0.39	0.39	0.43	0.97
Proteolytic activity	254.8	284.2	234.4	239.7	30.3	0.28	0.19	0.53

Values presented as means (n = 9), and standard error of the mean (SEM). Different lower-case letters denote significant differences between dietary P/CH ratios. In the case of interaction between factors, one-way ANOVA was performed for the P/CH ratio within each protein source, and protein source within each P/CH ratio. The significant interactions between the factors are presented in the upper part of the table. CH: carbohydrate; FM: fishmeal; I: interaction; PF: plant feedstuffs; P: protein; PS: protein source.

Alkaline protease zymograms, from both PC and midgut, revealed the presence of six bands with proteolytic activity against casein, three identified as trypsin-like proteases (90, 60, and 55 KDa), and the other three as chymotrypsin-like proteases (50, 30, and 25 KDa). All treatments presented the same number of proteolytic bands (Figure 5).

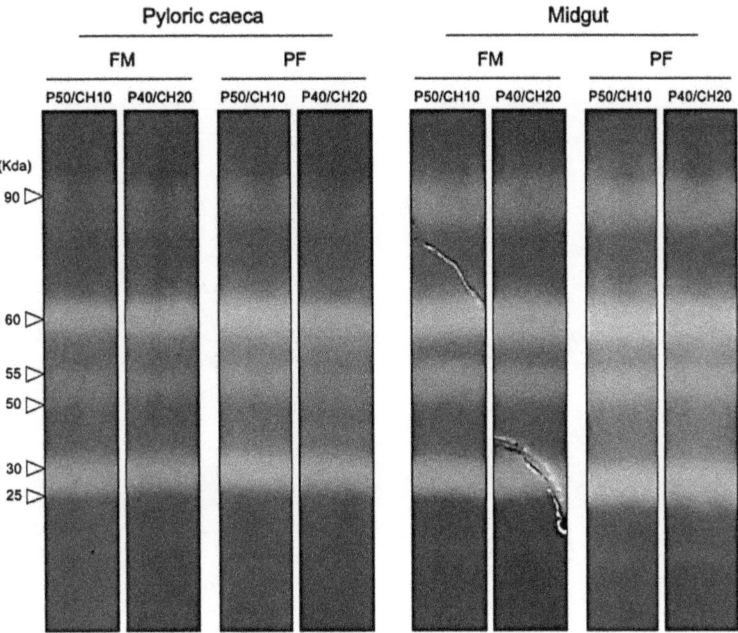

Figure 5. Representative model zymogram of alkaline proteases in pyloric caeca and midgut extracts. The molecular weight of each band with proteolytic activity is indicated. All samples were analyzed individually.

Concerning immune-related gene expressions, only *cyclooxygenase-2* (*cox2*) presented significant changes, being higher in fish fed the P50/CH10 diet (Table 7). Gene expression of *immunoglobulin M heavy chain* (*igM*), *interleukin-1β* (*il1β*), and *tumor necrosis factor-α* (*tnf-α*) was not affected by dietary composition.

Table 7. Normalized gene expression [1] of immunology and oxidative stress-related genes in midgut of gilthead seabream fed the experimental diets.

PS	FM		PF		SEM	Two-Way ANOVA		
P/CH Ratio	50/10	40/20	50/10	40/20		PS	P/CH Ratio	I
Immunology								
cox2	0.20	0.13	0.19	0.13	0.01	0.75	0.01	0.64
igM	19.5	16.0	11.1	18.0	1.5	0.28	0.58	0.09
il1β	0.15	0.11	0.14	0.10	0.02	0.78	0.44	0.48
tnf-α	0.13	0.09	0.10	0.11	0.01	0.61	0.37	0.22
Oxidative Stress								
hsp70	195.1	178.1	168.2	171.5	10.8	0.78	0.50	0.42
cat	61.5	46.1	47.1	44.0	4.1	0.33	0.26	0.77
gr	8.9	4.6	3.8	4.7	0.6	0.01	0.29	0.06
gpx1	13.3	9.3	8.7	8.4	0.6	0.02	0.05	0.09
sod	69.3	32.4	42.7	20.1	6.9	0.14	0.01	0.97

[1] All values expressed as arbitrary unit $\times 10^2$. Values presented as means (*n* = 9), and standard error of the mean (SEM). cat: catalase; CH: Carbohydrate; cox2: cyclooxygenase 2; FM: fishmeal; gpx1: glutathione peroxidase; gr: glutathione reductase; hsp70: 70 kilodalton heat shock proteins; igM: immunoglobulin M heavy chain; I: interaction; il1β: interleukin 1β; PF: plant feedstuffs; P: protein; PS: protein source; sod: superoxide dismutase; tnf-α: tumor necrosis factor α.

Regarding the oxidative stress-related genes, PF-based diets led to a lower expression of *glutathione reductase* (*gr*) and *glutathione peroxidase* (*gpx1*), while *superoxide dismutase* (*sod*) expression was lower in fish fed the P40/CH20 diet. The gene expression of *70 kilodalton heat shock proteins* (*hsp70*) and *catalase* (*cat*) was not affected by dietary composition.

4. Discussion

The presence of antinutritional factors on PF, namely, in soybean products, was reported as leading to gut inflammation in gilthead seabream [15,18,20,21,39]. Among the observed gut morphological alterations caused by soybean meal were a decrease in gut fold height, an enlargement of submucosa and lamina propria, an increased number of inflammatory cells on tissues, and modifications on enterocytes vacuolization [15,18,20,21]. Although we have assessed the midgut and PC, and previous studies analyzed the distal gut, the present results agree with the reported observations in this species, since fish fed the PF-P50/CH10 diet, which has a higher soybean meal content (25% compared with 19% for PF-P40/CH20, and no soybean meal content for FM-based diets), also presented more histological alterations when compared with fish fed the other diets. The histological modifications observed in the midgut and PC were mainly in gut fold height, width and cellularity of lamina propria, width of the submucosa, and/or in enterocytes vacuolization. Similarly, gilthead seabream juveniles fed 30% soybean meal presented a moderately and diffusely expanded distal gut lamina propria [14], while juveniles fed soy saponins and phytosterols presented histomorphological alterations of the intestinal mucosal structure [17]. Nonetheless, during the on-growing period (fish of similar sizes to those of the present study) gilthead seabream showed a high tolerance to soy saponins and phytosterols [29]. This indicates that fish responses can be different, depending on the life stage, dietary ingredients/antinutrients combinations, and intestine portions.

Moreover, in the present study, PC seemed to be more sensitive to dietary composition changes than midgut, where fewer histomorphological alterations were observed. This agreed with the study of Couto et al. [29], which observed that dietary soy saponins and phytosterols affected PC histomorphology but not the distal gut of on-growing gilthead seabream.

However, it is important to add that the observed histomorphology modifications were not enough to consistently affect gilthead seabream growth [28]. Nonetheless, a longer experimental trial could have exposed those differences.

The composition of gut microbiota also affects gut functionality since, for instance, bacteria might have a role in nutrients' digestion and immune functions, being affected by diet composition [39]. In the present study, protein source was the single factor affecting gut microbiota. The only detectable effect on digesta microbiota was an increase of the similarity index in fish fed the PF-P50/CH10 diet, indicating that this diet might modulate gut bacteria populations towards a higher similarity between samples. The absence of any other major effect on digesta microbiota in fish fed different dietary compositions was previously observed in gilthead seabream [25]. This lack of effect could be expected, since digesta microbiota comprises transient (allochthonous) microorganisms, which are often surrounded by the resident microbiota to the gut wall and, thus, do not last a long time in the gut [40].

The higher number of OTUs, richness, and diversity indices observed in the mucosa microbiota of fish fed the PF-based diets agree with what was previously reported for this species, at the juvenile stage, fed soybean meal-based diets compared with FM-based diets [16], and for other species also fed PF-based diets, such as Senegalese sole (*Solea senegalensis*) and Atlantic salmon [41–43]. These results could be explained by the presence of non-digestible carbohydrates on PF, which provide the required substrate for gut bacteria proliferation [44,45]. It should be noticed that higher richness and diversity indices, as in fish fed the PF-based diets, can be undesirable since they can be associated with the presence of pathogenic bacteria in gut microbiota [18,46]. On the other hand, a diverse gut microbiome, with the increase of microorganisms from the Firmicutes phylum, can stimulate a fish's innate immunity and reduce the gut surface area for the establishment of pathogenic bacteria, improving the fish's health [47–49]. Although, in the present study, none of the immune-related genes measured were affected by the use of PF, the dominant allochthonous and autochthonous bacteria detected were indeed the most closely related to the Firmicutes and Proteobacteria phyla, as already described in gilthead seabream fed different dietary compositions [18,47]. However, in future studies, a higher-resolution method, such as next-generation sequencing and FISH, could improve the characterization of the bacterial communities under different dietary feeding regimes, providing not only the full identification of the species and/or subspecies of the bacteria, but also allowing for their quantification. This more in-depth characterization and quantification of the bacterial species and/or subspecies will possibly allow for a clearer connection between microbiota and gut functionality.

Both *Pseudomonas* sp. and *Lactobacillus* sp. can produce α-amylase [50]; however, as their presence was detected in fish fed all experimental diets, no link can be made between the presence of α-amylase-producing bacteria, the dietary ratios, and α-amylase activity measured. Indeed, the lack of differences in the gut microbiota of fish fed different dietary P/CH ratios could be partially explained by the use of pregelatinized maize starch as the main carbohydrate source. Gilthead seabream presents almost 100% starch digestibility of diets including 10 to 30% of this ingredient [26]; thus, pregelatinized maize starch does not seem to provide a substantial substrate for microbial fermentation and development. A similar lack of changes in gut microbiota was reported for gilthead seabream and other fish species fed also with highly digestible starch [26,51,52].

For diets' digestion, several enzymes are needed, with each enzyme presenting a specific role. α-amylase, proteases, and lipase are, respectively, responsible for the enzymatic hydrolysis of starch, proteins, and lipids [51–53]. Despite that we did not observe any major effect on the feed intake of fish fed the different diets [28], in the present study, α-amylase activity in PC and midgut and proteolytic activity in PC were affected by the dietary composition. The α-amylase activity was lower in the PC and midgut of fish fed the P40/CH20 diet, and in the PC, it was also lower in fish fed the PF-based diets than those fed the FM-based diets. The influence of dietary P/CH ratio can be related to the adsorption of α-amylase by starch, as suggested by Spannhof and Plantikow [54], who observed that α-amylase secreted by fish during the digestive process was adsorbed by the starch present in the diets [55,56]. This lower α-amylase activity observed in fish fed the P40/CH20 diet can partially explain the lower feed efficiency observed in our previous study in fish fed the P40/CH20 diet, in comparison with those fed the P50/CH10 diet [28]. The effects of dietary protein sources may be related to the ingredients used, namely, wheat gluten, which is a source of α-amylase inhibitors [55,56].

According to Hidalgo et al. [57] and Fernández et al. [58], α-amylase activity is more dependent on fish nutritional habits than the proteolytic activity, and this is further supported by the lack of effects on the proteolytic activity reported in gilthead seabream fed diets with different P/CH ratios [25–27]. However, studies in other fish species showed that higher dietary protein levels increased proteolytic activity [59–62]. In the present study, higher proteolytic activity in fish fed the diet with a higher protein content was also observed in the PC, but only in fish fed the PF-based diets. Moreover, no differences were found regarding the alkaline protease pattern, as observed in the zymograms of the different dietary treatments, suggesting the proteases present are the same independently of the diet offered. Differently, García-Meilán et al. [24] observed that, in gilthead seabream fed FM-based diets, PC proteolytic activity was higher in fish fed lower dietary protein-content diets (P35 and P38), while in the midgut, the proteolytic activity increased progressively as dietary protein increased, stabilizing at 41% to 47% of protein. Thus, more studies should be conducted to clarify the effects of dietary protein level and source on proteolytic activity in the gut.

In the present study, fish fed the PF-based diets presented lower *gr* and *gpx1* gene expression than those fed FM-based diets, which may indicate that the former were more vulnerable to oxidative stress [63]. This evidence seems to be in agreement with the presence of soybean meal antinutritional factors, such as the β-conglycinin, which has been identified as one of the major feed allergens [64,65]. This allergen has an N-glycan structure, essential for the formation of di-tyrosine bridges, which trigger the process responsible for oxidative stress, increasing the malondialdehyde content, and causing oxidative damages [64].

Regarding dietary P/CH ratio effects on oxidative stress, the decrease of *sod* gene expression in fish fed the P40/CH20 diet may indicate that those fish were also more susceptible to oxidative stress. Nevertheless, Castro et al. [23] observed in gilthead seabream that the intestinal sod activity was not affected by the use of different dietary P/CH ratios. Indeed, a disconnection between the gene expression and enzymatic activity results was previously reported by other studies [22,66]. Thus, we might not disregard that the response at the biochemical level might be different of the one obtained at molecular level. Hence, future studies should also include enzymatic activities which, together with the gene expression analyses, will allow for a more complete conclusion.

Sitjà-Bobadilla et al. [13] and Kokou et al. [19,20] reported, in gilthead seabream fed PF-based diets, a synchronism between the immune and stress responses and the gut histomorphological alterations. A similar relationship was observed in the present study, although no effects were observed in the immune-related genes analyzed, except for *cox2* expression, which was higher in fish fed the high-protein diets. Cox2 is linked mainly to inflammation [67,68], so it might be expected that an increase of *cox2* gene expression would be accompanied by higher histomorphological scores in this group, which did not happen. The absence of effects on immune-related responses seems to agree with the lack of mortality or diseases observed in our previous study [28].

5. Conclusions

The present study aimed to provide an integrated view of the effects on gut health and functionality of gilthead seabream when fed diets with FM or PF as the main dietary protein sources and different P/CH ratios. However, no major statistical interactions between those two factors were observed, and in general, only independent effects were reported, which did not allow us to conclude on the cumulative effect of both factors. Dietary P/CH ratio has little effects on gut health or functionality; only a decrease of α-amylase activity and gut *cox2* and *sod* gene expression were observed.

PF-based diets are more prone to compromise CH digestibility, induce gut histomorphological changes and modifications of gut mucosa microbiota profile, and decrease expression of oxidative stress-related genes. Overall, the present data demonstrates the need of fine-tunning fish feed formulations with PF to properly preserve fish intestinal physiology.

Supplementary Materials: The following supporting information can be downloaded at: https://www.mdpi.com/article/10.3390/fishes7020059/s1, Table S1: Details of diets, ingredient composition, and proximate analysis published in Basto-Silva, et al. [28].

Author Contributions: Conceptualization: A.O.-T. and I.G.; Experimental analyses, data analyses, and experimental design methodology: C.B.-S. and I.G.-M.; Supervision team: P.E. and I.G. (experimental trial and digestive enzymes activity), A.C. (histological evaluation), C.R.S. (microbial diversity analysis), and E.C. (gene expression analysis). The first manuscript draft was written by C.B.-S., and all authors commented on previous versions of the manuscript. All authors have read and agreed to the published version of the manuscript.

Funding: This research was partially supported by the Strategic Funding to UID/Multi/04423/2019 (POCI-480 01-0145-FEDER-007621) through national funds provided by Foundation for Science and Technology (FCT) and European Regional Development Fund (ERDF), in the framework of the programme Portugal2020; also partially supported by the structured program of R&D&I ATLANTIDA—Platform for the monitoring of the North Atlantic Ocean and tools for the sustainable exploitation of the marine resources (NORTE-01-0145-FEDER-000040), supported by the North Portugal Regional Operational Programme (NORTE2020), through the ERDF; and by the Ministerio de Ciencia, Innovación y Universidades (MICIUN) (AGL2017-89436-R project). Catarina Basto-Silva and Inês Guerreiro were supported by the European Social Fund (ESF) and FCT (SFRH/BD/130171/2017 and SFRH/BPD/114959/2016, respectively). Irene García-Meilán was supported by the MICIUN (Programa Jose Castillejo CAS18/00436). Ana Couto, Cláudia Serra, and Paula Enes had scientific employment contracts supported by national funds through FCT.

Institutional Review Board Statement: The experiment was performed by accredited scientists (following FELASA category C recommendations) and approved by the General Directorate of Food and Veterinary from Portugal (Certification number ORBEA-CIIMAR 30-2019), according to the European Union directive 2010/63/EU on the protection of animals for scientific purposes.

Data Availability Statement: The data used to generate the results in this manuscript can be made available if requested to the corresponding author.

Conflicts of Interest: The authors declare no conflict of interest.

References

1. Tacon, A.G.J.; Metian, M. Global overview on the use of fish meal and fish oil in industrially compounded aquafeeds: Trends and future prospects. *Aquaculture* **2008**, *285*, 146–158. [CrossRef]
2. Tacon, A.G.J.; Metian, M. Feed matters: Satisfying the feed demand of aquaculture. *Rev. Fish. Sci. Aquac.* **2015**, *23*, 1–10. [CrossRef]
3. Olsen, R.L.; Hasan, M.R. A limited supply of fishmeal: Impact on future increases in global aquaculture production. *Trends Food Sci. Technol.* **2012**, *27*, 120–128. [CrossRef]
4. FAO. *The State of World Fisheries and Aquaculture 2020. Sustainability in Action*; FAO: Rome, Italy, 2020; p. 224.
5. Kissil, G.W.; Lupatsch, I. Successful replacement of fishmeal by plant proteins in diets for the gilthead seabream, *Sparus aurata* L. *Isr. J. Aquac.-Bamidgeh* **2004**, *56*, 20378. [CrossRef]
6. Fernández, F.; Miquel, A.G.; Córdoba, M.; Varas, M.; Metón, I.; Caseras, A.; Baanante, I.V. Effects of diets with distinct protein-to-carbohydrate ratios on nutrient digestibility, growth performance, body composition and liver intermediary enzyme activities in gilthead sea bream (*Sparus aurata*, L.) fingerlings. *J. Exp. Mar. Biol. Ecol.* **2007**, *343*, 1–10. [CrossRef]
7. Enes, P.; Panserat, S.; Kaushik, S.; Oliva-Teles, A. Dietary carbohydrate utilization by European sea bass (*Dicentrarchus labrax* L.) and gilthead sea bream (*Sparus aurata* L.) juveniles. *Rev. Fish. Sci.* **2011**, *19*, 201–215. [CrossRef]
8. Cabral, E.M.; Fernandes, T.J.R.; Campos, S.D.; Castro-Cunha, M.; Oliveira, M.; Cunha, L.M.; Valente, L.M.P. Replacement of fish meal by plant protein sources up to 75% induces good growth performance without affecting flesh quality in ongrowing Senegalese sole. *Aquaculture* **2013**, *380*, 130–138. [CrossRef]
9. Monge-Ortiz, R.; Martínez-Llorens, S.; Márquez, L.; Moyano, F.J.; Jover-Cerdá, M.; Tomás-Vidal, A. Potential use of high levels of vegetal proteins in diets for market-sized gilthead sea bream (*Sparus aurata*). *Arch. Anim. Nutr.* **2016**, *70*, 155–172. [CrossRef]
10. Wang, X.X.; Chen, M.Y.; Wang, K.; Ye, J.D. Growth and metabolic responses in Nile tilapia (*Oreochromis niloticus*) subjected to varied starch and protein levels of diets. *Ital. J. Anim. Sci.* **2017**, *16*, 308–316. [CrossRef]
11. Lupatsch, I.; Kissil, G.W.; Sklan, D. Defining energy and protein requirements of gilthead seabream (*Sparus aurata*) to optimize feeds and feeding regimes. *Isr. J. Aquac.-Bamidgeh* **2003**, *55*, 243–257. [CrossRef]
12. García-Meilán, I.; Ordóñez-Grande, B.; Valentín, J.M.; Fontanillas, R.; Gallardo, Á. High dietary carbohydrate inclusion by both protein and lipid replacement in gilthead sea bream. Changes in digestive and absorptive processes. *Aquaculture* **2020**, *520*, 734977. [CrossRef]
13. Sitjà-Bobadilla, A.; Peña-Llopis, S.; Gómez-Requeni, P.; Médale, F.; Kaushik, S.; Pérez-Sánchez, J. Effect of fish meal replacement by plant protein sources on non-specific defence mechanisms and oxidative stress in gilthead sea bream (*Sparus aurata*). *Aquaculture* **2005**, *249*, 387–400. [CrossRef]
14. Bonaldo, A.; Roem, A.J.; Fagioli, P.; Pecchini, A.; Cipollini, I.; Gatta, P.P. Influence of dietary levels of soybean meal on the performance and gut histology of gilthead sea bream (*Sparus aurata* L.) and European sea bass (*Dicentrarchus labrax* L.). *Aquac. Res.* **2008**, *39*, 970–978. [CrossRef]
15. Santigosa, E.; Sánchez, J.; Médale, F.; Kaushik, S.; Pérez-Sánchez, J.; Gallardo, M.A. Modifications of digestive enzymes in trout (*Oncorhynchus mykiss*) and sea bream (*Sparus aurata*) in response to dietary fish meal replacement by plant protein sources. *Aquaculture* **2008**, *282*, 68–74. [CrossRef]
16. Dimitroglou, A.; Merrifield, D.L.; Spring, P.; Sweetman, J.; Moate, R.; Davies, S.J. Effects of mannan oligosaccharide (MOS) supplementation on growth performance, feed utilisation, intestinal histology and gut microbiota of gilthead sea bream (*Sparus aurata*). *Aquaculture* **2010**, *300*, 182–188. [CrossRef]
17. Couto, A.; Kortner, T.M.; Penn, M.; Bakke, A.M.; Krogdahl, A.; Oliva-Teles, A. Effects of dietary phytosterols and soy saponins on growth, feed utilization efficiency and intestinal integrity of gilthead sea bream (*Sparus aurata*) juveniles. *Aquaculture* **2014**, *432*, 295–303. [CrossRef]
18. Estruch, G.; Collado, M.C.; Peñaranda, D.S.; Tomás-Vidal, A.; Jover Cerdá, M.; Pérez Martínez, G.; Martinez-Llorens, S. Impact of fishmeal replacement in diets for gilthead sea bream (*Sparus aurata*) on the gastrointestinal microbiota determined by pyrosequencing the 16S rRNA gene. *PLoS ONE* **2015**, *10*, e0136389. [CrossRef]

19. Kokou, F.; Sarropoulou, E.; Cotou, E.; Rigos, G.; Henry, M.; Alexis, M.; Kentouri, M. Effects of fish meal replacement by a soybean protein on growth, histology, selected immune and oxidative status markers of gilthead sea bream, *Sparus aurata*. *J. World Aquac. Soc.* **2015**, *46*, 115–128. [CrossRef]

20. Kokou, F.; Sarropoulou, E.; Cotou, E.; Kentouri, M.; Alexis, M.; Rigos, G. Effects of graded dietary levels of soy protein concentrate supplemented with methionine and phosphate on the immune and antioxidant responses of gilthead sea bream (*Sparus aurata* L.). *Fish Shellfish Immunol.* **2017**, *64*, 111–121. [CrossRef]

21. Estruch, G.; Collado, M.C.; Monge-Ortiz, R.; Tomás-Vidal, A.; Jover-Cerdá, M.; Peñaranda, D.S.; Pérez Martínez, G.; Martínez-Llorens, S. Long-term feeding with high plant protein based diets in gilthead seabream (*Sparus aurata*, L.) leads to changes in the inflammatory and immune related gene expression at intestinal level. *BMC Vet. Res.* **2018**, *14*, 302. [CrossRef]

22. Castro, C.; Corraze, G.; Firmino-Diógenes, A.; Larroquet, L.; Panserat, S.; Oliva-Teles, A. Regulation of glucose and lipid metabolism by dietary carbohydrate levels and lipid sources in gilthead sea bream juveniles. *Br. J. Nutr.* **2016**, *116*, 19–34. [CrossRef] [PubMed]

23. Castro, C.; Diógenes, A.F.; Coutinho, F.; Panserat, S.; Corraze, G.; Pérez-Jiménez, A.; Peres, H.; Oliva-Teles, A. Liver and intestine oxidative status of gilthead sea bream fed vegetable oil and carbohydrate rich diets. *Aquaculture* **2016**, *464*, 665–672. [CrossRef]

24. García-Meilán, I.; Valentín, J.M.; Fontanillas, R.; Gallardo, M.A. Different protein to energy ratio diets for gilthead sea bream (*Sparus aurata*): Effects on digestive and absorptive processes. *Aquaculture* **2013**, *412*, 1–7. [CrossRef]

25. Castro, C.; Couto, A.; Diógenes, A.F.; Corraze, G.; Panserat, S.; Serra, C.R.; Oliva-Teles, A. Vegetable oil and carbohydrate-rich diets marginally affected intestine histomorphology, digestive enzymes activities, and gut microbiota of gilthead sea bream juveniles. *Fish Physiol. Biochem.* **2019**, *45*, 681–695. [CrossRef] [PubMed]

26. Couto, A.; Enes, P.; Peres, H.; Oliva-Teles, A. Temperature and dietary starch level affected protein but not starch digestibility in gilthead sea bream juveniles. *Fish Physiol. Biochem.* **2012**, *38*, 595–601. [CrossRef] [PubMed]

27. Fountoulaki, E.; Alexis, M.N.; Nengas, I.; Venou, B. Effect of diet composition on nutrient digestibility and digestive enzyme levels of gilthead sea bream (*Sparus aurata* L.). *Aquac. Res.* **2005**, *36*, 1243–1251. [CrossRef]

28. Basto-Silva, C.; Enes, P.; Oliva-Teles, A.; Balbuena-Pecino, S.; Navarro, I.; Capilla, E.; Guerreiro, I. Dietary protein source and protein/carbohydrate ratio affects appetite regulation-related genes expression in gilthead seabream (*Sparus aurata*). *Aquaculture* **2021**, *533*, 736142. [CrossRef]

29. Couto, A.; Kortner, T.M.; Penn, M.; Bakke, A.M.; Krogdahl, A.; Oliva-Teles, A. Effects of dietary soy saponins and phytosterols on gilthead sea bream (*Sparus aurata*) during the on-growing period. *Anim. Feed Sci. Technol.* **2014**, *198*, 203–214. [CrossRef]

30. Khojasteh, S.M. The morphology of the post-gastric alimentary canal in teleost fishes: A brief review. *Int. J. Aquatic Sci.* **2012**, *3*, 71–88.

31. Penn, M.H.; Bendiksen, E.Å.; Campbell, P.; Krogdahl, Å. High level of dietary pea protein concentrate induces enteropathy in Atlantic salmon (*Salmo salar* L.). *Aquaculture* **2011**, *310*, 267–273. [CrossRef]

32. Alarcón, F.J.; Díaz, M.; Moyano, F.J.; Abellán, E. Characterization and functional properties of digestive proteases in two sparids; gilthead seabream (*Sparus aurata*) and common dentex (*Dentex dentex*). *Fish Physiol. Biochem.* **1998**, *19*, 257–267. [CrossRef]

33. Moyano, F.J.; Díaz, M.; Alarcón, F.J.; Sarasquete, M.C. Characterization of digestive enzyme activity during larval development of gilthead seabream (*Sparus aurata*). *Fish Physiol. Biochem.* **1996**, *15*, 121–130. [CrossRef] [PubMed]

34. Bradford, M.M. A rapid and sensitive method for the quantitation of microgram quantities of protein utilizing the principle of protein-dye binding. *Anal. Biochem.* **1976**, *72*, 248–254. [CrossRef]

35. Castro, C.; Couto, A.; Pérez-Jiménez, A.; Serra, C.R.; Díaz-Rosales, P.; Fernandes, R.; Corraze, G.; Panserat, S.; Oliva-Teles, A. Effects of fish oil replacement by vegetable oil blend on digestive enzymes and tissue histomorphology of European sea bass (*Dicentrarchus labrax*) juveniles. *Fish Physiol. Biochem.* **2016**, *42*, 203–217. [CrossRef]

36. National Center for Biotechnology Information. Available online: https://www.ncbi.nlm.nih.gov/ (accessed on 14 January 2022).

37. Taylor, S.; Wakem, M.; Dijkman, G.; Alsarraj, M.; Nguyen, M. A practical approach to RT-qPCR—Publishing data that conform to the MIQE guidelines. *Methods* **2010**, *50*, S1–S5. [CrossRef]

38. Pfaffl, M.W. A new mathematical model for relative quantification in real-time RT-PCR. *Nucleic Acids Res.* **2001**, *29*, e45. [CrossRef]

39. Ringø, E.; Zhou, Z.; Vecino, J.L.G.; Wadsworth, S.; Romero, J.; Krogdahl, A.; Olsen, R.E.; Dimitroglou, A.; Foey, A.; Davies, S.; et al. Effect of dietary components on the gut microbiota of aquatic animals. A never-ending story? *Aquac. Nutr.* **2016**, *22*, 219–282. [CrossRef]

40. Yukgehnaish, K.; Kumar, P.; Sivachandran, P.; Marimuthu, K.; Arshad, A.; Paray, B.A.; Arockiaraj, J. Gut microbiota metagenomics in aquaculture: Factors influencing gut microbiome and its physiological role in fish. *Rev. Aquac.* **2020**, *12*, 1903–1927. [CrossRef]

41. Bakke-McKellep, A.M.; Penn, M.H.; Salas, P.M.; Refstie, S.; Sperstad, S.; Landsverk, T.; Ringo, E.; Krogdahl, A. Effects of dietary soyabean meal, inulin and oxytetracycline on intestinal microbiota and epithelial cell stress, apoptosis and proliferation in the teleost Atlantic salmon (*Salmo salar* L.). *Br. J. Nutr.* **2007**, *97*, 699–713. [CrossRef]

42. Green, T.J.; Smullen, R.; Barnes, A.C. Dietary soybean protein concentrate-induced intestinal disorder in marine farmed Atlantic salmon, *Salmo salar* is associated with alterations in gut microbiota. *Vet. Microbiol.* **2013**, *166*, 286–292. [CrossRef]

43. Batista, S.; Ozório, R.O.A.; Kollias, S.; Dhanasiri, A.K.; Lokesh, J.; Kiron, V.; Valente, L.M.P.; Fernandes, J.M.O. Changes in intestinal microbiota, immune- and stress-related transcript levels in Senegalese sole (*Solea senegalensis*) fed plant ingredient diets intercropped with probiotics or immunostimulants. *Aquaculture* **2016**, *458*, 149–157. [CrossRef]

44. Scott, K.P.; Gratz, S.W.; Sheridan, P.O.; Flint, H.J.; Duncan, S.H. The influence of diet on the gut microbiota. *Pharmacol. Res.* **2013**, *69*, 52–60. [CrossRef] [PubMed]

45. Villasante, A.; Ramírez, C.; Catalán, N.; Opazo, R.; Dantagnan, P.; Romero, J. Effect of dietary carbohydrate-to-protein ratio on gut microbiota in Atlantic salmon (*Salmo salar*). *Animals* **2019**, *9*, 89–106. [CrossRef] [PubMed]

46. Miao, S.; Zhao, C.; Zhu, J.; Hu, J.; Dong, X.; Sun, L. Dietary soybean meal affects intestinal homoeostasis by altering the microbiota, morphology and inflammatory cytokine gene expression in northern snakehead. *Sci. Rep.* **2018**, *8*, 113. [CrossRef] [PubMed]

47. Parma, L.; Candela, M.; Soverini, M.; Turroni, S.; Consolandi, C.; Brigidi, P.; Mandrioli, L.; Sirri, R.; Fontanillas, R.; Gatta, P.P.; et al. Next-generation sequencing characterization of the gut bacterial community of gilthead sea bream (*Sparus aurata*, L.) fed low fishmeal based diets with increasing soybean meal levels. *Anim. Feed Sci. Technol.* **2016**, *222*, 204–216. [CrossRef]

48. Parma, L.; Yúfera, M.; Navarro-Guillén, C.; Moyano, F.J.; Soverini, M.; D'Amico, F.; Candela, M.; Fontanillas, R.; Gatta, P.P.; Bonaldo, A. Effects of calcium carbonate inclusion in low fishmeal diets on growth, gastrointestinal pH, digestive enzyme activity and gut bacterial community of European sea bass (*Dicentrarchus labrax* L.) juveniles. *Aquaculture* **2019**, *510*, 283–292. [CrossRef]

49. Banerjee, G.; Ray, A.K. Bacterial symbiosis in the fish gut and its role in health and metabolism. *Symbiosis* **2016**, *72*, 1–11. [CrossRef]

50. Ray, A.K.; Ghosh, K.; Ringø, E. Enzyme-producing bacteria isolated from fish gut: A review. *Aquac. Nutr.* **2012**, *18*, 465–492. [CrossRef]

51. Munilla-Morán, R.; Saborido-Rey, F. Digestive enzymes in marine species. II. Amylase activities in gut from seabream (*Sparus aurata*), turbot (*Scophthalmus maximus*) and redfish (*Sebastes mentella*). *Comp. Biochem. Physiol. B Biochem. Mol. Biol.* **1996**, *113*, 827–834. [CrossRef]

52. Munilla-Morán, R.; Saborido-Rey, F. Digestive enzymes in marine species. I. Proteinase activities in gut from redfish (*Sebastes mentella*), seabream (*Sparus aurata*) and turbot (*Scophthalmus maximus*). *Comp. Biochem. Physiol. B Biochem. Mol. Biol.* **1996**, *113*, 395–402. [CrossRef]

53. Rust, M.B. Nutritional physiology. In *Fish Nutrition*, 3rd ed.; Elsevier: Amsterdam, The Netherlands, 2002; p. 367.

54. Spannhof, L.; Plantikow, H. Studies on carbohydrate digestion in rainbow trout. *Aquaculture* **1983**, *30*, 95–108. [CrossRef]

55. Storebakken, T.; Shearer, K.D.; Baeverfjord, G.; Nielsen, B.G.; Asgard, T.; Scott, T.; De Laporte, A. Digestibility of macronutrients, energy and amino acids, absorption of elements and absence of intestinal enteritis in Atlantic salmon, *Salmo salar*, fed diets with wheat gluten. *Aquaculture* **2000**, *184*, 115–132. [CrossRef]

56. Bakke-McKellep, A.M.; Refstie, S. Alternative protein sources and digestive function alterations in teleost fishes. In *Feeding and Digestive Functions in Fish*; Cyrino, J.E.P., Bureau, D.P., Kapoor, R.G., Eds.; CRC Press: Boca Raton, FL, USA, 2008.

57. Hidalgo, M.C.; Urea, E.; Sanz, A. Comparative study of digestive enzymes in fish with different nutritional habits. Proteolytic and amylase activities. *Aquaculture* **1999**, *170*, 267–283. [CrossRef]

58. Fernández, I.; Moyano, F.J.; Díaz, M.; Martínez, T. Characterization of alpha-amylase activity in five species of Mediterranean sparid fishes (Sparidae, Teleostei). *J. Exp. Mar. Biol. Ecol.* **2001**, *262*, 1–12. [CrossRef]

59. Giri, S.S.; Sahoo, S.K.; Sahu, A.K.; Meher, P.K. Effect of dietary protein level on growth, survival, feed utilisation and body composition of hybrid Clarias catfish (*Clarias batrachus* × *Clarias gariepinus*). *Anim. Feed Sci. Technol.* **2003**, *104*, 169–178. [CrossRef]

60. Debnath, D.; Pal, A.K.; Sahu, N.P.; Yengkokpam, S.; Baruah, K.; Choudhury, D.; Venkateshwarlu, G. Digestive enzymes and metabolic profile of *Labeo rohita* fingerlings fed diets with different crude protein levels. *Comp. Biochem. Physiol. B Biochem. Mol. Biol.* **2007**, *146*, 107–114. [CrossRef]

61. Habte-Tsion, H.M.; Liu, B.; Ge, X.P.; Xie, J.; Xu, P.; Ren, M.C.; Zhou, Q.L.; Pan, L.K.; Chen, R.L. Effects of dietary protein level on growth performance, muscle composition, blood composition, and digestive enzyme activity of Wuchang bream (*Megalobrama amblycephala*) fry. *Isr. J. Aquac.-Bamidgeh* **2013**, *65*, 9.

62. Yan, X.; Yang, J.; Dong, X.; Tan, B.; Zhang, S.; Chi, S.; QihuiYang; Liu, H.; Yang, Y. Optimum protein requirement of juvenile orange-spotted grouper (*Epinephelus coioides*). *Sci. Rep.* **2021**, *11*, 6230. [CrossRef]

63. Kortner, T.M.; Skugor, S.; Penn, M.H.; Mydland, L.T.; Djordjevic, B.; Hillestad, M.; Krasnov, A.; Krogdahl, A. Dietary soyasaponin supplementation to pea protein concentrate reveals nutrigenomic interactions underlying enteropathy in Atlantic salmon (*Salmo salar*). *BMC Vet. Res.* **2012**, *8*, 101. [CrossRef]

64. Zhang, J.-X.; Guo, L.-Y.; Feng, L.; Jiang, W.-D.; Kuang, S.-Y.; Liu, Y.; Hu, K.; Jiang, J.; Li, S.-H.; Tang, L.; et al. Soybean β-conglycinin induces inflammation and oxidation and causes dysfunction of intestinal digestion and absorption in fish. *PLoS ONE* **2013**, *8*, e58115. [CrossRef]

65. Tan, C.; Zhou, H.; Wang, X.; Mai, K.; He, G. Resveratrol attenuates oxidative stress and inflammatory response in turbot fed with soybean meal based diet. *Fish Shellfish Immunol.* **2019**, *91*, 130–135. [CrossRef] [PubMed]

66. Zhang, G.; Mao, J.; Liang, F.; Chen, J.; Zhao, C.; Yin, S.; Wang, L.; Tang, Z.; Chen, S. Modulated expression and enzymatic activities of Darkbarbel catfish, *Pelteobagrus vachelli* for oxidative stress induced by acute hypoxia and reoxygenation. *Chemosphere* **2016**, *151*, 271–279. [CrossRef] [PubMed]

67. Olsen, R.E.; Svardal, A.; Eide, T.; Wargelius, A. Stress and expression of cyclooxygenases (cox1, cox2a, cox2b) and intestinal eicosanoids, in Atlantic salmon, *Salmo salar* L. *Fish Physiol. Biochem.* **2012**, *38*, 951–962. [CrossRef] [PubMed]

68. Wang, T.; Yan, J.; Xu, W.; Ai, Q.; Mai, K. Characterization of cyclooxygenase-2 and its induction pathways in response to high lipid diet-induced inflammation in *Larmichthys crocea*. *Sci. Rep.* **2016**, *6*, 19921. [CrossRef] [PubMed]

Article

Solid-State Hydrolysis (SSH) Improves the Nutritional Value of Plant Ingredients in the Diet of *Mugil cephalus*

Francisca P. Martínez-Antequera [1],*, Isabel Barranco-Ávila [2], Juan A. Martos-Sitcha [2] and Francisco J. Moyano [1]

[1] Department of Biology and Geology, Faculty of Experimental Sciences, Campus de Excelencia Internacional del Mar (CEI·MAR), University of Almería, 04120 Almeria, Spain; fjmoyano@ual.es

[2] Department of Biology, Faculty of Marine and Environmental Sciences, Instituto Universitario de Investigación Marina (INMAR), Campus de Excelencia Internacional del Mar (CEI·MAR), University of Cádiz, 11519 Puerto Real, Spain; isabelbavila@gmail.com (I.B.-Á.); juanantonio.sitcha@uca.es (J.A.M.-S.)

* Correspondence: fma996@ual.es

Abstract: The possibility of improving the nutritional quality of plant byproducts (brewers' spent grain and rice bran) through an enzyme treatment was tested in a formulated feed for grey mullet (*Mugil cephalus*). The enzyme treatment was carried out by Solid-State Hydrolysis (SSH) using a commercial preparation including carbohydrases and phytase. A feed prepared without the treatment and a commercial feed for carp were used as controls. In a preliminary short-term trial carried out at laboratory facilities, fish receiving the enzyme-treated feed showed significant improvement in both FCR and SGR when compared to those obtained with the untreated diet, although both experimental diets presented worse values than those obtained with the commercial feed. Different metabolic indicators including higher values of muscle glycogen and plasmatic triglycerides supported the positive effect of the enzyme treatment on the nutritional condition of the fish over those fed on the diet containing non-treated ingredients. Results of growth and feed efficiency that were obtained in a second long-term trial developed for 148 days under real production conditions evidenced the equivalence among the experimental and commercial diets and confirmed that enzyme pretreatment of plant ingredients by SSH may be a useful procedure to improve the nutritive value of high fiber plant byproducts when included in practical diets for this species and others with similar nutritional features.

Keywords: aquaculture feeds; plant byproducts; enzymatic pretreatment

Citation: Martínez-Antequera, F.P.; Barranco-Ávila, I.; Martos-Sitcha, J.A.; Moyano, F.J. Solid-State Hydrolysis (SSH) Improves the Nutritional Value of Plant Ingredients in the Diet of *Mugil cephalus*. *Fishes* **2022**, *7*, 4. https://doi.org/10.3390/fishes7010004

Academic Editors: Maria Angeles Esteban, Bernardo Baldisserotto and Eric Hallerman

Received: 5 December 2021
Accepted: 23 December 2021
Published: 25 December 2021

Publisher's Note: MDPI stays neutral with regard to jurisdictional claims in published maps and institutional affiliations.

1. Introduction

Although in the last years important efforts have been carried out to reduce the levels of the conventional marine resources, i.e., fish meal and fish oil, in the diets of cultured fish, the sustainable development of marine aquaculture requires species that can be produced without the need to use high amounts of such ingredients, the availability of which is growing progressively more limited. Mugilidae (mullets) are a group of fish gaining increasing interest for aquaculture due to their rapid growth, resistance to a wide range of environmental conditions, and omnivorous profile. Over the past few years, the culture of these species, particularly of the grey mullet (*Mugil cephalus*), is considered a priority within the current strategies of aquaculture diversification in different parts of the world, with particular interest in some Mediterranean countries [1]. In addition to specific research aimed at completing its reproduction in captivity, the culture of grey mullet requires the development of suitable species-specific diets, the availability of which represents a bottleneck for their production under intensive systems. Recent studies show good results when testing highly nutritive diets based in the use of zooplankton species [2,3], and several others support the possibility of using high amounts of plant byproducts in feeds for these species, even during early stages of their development [4–8]. Nevertheless, in this latter case the selected products presented a high nutritional quality (high protein, low fibre contents) and hence the potential of using other vegetable ingredients with a more limited nutritional value has not been properly tested.

As indicated previously, due to their low trophic level and opportunistic nature, mullets are ideal candidates to take advantage of feeds including high percentages of alternative products and byproducts, many of which have high interest for local use within the framework of the circular economy. Nevertheless, from a nutritional point of view, such plant byproducts may present important limitations linked to both their amino acid imbalances and reduced digestibility due to the presence of a wide variety of antinutritional compounds including alkaloids, lectins, digestive enzyme inhibitors, indigestible carbohydrates (mainly non-starch polysaccharides, NSPs) and phytate [9]. In this sense, the use of enzyme additives may be a powerful tool to counteract the potential negative effects derived from the presence of some of these compounds, such as phytate and NSP, thus increasing the whole nutritional value of the ingredients. A number of commercially available multienzyme complexes have been developed to improve the use of carbohydrates and phytate present in plant ingredients used in feeds for terrestrial animals. Nevertheless, they have been designed for optimal functioning under the body temperature and pH conditions existing in the digestive systems of pigs and poultry, which are notably different from those present in aquatic species. This may explain the limited effectiveness and somewhat contradictory results obtained when such products are tested in some fish species, such as Japanese sea bass, carp or rainbow trout [10–13]. The efficiency of hydrolysis produced by such enzymes is greatly conditioned by several aspects. As an example, there may be interactions between gastric and intestinal proteases produced by fish and the exogenous enzymes that can negatively affect their potential beneficial effects [14]. Furthermore, the effectiveness of the exogenous enzymes may be greatly reduced by the high temperatures reached during feed preparation or, in the case of being applied post-extrusion via oil top coating or spraying, by the time available for the enzymatic action inside the digestive system of the species, which is closely related to gut transit rates linked to water temperature.

An interesting alternative to overcome the aforementioned limitations is the pretreatment of plant ingredients with the enzyme compound before the preparation of the feed pellets, using Solid-State Hydrolysis (SSH). SSH operates with a percentage of solid substrate greater than 15%, so there is little or no free water [15]. This process is routinely used to obtain specific products such as glucose or other sugars, or directly to increase the nutritional value of plant ingredients by reducing the content of NSP [16]. By using SSH, hydrolysis can be carried out under optimal conditions for the enzymes, and their activity is not affected either by the high temperatures reached during feed preparation or by the biochemical conditions present in the guts of fish.

Considering all the above, the aim of the present work was to assess whether pretreatment of plant ingredients containing high proportions of NSP and phytate using a mixture of enzymes applied under an SSH protocol could improve nutritional value when used in the diet of *M. cephalus*. To achieve this objective two different trials were carried out: an initial short-term trial aimed at preliminary evaluation of the growth performance, feed efficiency and energy metabolism of fish fed on such diet, and a second long-term trial focused only on the evaluation of differences in growth and feed efficiency produced by the diet when evaluated under field production conditions.

2. Materials and Methods

2.1. Experiment 1. Short-Term Trial

2.1.1. Ingredients and Experimental Feeds

The experimental (EXP) diet was formulated taking the proximate composition (Table 1) of a commercial diet as a reference (AQUASOJA, SORGAL, Ovar, Portugal). This commercial diet (COM) was routinely used as maintenance feed in the culture of *M. cephalus* by the company providing the fish. It contains several animal ingredients (fishmeal, fish hydrolysate, feather meal, meat and bone meal, poultry fat) and plant ingredients (soybean, wheat, bean and sunflower meals) to give a total amount of 35 g/100 g crude protein and 9 g/100 g crude fat. The EXP diet was formulated including more than 75% plant ingredients, of which 30% were high-fibre by-products such as rice bran and brewer's spent grain. To prepare the enzyme treated diet (EXP/enz), all plant ingredients were milled to a mesh size of 0.5 mm and mixed with citrate buffer (pH 5.0, 0.1 M) to obtain a moist mass (1:2 *w/v*), providing the optimal conditions for the action of the multienzyme complex under SSH. The product used was Rovabio®, a mixture of xylanases, glucanases, arabinofuranosidases and phytase produced by Adisseo (Auvergne, France). It was added to the mixture of plant meals by dissolving the dose recommended for terrestrial species by the manufacturer (0.2 mL/kg) in a certain amount of citrate buffer (0.1 M, pH 5.5) that was then carefully sprayed and mixed. The enzymes were allowed to act, keeping the mixture at 45 °C for 6 h with manual stirring every hour to ensure the homogeneity of the reaction. After this time, the reaction was stopped by placing the mixture in a cold chamber at 4 °C until addition of the rest of the diet ingredients and preparation of feed pellets. The feeds were prepared using an extrusion machine with a mesh size of 2 mm, dried, and stored at 4 °C until use.

Table 1. Ingredients and proximate composition of the experimental feed used in the experiments.

Ingredient (in g/100 g d.w.)	EXP	COM
Fishmeal 67/10	10.00	
Soybean meal 47	18.83	
Defatted rice bran	10.00	
Soybean protein concentrate	8.00	
Corn gluten meal 60	8.00	
Guar meal (Korma)	11.16	
Brewer's spent grain	20.00	
Fish oil	3.25	
Sunflower oil	2.60	
Soy lecithin	0.65	
Vitamin/mineral premix	0.05	
Taurine	0.30	
Yeast	3.00	
Squid hydrolysate	1.50	
Proximate composition (in g/100 g)		
Crude protein	35.60	35.00
Crude fat	9.03	9.00
Digestible carbohydrates (starch + oligosaccharides)	10.56	4.00
NSP	27.73	
Ash	6.21	8.00
Phosphorus	0.85	1.30
Phytate P	0.35	
Gross Energy (MJ kg^{-1})	17.80	17.10

Declared ingredients in the commercial diet: Meat and bone meal, feather meal, fishmeal, fish hydrolysate, wheat meal, horse bean meal, sunflower meal, dehulled soybean meal, rice bran, fish oil, poultry fat, brewers' yeast. EXP: experimental; COM: commercial; NSP: Non-starch polysaccharides; P: phosphorus. Vitamins and mineral premix (IU or mg kg^{-1} diet);DL-alpha tocopherol acetate, 200 mg; sodium menadione bisulphate, 10 mg; retinyl acetate, 16,650 IU; DL-cholecalciferol, 2000 IU; thiamine, 25 mg; riboflavin, 25 mg; pyridoxine, 25 mg; cyanocobalamin, 0.1 mg; niacin, 150 mg; folic acid, 15 mg; L-ascorbic acid monophosphate, 750 mg; inositol, 500 mg; biotin, 0.75 mg; calcium panthotenate, 100 mg; choline chloride, 1000 mg; copper sulphate heptahydrate, 25 mg; ferric sulphate monohydrate,100 mg; potassium iodide, 2 mg; manganese sulphate monohydrate,100 mg; sodium selenite, 0.05 mg; zinc sulphate monohydrate, 200 mg.

The effect of SSH on the feeds was assessed by chemical analysis of some specific compounds of which the relative concentrations were expected to be modified as a result of enzyme treatment: soluble protein, reducing sugars, pentoses, total phosphorus and phytate phosphorus. Soluble protein was analysed by the Bradford method [17] using a SIGMA Total Protein Kit (TP0100). Reducing sugars were measured using 3,5-dinitrosalicylic acid

(DNS) following the method described by Miller [18]. Pentoses were measured by the phloroglucinol method described by Douglas [19]. Total phosphorus was determined by the molybdovanadate method after total digestion of the organic matter with concentrated nitric acid. Phytic acid was determined following the bipyridine method described by Haug and Lantzsch [20]. All the analyses were performed in triplicate on samples from each diet. The rest of compounds (total crude protein and lipids, moisture, ash) were analysed using AOAC protocols [21]. In brief, crude protein (N × 6.25) was evaluated using the Kjeldahl method, lipid content was determined by petroleum ether extraction (40–60 °C) using a Soxhlet System, moisture content was calculated by drying at 105 °C for 24 h, and ash content was determined using a muffle furnace at 550 °C for 5 h. In addition to these chemical analyses, change in the water retention capacity of feed pellets as a result of partial hydrolysis of the carbohydrate fraction was evaluated as described in Heywood et al. [22].

2.1.2. Feeding Trial, Samples Collection and Data Recording

Juvenile grey mullets (*Mugil cephalus*) were provided by PIMSL (Sevilla, Spain), transferred to the experimental facility (CTAQUA, Centro Tecnológico de la Acuicultura de Andalucía, El Puerto de Santa María, Cádiz, Spain) and acclimated to laboratory conditions for two weeks. Then, the fish (12.24 ± 1.05 g body weight) were randomly distributed in triplicate groups in nine 400 L tanks (n = 100 fish per tank, 300 fish per experimental diet) coupled to a recirculation aquaculture system (RAS) and equipped with physical and biological filters and programmable temperature and O_2 suppliers. Water flow of 5–6 L/tank/min ensured a daily renewal of ten times total volume, and was maintained at 20.3 ± 1.0 °C during the experiment. Experimental diets were offered to apparent satiation three times per day, with the orientative daily ration adapted according to weight controls carried out every 14 days. The experiment lasted six weeks. Total feed intake was recorded for each experimental unit to calculate growth performance parameters. At the end of the trial, overnight fasted fish (four fish per tank, twelve per experimental conditions) were randomly sampled and deeply anaesthetized with 2-fenoxiethanol in a lethal dose (1 mL/L SW) to obtain blood and tissue samples. Blood was drawn from caudal vessels with heparinized syringes and centrifuged at 3000× g for 15 min at 4 °C to separate plasma, which was then snap-frozen in liquid nitrogen and stored at −80 °C until used for biochemical analysis. Fish were cervically sectioned in order to obtain biopsies of different tissues; samples of liver were rapidly taken and weighed to calculate the hepatosomatic index (HSI) and, together with samples of white skeletal muscle, were snap-frozen in liquid nitrogen and stored at −80 °C for subsequent biochemical analysis. Maintenance and sampling of the fish was carried out in compliance with the Guidelines of the European Union Council (86/609/EU) for the use of laboratory animals.

2.1.3. Growth Performance and Biometric Parameters

The following growth parameters were evaluated:

$$\text{specific growth rate (SGR)} = (100 \times (\ln \text{final body weight} - \ln \text{initial body weight})/\text{days} \tag{1}$$

$$\text{weight gain percent (WG)} = (100 \times (\text{body weigh increase})/\text{initial body weight} \tag{2}$$

$$\text{feed conversion ratio (FCR)} = \text{total feed intake}/\text{weight gain} \tag{3}$$

$$\text{condition factor} = (100 \times \text{body weight})/\text{fork length}^3 \tag{4}$$

$$\text{hepatosomatic index (HSI)} = (100 \times \text{liver weight})/\text{fish weight} \tag{5}$$

2.1.4. Biochemical Parameters

Glucose (Ref. 1001200), lactate (Ref. 1001330) and triglycerides (Ref. 1001311) in plasma and tissues were measured using commercial kits (Spinreact, St. Esteve d'en Bas, Girona, Spain) adapted to 96-well microplates. Plasma total protein concentration was determined with a BCA Protein Assay Kit (Ref. 23225, Thermo Fisher Scientific Pierce, Waltham, MA, USA,) using BSA as the standard. Glycogen concentration was quantified using the method described by Decker and Keppler [23], while glucose obtained after glycogen breakdown with amyloglucosidase (Ref. A7420; Sigma-Aldrich, St. Louis, MO, USA) was determined using the same commercial kit described above. To analyse biochemical parameters in liver and muscle, frozen tissues were homogenized by ultrasonic disruption in 7.5 volumes ice-cold 0.6 N perchloric acid, neutralized using 1 M KCO_3, and centrifuged (30 min, $3220 \times g$ and 4 °C); the supernatants were then isolated to determine tissue metabolites. All assays were performed using a PowerWaveTM 340 microplate spectrophotometer (Bio-Tek Instruments, Winooski, VT, USA) using the KCjunior^{TM} data analysis software (Bio-Tek Instruments, Winooski, VT, USA) for Microsoft$^{®}$

2.2. Experiment 2. Field Trial

Experimental Feeds, Feeding Trial, Samples Collection and Data Recording

The same three experimental feeds described and evaluated in Experiment 1 were used in this trial. In this case, the experiment was carried out in the facilities of PIMSL (Sevilla, Spain); 1200 juvenile fish with a 39.63 ± 1.14 g initial mean body weight were randomly distributed in six 8 m^3 concrete outdoor tanks in duplicate groups (n = 200 fish per tank) coupled to a recirculation aquaculture system (RAS) equipped with physical and biological filters. Water temperature varied within a wide range during the 148 days of the experimental period according to season, from June to December 2020 (19.5 ± 3.14 °C). Fish were fed twice a day at an initial ration of 2.0%bw that was adjusted after one weight control carried out at an intermediate point during the experiment. Total feed intake and weight increase were recorded for each experimental group in order to calculate FCR and SGR at the end of the experiment, as indicated in Section 2.1.4.

2.3. Statistical Analysis

After a preliminary evaluation to determine the normality of the data using the Shapiro–Wilk test, homoscedasticity analysis was conducted using the Brown–Forsythe test. Due to the different composition of the two diets used as a reference (commercial and experimental without enzyme), separate comparisons were carried out among the diets, on one hand comparing the commercial diet with each of the two experimental diets, and on the other comparing only these latter between themselves. The objective was to assess whether any of the two experimental feeds were equivalent to the commercial diet, while in the second case the objective was to check whether the enzyme treatment could improve on the performance obtained with the experimental diet. The first was carried out by one-way ANOVA followed by the Bonferroni test, and the second was performed using Student *T*-test. The significance level was established at $p < 0.05$. Data expressed in percentage were previously arc-sin transformed. All analyses were carried out using Statgraphics Centurion software (Statgraphics Technologies, Inc., The Plains, VA, USA).

3. Results

3.1. Experiment 1

The analysis of the diets showed significantly higher values of some compounds (soluble protein, reducing sugars and phytate) and lower values of total phosphorus in EXP feed when compared to COM feed, reflecting differences in the ingredients used in their elaboration. The Solid-State enzymatic Hydrolysis significantly increased the amount of potentially available reducing sugars and pentoses, and reduced the amount of phytate (Table 2). Furthermore, physical transformation resulting from the enzyme treatment determined a significant reduction in water retention capacity in the EXP/enz diet (Table 2).

Table 2. Differences in the nutrient content and water retention of experimental feeds (g/100 d.m.).

	Soluble Protein	Reducing Sugars	Pentoses	Phosphorus	Phytate	Water Retention
COM	3.13 ± 0.05 [A]	0.45 ± 0.00 [A]	0.29 ± 0.01 [A]	1.31 ± 0.02 [A]	0.22 ± 0.00 [A]	305.77 ± 6.89 [A]
EXP	6.83 ± 1.13 [Ba]	2.08 ± 0.02 [Ba]	0.28 ± 0.00 [Aa]	0.83 ± 0.10 [Ba]	0.34 ± 0.01 [Ba]	305.45 ± 3.87 [Aa]
EXP/enz	5.58 ± 0.24 [Ba]	2.83 ± 0.03 [Bb]	0.36 ± 0.05 [Bb]	0.76 ± 0.03 [Ba]	0.21 ± 0.02 [Bb]	280.94 ± 9.29 [Bb]

Statistical comparisons between COM and any of the EXP diets is noted in capital letters, while comparisons between EXP and EXP/enz are detailed in lowercase. Each assay was performed in triplicate. Values not sharing the same letter differ significantly with $p < 0.05$. COM: commercial; EXP: experimental; EXP/enz: experimental enzyme treated.

3.1.1. Growth Performance

Results of growth performance and feed utilisation are presented in Table 3. No mortality occurred during the experiment, and all groups presented an increase in body mass, accounting for 11% to 33% of their initial mean body weight. Fish presented a normal condition factor for the species (K = 1.07–1.12). The results presented in Table 4 show that the experimental diet, irrespective of being enzymatically treated or not, clearly offered worse results in growth and feed efficiency than the control diet. Nevertheless, it was also clear that the Solid-State enzymatic Hydrolysis significantly improved the same parameters when comparing fish fed on EXP/enz to those obtained with EXP.

Table 3. Growth and feed efficiency measured in fish fed on the experimental diets.

Parameter	COM	EXP	EXP/enz
Initial body mass (g/fish)	12.02 ± 0.33 [A]	11.89 ± 0.55 [Aa]	12.38 ± 0.16 [Aa]
Final body mass (g/fish)	16.60 ± 0.60 [A]	13.21 ± 0.49 [Ba]	15.36 ± 0.04 [Ab]
Feed consumption (g/fish)	10.02 ± 0.17 [A]	8.65 ± 0.45 [Ba]	9.23 ± 0.17 [Ab]
FCR (g feed/g fish)	2.19 ± 0.11 [A]	5.89 ± 0.60 [Ba]	3.11 ± 0.19 [Bb]
SGR (%/day)	0.75 ± 0.02 [A]	0.25 ± 0.05 [Ba]	0.50 ± 0.03 [Bb]
HIS (%)	1.27 ± 0.07 [A]	0.93 ± 0.18 [Ba]	0.83 ± 0.20 [Bb]
Condition factor (K)	1.11 ± 0.12 [A]	1.07 ± 0.12 [Ba]	1.07 ± 0.17 [Ba]

Statistical comparisons between COM and any of the EXP diets is noted in capital letters, while comparisons between EXP and EXP/enz are detailed in lowercase. COM: commercial; EXP: experimental; EXP/enz: experimental enzyme treated; FCR: feed conversion ratio; SGR: specific growth rate; HIS: hepatosomatic index.

Table 4. Metabolites measured in plasma and tissues of fish fed on the different experimental diets.

Parameter	COM	EXP	EXP/enz
In plasma (mg/dL)			
Glucose	80.54 ± 26.21 [A]	92.12 ± 29.61 [Aa]	76.76 ± 14.81 [Aa]
Lactate	67.28 ± 32.89 [A]	56.71 ± 11.07 [Ba]	47.13 ± 22.73 [Bb]
Protein	32.60 ± 5.46 [A]	34.62 ± 10.08 [Aa]	38.23 ± 4.35 [Aa]
TAG	38.09 ± 7.60 [A]	34.13 ± 4.76 [Aa]	42.53 ± 6.48 [Bb]
In liver (mg/g w/w)			
Glucose	1.41 ± 0.48 [A]	1.37 ± 0.56 [Aa]	1.34 ± 0.29 [Aa]
Glycogen	7.26 ± 1.26 [A]	2.61 ± 1.52 [Ba]	3.31 ± 1.32 [Ba]
In muscle (mg/g w/w)			
Glucose	0.77 ± 0.30 [A]	0.84 ± 0.28 [Aa]	0.72 ± 0.30 [Aa]
Glycogen	0.78 ± 0.45 [A]	0.38 ± 0.14 [Ba]	0.79 ± 0.31 [Ab]
Lactate	24.57 ± 6.63 [A]	26.35 ± 7.62 [Ba]	24.96 ± 3.90 [Aa]
TAG	10.48 ± 5.14 [A]	10.79 ± 3.83 [Aa]	11.17 ± 7.15 [Aa]

Statistical comparisons between COM and any of the EXP diets is noted in capital letters, while comparisons between EXP and EXP/enz are detailed in lowercase.

3.1.2. Biochemical Parameters

Data on blood and tissue biochemistry are detailed in Table 4. The parameters measured in plasma show significantly lower values of lactate in fish fed any of the experimental diets when compared to those obtained in fish fed the COM diet. Additionally, the amount of TAG measured in fish fed the EXP/enz diet was significantly higher than in fish fed the EXP diet. Liver glycogen measured in fish fed any of the experimental diets was also significantly lower than in fish fed the COM diet. Indeed, muscle glycogen measured in fish fed on the EXP diet was significantly reduced compared to the other two groups.

3.2. Experiment 2. Field Trial

The results on growth performance and feed efficiency in the field trial are presented in Table 5. Average fish growth during the whole experimental period ranged from 0.32 to 0.43%, although it was low or even absent during a great part of the period due to low winter temperatures. Nevertheless, all of the experimental groups doubled their initial weight by the end of the experiment. Moreover, all groups presented reasonably good FCR values, ranging from 2.33 to 2.67, and in contrast to those obtained in experiment 1, the values of FCR were significantly enhanced when fish were fed on the EXP/enz diet.

Table 5. Growth and feed efficiency in fish fed on the experimental diets in the field trial.

Parameter	COM	EXP	EXP/enz
Initial body mass (g/fish)	40.13 ± 0.18 [A]	38.25 ± 0.35 [Ba]	40.50 ± 0.71 [Ab]
Final body mass (g/fish)	88.05 ± 11.53 [A]	94.65 ± 13.08 [Aa]	104.70 ± 13.86 [Aa]
Feed consumption (g/fish)	18.58 ± 0.96 [A]	17.83 ± 1.68 [Aa]	17.76 ± 0.01 [Aa]
FCR (g feed/g fish)	2.65 ± 0.12 [A]	2.67 ± 0.29 [Aa]	2.33 ± 0.58 [Bb]
SGR (%/day)	0.32 ± 0.08 [A]	0.38 ± 0.09 [Aa]	0.43 ± 0.09 [Aa]

Statistical comparisons between COM and any of the EXP diets is noted in capital letters, while comparisons between EXP and EXP/enz are detailed in lowercase.

4. Discussion

The "extreme" experimental feed (EXP) used in the present work was designed to assess the limits of including a high amount of plant ingredients of limited nutritional value in practical diets for *M. cephalus*, as well as the potential benefits derived from the enzymatic treatment of such plant ingredients. For this reason, it included a low proportion of fishmeal (10%), a high amount of fibrous byproducts containing significant levels of phytate and NSP (10% rice bran, 20% brewer's spent grain) and no supplementation with lysine or methionine. Although experiment 1, developed with small juveniles, was not maintained for enough time to allow the fish to at least double their weight, it offered preliminary insight on the potential performance of the diets and two clear results: on the one hand, it demonstrated that the EXP diet did not fulfill the nutritional requirements of the species at this early stage of development; on the other, it showed that enzyme pretreatment of the plant ingredients had a significant positive effect on the nutritional value of such a diet.

Related to the first point, despite the values of FCR and SGR obtained with diets EXP being not directly comparable to those obtained with diet COM (as both types of diets were differently formulated and contained quite different ingredients in terms of acceptance, digestibility and nutritional value), it was clear that the use of the EXP diet impaired the growth of the fish. Similarly, El-Gendy et al. [24] reported a clear reduction in FCR and SGR when feeding juveniles of *M. cephalus* with diets prepared with increasing amounts of plant ingredients (from 20 to 100%), in that case including a significant proportion of cereal bran. Certainly, a different selection of plant ingredients with a higher nutritional value should have produced much better results; in this way, Gisbert et al. [8] reported good results in terms of SGR, digestive physiology and fish condition when feeding small (0.2 g) juveniles of *M. cephalus* on feeds including a blend of corn and wheat gluten and soy protein concentrate, supplemented with crystalline L-lysine and DL-methionine. Nevertheless, as mentioned previously, the objective of the present study was to assess to what extent feeds for *M. cephalus* could include high amounts of plant byproducts and no specific supplementation to minimize the final cost of the diet without compromising growth.

Regarding the second point, it is clear that Solid-State enzymatic Hydrolysis of the EXP diet with a commercial mixture of xylanases, β-glucanases and pectinases resulted in a significant improvement in both FCR and SGR compared to the values obtained with the untreated diet, suggesting a clear positive effect of the enzyme complex. Similar positive effects have been reported in terrestrial species such as pigs [25] and poultry [26], being associated to a great extent with the modification of the structure of carbohydrates present in plant ingredients, which increases the bioaccessibility of nutrients to the action of digestive enzymes. In the present study, such modification was indirectly evidenced by the significant reduction in the water retention capacity, as well as by the significant increase in bioavailable monosaccharides and the reduction in phytate measured in the EXP/enz feed (Table 2). The improvement observed in the FCR and SGR over those obtained in fish fed the EXP diet was higher than 50%, suggesting that partial hydrolysis of the antinutritive factors (NSP and phytate) exerted a positive impact on the nutritional value of the feed, and hence on the performance indicators.

In the present study, the measurement of different metabolites was intended to evaluate the potential impact of the EXP diet on fish energy orchestration, as well as whether enzyme treatment yielded significant effects on metabolic homeostasis. As presented herein, no significant differences were observed in glucose levels measured in plasma, muscle or liver irrespective of the diet or enzyme treatment, even when this latter treatment significantly increased the amount of available dietary sugars measured in the fish receiving the EXP/enz diet. In contrast to what has been described for carnivorous fish [27], this suggests a good ability of the grey mullet to use carbohydrates as a source of energy, demonstrating a homeostatic load of this metabolite that might be considered sufficient for the potential growth of this species. The unfavourable nutritional status of fish fed on the EXP diet was reflected by several indicators. First, a significantly lower accumulation of liver glycogen was correlated with the lower weight of this organ, which determined decreased values of HSI as well as lower plasma levels of lactate and could reflect an impairment between the total energy incorporated through the feed and the demand for physiological processes such as growth. Furthermore, a significantly lower concentration of muscle glycogen was measured in fish receiving the EXP diet. Under normal conditions, if the feed is able to cover nutritional needs excess glucose may be stored as glycogen (glycogenesis) or converted into lipids (lipogenesis) instead of being oxidized for energy; however, under conditions of food deprivation or nutritionally unbalanced feeding, glucose requirements are satisfied either by glycogen depletion to glucose (glycogenolysis) or by de novo glucose synthesis through gluconeogenesis from lactate, glycerol or certain amino acids [28]. In contrast, the comparatively improved nutritional status associated with consumption of the EXP/enz feed was supported by significantly higher values of muscle glycogen and plasma TAG. Moreover, the results also suggest that de novo gluconeogenesis in the muscle may contribute to higher glycogen content, although to elucidate whether this fact is a cause or a consequence of better feed utilisation would require further investigation related to the key role of several metabolic enzymes in this and other tissues.

The results obtained in experiment 2 were noticeably different, for two main reasons: the initial size of the juvenile fish was much higher (nearly 40 g average weight) and the experimental period was long enough that the fish were able to double their initial weight (Table 5). In addition, the environmental conditions were different, as the experiment was carried on outdoors in standard facilities used for rearing the intermediate stages of growing fish. Under such conditions, both the EXP and EXP/enz diets appeared to be equivalent to the COM diet, and Solid-State enzymatic Hydrolysis confirmed its positive effect on the nutritional value of the ingredients. Although the growth rates of *M. cephalus* were relatively low at the end of the 28-week feeding period, they were comparable to those reported in previous studies carried out with this and other similar species that evidence the slow growth rate of mugilids [29–31]. Legarda et al. [32] reported an SGR of 0.56%/day for *Mugil cephalus* juveniles fed 1%bw daily in a biofloc system maintained at 28 ± 1 °C. Nengas et al. [33] also recorded low SGR values (around 0.45%/day) in juveniles of *Liza aurata* fed once per day at 2% of their biomass at temperatures ranging from 12 to 26 °C.

Values of FCR obtained with any of the EXP diets were clearly improved in relation to those obtained in experiment 1, being in this case equivalent or even better than obtained with the COM diet. This could be thanks to for two main reasons: on the one hand, as indicated previously, the experiment was carried out using older fish, and presumably their ability to digest fibrous materials improved greatly with age, this being associated with the maturation and development of the digestive function. These changes in digestive capability with development have been previously reported in other mullet species including the thick-lipped grey mullet, *Chelon labrosus* [34,35]. This suggests that the possibility of using a high amount of fibrous plant byproducts in diets for this species is strongly conditioned by the age of the fish, and probably linked to the ability, acquired with development, to properly digest such ingredients. In addition, it must be considered that the conditions in which the fish were maintained in this latter experiment, mainly a lower stocking density, could result in lower stress and a positive impact on feeding behaviour, resulting in better food utilization. Although there are few studies on the nutrition of mullets under field conditions, the study carried out by [33] on *Liza aurata* showed that growth and feed utilization of the fish was not significantly affected by variations in the dietary protein level. Such absence of a clear effect could be due to the possible complementary effect of natural food present in the water mass. In the present study, the only source of nutrients was the artificial feeds, and no other food source was available; hence, the observed response was representative of the nutritional value of the feeds.

The results further confirmed the positive effect of the enzyme treatment already pointed out in experiment 1, which is in line with other previous studies. A 10% improvement in FCR was reported by Maas et al. [36] when including enzymes in diets for tilapia formulated solely using vegetable ingredients, and a 14% improvement in FCR was obtained after addition of multi-enzyme complexes (Natuzyme® or Hemicell®) in diets for Caspian salmon [37]. As previously indicated, a number of beneficial effects have been reported resulting from the hydrolysis of NSP [38,39]. In addition, the significant reduction in phytate derived from Solid-State enzymatic Hydrolysis (Table 2) could result in not only an increased availability of phosphorus, but also in a decrease in some other negative effects associated with the presence of phytate on the digestive bioavailability of proteins and some minerals, as described by several authors [40,41]. Furthermore, the citrate buffer used to develop the process of SSH could enhance the solubilisation of certain minerals, such as Fe or Mn [42].

The results support the suitability of SSH as a method for proper application of exogenous enzymes, as the hydrolysis is performed under optimal conditions and the enzyme activity is not affected by either the high temperatures reached during feed preparation or by the biochemical conditions present in the gut of the fish. This overcomes most of the physiological and technical limitations described when enzymes are included in the feed. To date, only one study, carried out by Denstadli et al. [43] on trout feeds, has tested this way of performing enzyme treatment. In that study, the feeds included 450 g/kg feed of plant ingredients and an estimated amount of NSPs accounted for 80 g/kg feed, values much lower than those used in the present study where the feed contained nearly 700 g plant ingredients per kg of feed and the estimated amount of NSP exceeded 200 g/kg. While in that work the enzyme treatment determined a 10–13% reduction of NSP content when using soybean meal as the main ingredient and of 4–6% when using rapeseed meal, in the present study the hydrolysis of NSP was not measured directly; however, the observed changes in the bioavailability of pentoses and reducing sugars suggested a modification of the nutritional value of the feed that impacted the results obtained on growth and nutritive utilization. From a practical point of view, performing such treatment by SSH allows adaptation of the more suitable operative conditions (dose, reaction time, etc.) to the specific features of different plant ingredients. In addition, because the enzyme mixture is used prior to pelleting, inactivation due to thermal processing should eliminate any further undesired effects. Although the potential application of such procedures requires further research, it offers interesting possibilities for a wider utilization of different inexpensive byproducts in feeds for herbivorous/omnivorous fish species such as *M. cephalus*, which has highly positive features from an environmental perspective.

5. Conclusions

Considering the aforementioned, the present study strongly suggests that enzyme pretreatment of highly fibrous plant ingredients by SSH using a commercial mixture of different carbohydrases and phytase may be a useful procedure to improve the nutritive value of high fiber plant byproducts for inclusion in practical diets for *M. cephalus* and other fish with similar nutritional features.

Author Contributions: Conceptualization, F.J.M.; methodology, F.J.M., F.P.M.-A., I.B.-Á. and J.A.M.-S.; software, F.J.M.; validation, F.J.M., F.P.M.-A., I.B.-Á. and J.A.M.-S.; formal analysis, F.P.M.-A., I.B.-Á. and J.A.M.-S.; investigation, F.J.M., F.P.M.-A. and J.A.M.-S.; resources, F.J.M. and J.A.M.-S.; data curation, F.J.M., F.P.M.-A., I.B.-Á. and J.A.M.-S.; writing—original draft preparation, F.P.M.-A.; writing—review and editing, F.J.M. and J.A.M.-S.; visualization, F.P.M.-A.; supervision, F.J.M.; project administration, F.J.M.; funding acquisition, F.J.M. All authors have read and agreed to the published version of the manuscript.

Funding: This work was funded by the Excellence Campus of Marine Science (CEIMAR) within the "II Call for Innovative Projects in the field of Blue Economy".

Institutional Review Board Statement: The study was conducted according to the guidelines of the Declaration of Helsinki and agree with the Directives of the Spanish Government (RD53/2013) and the European Union Council (2010/63/EU) for the use of animals in research. All experiments were carried out at the Centro Tecnológico de la Acuicultura de Andalucía (CTAQUA) which is authorized for animal experimentation (Spanish Operational Code REGA ES110270000411).

Data Availability Statement: Data available on request from the authors.

Acknowledgments: Thanks to "Origen" brewery (Huércal de Almería, Almería, Spain) for providing the brewer's spent grain used in the elaboration of the feeds. Thanks also to PIMSL for providing the fish and facilities required for the field experiment, and especially to its production manager, Narciso Mazuelos.

Conflicts of Interest: The authors declare that there is no conflict of interest that could be perceived as prejudicing the impartiality of the research reported.

References

1. Dror, A. Marine Aquaculture in the Mediterranean. In *Sustainable Food Production*; Christou, P., Savin, E., Costa-Pierce, B., Eds.; Springer Science & Business Media: Berlin, Germany, 2013; pp. 1121–1138.
2. Abo-Taleb, H.A.; El-feky, M.M.; Azab, A.M.; Mabrouk, M.M.; Elokaby, M.A.; Ashour, M.; Mansour, A.T.; Abdelzaher, O.F.; Abualnaja, K.M.; Sallam, A.E. Growth performance, feed utilization, gut integrity, and economic revenue of grey mullet, *Mugil cephalus*, fed an increasing level of dried zooplankton biomass meal as fishmeal substitutions. *Fishes* **2021**, *6*, 38. [CrossRef]
3. Abo-Taleb, H.A.; Ashour, M.; Elokaby, M.A.; Mabrouk, M.M.; El-feky, M.M.; Abdelzaher, O.F.; Gaber, A.; Alsanie, W.F.; Mansour, A.T. Effect of a new feed *Daphnia magna* (Straus, 1820), as a fish meal substitute on growth, feed utilization, histological status, and economic revenue of grey mullet, *Mugil cephalus* (Linnaeus 1758). *Sustainability* **2021**, *13*, 7093. [CrossRef]
4. Wassef, E.A.; El Masry, M.H.; Mikhail, F.R. Growth enhancement and muscle structure of striped mullet, *Mugil cephalus* L., fingerlings by feeding algal meal-based diets. *Aquac. Res.* **2001**, *32*, 315–322. [CrossRef]
5. Kalla, A.; Garg, S.K.; Kaushik, C.P.; Arasu, A.R.T.; Dinodia, G.S. Effect of replacement of fish meal with processed soybean on growth, digestibility and nutrient retention in *Mugil cephalus* (Linn.) fry. *Ind. J. Fisher.* **2003**, *50*, 509–518.
6. Jana, N.S.; Sudesh; Garg, S.K.; Sabhlok, V.P.; Bhatnagar, A. Nutritive evaluation of lysine- and methionine-supplemented raw vs heat-processed soybean to replace fishmeal as a dietary protein source for grey mullet, *Mugil cephalus*, and milkfish, *Chanos chanos*. *J. Appl. Aquac.* **2012**, *24*, 69–80. [CrossRef]
7. El-Dahhar, A.A.; Salama, M.A.; Moustafa, Y.T.; Elmorshedy, E.M. Effect of using equal mixture of seaweeds and marine algae in striped mullet (*Mugil cephalus*) larval diets on growth performance and feed utilization. *J. Arab. Aquacult. Soc.* **2014**, *9*, 145–158.

8. Gisbert, E.; Mozanzadeh, M.T.; Kotzamanis, Y.; Estévez, A. Weaning wild flathead grey mullet (*Mugil cephalus*) fry with diets with different levels of fish meal substitution. *Aquaculture* **2016**, *462*, 92–100. [CrossRef]

9. Gatlin, D.M.; Barrows, F.T.; Brown, P.; Dabrowski, K.; Gaylord, G.; Hardy, R.; Herman, E.; Hu, G.; Krogdahl, A.; Nelson, R. Expanding the utilization of sustainable plant products in aquafeeds: A review. *Aquac. Res.* **2007**, *38*, 551–579. [CrossRef]

10. Ai, Q.; Mai, K.; Zhang, W.; Xu, W.; Tan, B.; Zhang, C.; Li, H. Effects of exogenous enzymes (phytase, non-starch polysaccharide enzyme) in diets on growth, feed utilization, nitrogen and phosphorus excretion of Japanese seabass, *Lateolabrax japonicus*. *Comp. Biochem. Physiol. A* **2007**, *147*, 502–508. [CrossRef]

11. Dalsgaard, J.; Verlhac, V.; Hjermitslev, N.H.; Ekmann, K.S.; Fischer, M.; Klausen, M.; Pedersen, P.B. Effects of exogenous enzymes on apparent nutrient digestibility in rainbow trout (*Oncorhynchus mykiss*) fed diets with high inclusion of plant-based protein. *Anim. Feed Sci. Technol.* **2012**, *171*, 181–191. [CrossRef]

12. Jiang, T.T.; Feng, L.; Liu, Y.; Jiang, W.D.; Jiang, J.; Li, S.H.; Tang, L.; Kuang, S.Y.; Zhou, X.Q. Effects of exogenous xylanase supplementation in plant protein-enriched diets on growth performance, intestinal enzyme activities and microflora of juvenile Jian carp (*Cyprinus carpio var. Jian*). *Aquac. Nutr.* **2014**, *20*, 632–645. [CrossRef]

13. Castillo, S.; Gatlin, D.M. Dietary supplementation of exogenous carbohydrase enzymes in fish nutrition: A review. *Aquaculture* **2015**, *435*, 286–292. [CrossRef]

14. Fernandes, H.; Moyano, F.; Castro, C.; Salgado, J.; Martínez, F.; Aznar, M.; Fernandes, N.; Ferreira, P.; Gonçalves, M.; Belo, I.; et al. Solid-state fermented brewer's spent grain enzymatic extract increases in vitro and in vivo feed digestibility in European seabass. *Sci. Rep.* **2021**, *11*, 1–15. [CrossRef]

15. Cheng, H.Z.; Liu, Z.H. Enzymatic hydrolysis of lignocellulosic biomass from low to high solids loading. *Eng. Life Sci.* **2016**, *4*, 1–11.

16. Opazo, R.; Ortuzar, F.; Navarrete, P.; Espejo, R.; Romero, J. Reduction of soybean meal non-starch polysaccharides and α-galactosides by solid-state fermentation using cellulolytic bacteria obtained from different environments. *PLoS ONE* **2012**, *7*, e44783. [CrossRef] [PubMed]

17. Bradford, M.M. A rapid and sensitive method for the quantitation of microgram quantities of protein utilizing the principle of protein-dye binding. *Anal. Biochem.* **1976**, *72*, 248–254. [CrossRef]

18. Miller, G.L. Use of dinitrosalicylic acid reagent for determination of reducing sugar. *Anal. Chem.* **1959**, *31*, 426–428. [CrossRef]

19. Douglas, S.G. A rapid method for the determination of pentosans in wheat flour. *Food Chem.* **1981**, *7*, 139–145. [CrossRef]

20. Haug, W.; Lantzsch, H.J. Sensitive method for the rapid determination of phytate in cereals and cereal products. *J. Sci. Food Agric.* **1983**, *34*, 1423–1426. [CrossRef]

21. Association of Official Agricultural Chemists. *Official Methods of Analysis*, 17th ed.; AOAC International: Rockville, MD, USA, 2000.

22. Heywood, A.A.; Myers, D.J.; Bailey, T.B.; Johnson, L.A. Functional properties of low-fat soy flour produced by an extrusion-expelling system. *J. Am. Oil Chem. Soc.* **2002**, *79*, 1249. [CrossRef]

23. Decker, K.; Keppler, D. Galactosamine hepatitis: Key role of the nucleotide deficiency period in the pathogenesis of cell injury and cell death. *Rev. Physiol. Biochem. Pharmacol.* **1974**, *71*, 77–106.

24. El-Gendy, M.; Shehab El-Din, M.; Tolan, A. Studies on growth performance and health status when substituting fish meal by a mixture of oil seeds meal in diets of Nile tilapia (*Oreochromis niloticus*) and grey mullet (*Mugil cephalus*). *Egypt. J. Aquat. Biol. Fish.* **2016**, *20*, 47–58. [CrossRef]

25. Sun, H.; Cozannet, P.; Ma, R.; Zhang, L.; Huang, Y.K.; Preynat, A.; Sun, L.H. Effect of concentration of arabinoxylans and a carbohydrase mixture on energy, amino acids and nutrients total tract and ileal digestibility in wheat and wheat by-product-based diet for pigs. *Anim. Feed. Sci. Technol.* **2020**, *262*, 114380. [CrossRef]

26. Poernama, F.; Wibowo, T.A.; Liu, Y.G. The effect of feeding phytase alone or in combination with nonstarch polysaccharides-degrading enzymes on broiler performance, bone mineralization, and carcass traits. *J. Appl. Poult. Res.* **2021**, *30*, 100134. [CrossRef]

27. Polakof, S.; Panserat, S.; Soengas, J.L.; Moon, T.W. Glucose metabolism in fish: A review. *J. Comp. Physiol. B* **2012**, *182*, 1015–1045. [CrossRef]

28. Enes, P.; Panserat, S.; Kaushik, S.; Oliva-Teles, A.A. Nutritional regulation of hepatic glucose metabolism in fish. *Fish Physiol. Biochem.* **2009**, *35*, 519–539. [CrossRef] [PubMed]

29. Vallet, F.; Berhaut, J.; Leray, C.; Bonnet, B.; Pic, P. Preliminary experiments on the artificial feeding of Mugilidae. *Helgol. Wiss Meer* **1970**, *20*, 610–619. [CrossRef]

30. Chervinski, J. Growth of the golden grey mullet (*Liza aurata* (Risso)) in saltwater ponds during 1974. *Aquaculture* **1976**, *7*, 51–57. [CrossRef]

31. Richard, M.; Maurice, J.T.; Anginot, A.; Paticat, F.; Verdegem, M.C.J.; Hussenot, J.M.E. Influence of periphyton substrates and rearing density on *Liza aurata* growth and production in marine nursery ponds. *Aquaculture* **2010**, *310*, 106–111. [CrossRef]

32. Legarda, E.C.; Poli, M.A.; Martins, M.A.; Pereira, S.A.; Martins, M.L.; Machado, C.; de Lorenzo, M.A.; do Nascimento Vieira, F. Integrated recirculating aquaculture system for mullet and shrimp using biofloc technology. *Aquaculture* **2019**, *512*, 734308. [CrossRef]

33. Nengas, I.; Karacostas, I.; Neofitou, C.; Tsiara, V.; Tsiamis, V.; Kougioumtzis, N.; Karapanagiotidis, I.T. Growth and feed utilization by golden grey mullet (*Liza aurata*) in a coastal lagoon ecosystem, fed compound feeds with varying protein levels. *Isr. J. Aquac.-Bamidgeh* **2014**, *66*, 20784. [CrossRef]

34. Pujante, I.M.; Díaz-López, M.; Mancera, J.M.; Moyano, F.J. Characterization of digestive enzymes protease and alpha-amylase activities in the thick-lipped grey mullet (*Chelon labrosus*, Risso 1827). *Aquac. Res.* **2017**, *48*, 367–376. [CrossRef]

35. Gilannejad, N.; de Las Heras, V.; Martos-Sitcha, J.A.; Moyano, F.J.; Yúfera, M.; Martínez-Rodríguez, G. Ontogeny of expression and activity of digestive enzymes and establishment of gh/igf1 axis in the omnivorous fish *Chelon labrosus*. *Animals* **2020**, *10*, 874. [CrossRef]

36. Maas, R.M.; Verdegem, M.C.J.; Schrama, J.W. Effect of non-starch polysaccharide composition and enzyme supplementation on growth performance and nutrient digestibility in Nile tilapia (*Oreochromis niloticus*). *Aquac. Nutr.* **2018**, *25*, 622–632. [CrossRef]

37. Zamini, A.; Kanani, H.G.; Esmaeili, A.; Ramezani, S.; Zoriezahra, S.J. Effects of two dietary exogenous multi-enzyme supplementation, Natuzyme® and beta-mannanase (Hemicell®), on growth and blood parameters of Caspian salmon (*Salmo trutta caspius*). *Comp. Clin. Pathol.* **2014**, *23*, 187–192. [CrossRef]

38. Adeola, O.; Bedford, M.R. Exogenous dietary xylanase ameliorates viscosity-induced anti-nutritional effects in wheat-based diets for White Pekin ducks (*Anas platyrinchos domesticus*). *Br. J. Nutr.* **2004**, *92*, 87–94. [CrossRef]

39. Sinha, A.K.; Kumar, V.; Makkar, H.P.; De Boeck, G.; Becker, K. Non-starch polysaccharides and their role in fish nutrition–A review. *Food Chem.* **2011**, *127*, 1409–1426. [CrossRef]

40. Sugiura, S.H.; Gabaudan, J.; Dong, F.M.; Hardy, R.W. Dietary microbial phytase supplementation and the utilization of phosphorus, trace minerals and protein by rainbow trout (*Oncorhynchus mykiss*) fed soybean meal based diets. *Aquac. Res.* **2001**, *32*, 583–592. [CrossRef]

41. Cao, L.; Wang, W.; Yang, C.; Yang, Y.; Diana, J.; Yakupitiyage, A.; Luo, Z.; Li, D. Application of microbial phytase in fish feed. *Enzym. Microb. Technol.* **2007**, *40*, 497–507. [CrossRef]

42. Sugiura, S.H.; Dong, F.M.; Hardy, R.W. Effects of dietary supplements on the availability of minerals in fish meal; preliminary observations. *Aquaculture* **1998**, *160*, 283–303. [CrossRef]

43. Denstadli, V.; Hillestad, M.; Verlhac, V.; Klausen, M.; Øverland, M. Enzyme pretreatment of fibrous ingredients for carnivorous fish: Effects on nutrient utilisation and technical feed quality in rainbow trout (*Oncurhynchus mykiss*). *Aquaculture* **2011**, *319*, 391–397. [CrossRef]

 *fishes*

Article

Aurantiochytrium sp. Meal Improved Body Fatty Acid Profile and Morphophysiology in Nile Tilapia Reared at Low Temperature

Rosana Oliveira Batista [1], **Renata Oselame Nobrega** [1], **Delano Dias Schleder** [2], **James Eugene Pettigrew** [3] and **Débora Machado Fracalossi** [1,*]

[1] Departamento de Aquicultura, Universidade Federal de Santa Catarina (UFSC), Florianópolis 88034-001, SC, Brazil; rosana.engpesca@gmail.com (R.O.B.); renata.oselamenobrega@gmail.com (R.O.N.)
[2] Departamento de Medicina Veterinária, Instituto Federal Catarinense, Campus Araquari, Araquari 89245-000, SC, Brazil; delano.schleder@ifc.edu.br
[3] Pettigrew Research Services, Tubac, AZ 85646, USA; jim@pettigrewconsulting.com
* Correspondence: debora.fracalossi@ufsc.br; Tel.: +55-48-3721-6300

Abstract: *Aurantiochytrium* sp. is a heterotrophic microorganism that produces docosahexaenoic acid (DHA), thus being considered as a possible replacement for fish oil in aquafeeds. We investigated the effect of *Aurantiochytrium* sp. meal (AM) dietary levels (0, 5, 10, 20, and 40 g kg^{-1}) on Nile tilapia body and hepatopancreas fatty acid (FA) profile, body FA retention, somatic indices, and morphophysiological changes in the intestine and hepatopancreas, after feeding Nile tilapia juveniles (average initial weight 8.47 g) for 87 days at 22 °C. The 10AM diet was compared to a control diet containing cod liver oil (CLO), since their DHA concentration was similar. Within fish fed diets containing increasing levels of AM, there was a linear increase in n-3 FA content, especially DHA, which varied in the body (0.02 to 0.41 g 100 g^{-1}) and hepatopancreas (0.15 to 1.05 g 100 g^{-1}). The morphology of the intestines and hepatopancreas was positively affected in AM-fed fish. Fish fed 10AM showed less accumulation of n-3 FAs in the body and hepatopancreas when compared to fish fed CLO. Therefore, AM is an adequate substitute for fish oil in winter diets for Nile tilapia, with the supplementation of 40AM promoting the best results regarding intestine and hepatopancreas morphophysiology.

Keywords: *Aurantiochytrium* sp.; docosahexaenoic acid; histology; *Oreochromis niloticus*; physiology; temperature

Citation: Batista, R.O.; Nobrega, R.O.; Schleder, D.D.; Pettigrew, J.E.; Fracalossi, D.M. *Aurantiochytrium* sp. Meal Improved Body Fatty Acid Profile and Morphophysiology in Nile Tilapia Reared at Low Temperature. *Fishes* **2021**, *6*, 45. https://doi.org/10.3390/fishes6040045

Academic Editor: Eric Hallerman

Received: 12 September 2021
Accepted: 29 September 2021
Published: 8 October 2021

Publisher's Note: MDPI stays neutral with regard to jurisdictional claims in published maps and institutional affiliations.

1. Introduction

Fish are ectothermic organisms; therefore, their body temperature is influenced by water temperature, which consequently affects their metabolic rate, feed consumption, feed conversion, and other physiological functions [1]. Water temperature is one of the most important abiotic factors in aquaculture, as it directly affects the growth and survival of fish [2]. Whereas in nature, fish may use behavioral responses to overcome low temperature, fish under farming conditions cannot use such natural responses. This is the case in pond or cage farming situations where fish may be subjected to extreme temperature fluctuations [3]. As a primary response to low temperature, fish increase serum cortisol and catecholamine levels; as a secondary response, metabolic changes may occur [3]. These adaptations are commonly used as indicators of short-term cold responses [4]. However, if low water temperatures persist for longer periods, the continued stress can trigger tertiary responses, causing physiological changes that can lead to mortality [5].

Nile tilapia, *Oreochromis niloticus*, is the third most cultivated fish species in the world [6]; despite its tropical origin, it has been introduced in many subtropical and temperate regions around the world. In Brazil, the fourth country in Nile tilapia production worldwide, the highest production occurs in the Southern states of Paraná, São Paulo, Minas Gerais, and Santa Catarina [7], where the climate is predominately subtropical, or altitude-tropical, with important thermal variations between summer and winter. Average water temperatures of 20–22 °C were reported in tilapia farms in Brazil during the winter [8,9]. The optimal temperature for Nile tilapia growth is around 26–30 °C [10]. Temperatures below 22 °C result in a drastic decrease in zootechnical performance [9,11]. Between 16 and 13 °C, feed intake ceases and, below 9 °C, voluntary movements cease [10].

The worldwide importance of Nile tilapia farming has led to an urgent need to develop new technologies to improve production in subtropical regions [12]. The availability of Nile tilapia strains that are more tolerant to low temperatures and the adoption of sustainable management practices can improve production in subtropical regions. However, research on winter diets for Nile tilapia, focusing on nutrients and biological responses, are also reported to improve performance [9,13–16].

Water temperature plays an important role in lipid metabolism in fish [17]. Polyunsaturated fatty acids (PUFA) of 18-carbon chains, such as α-linolenic acid (α-LNA,18:3 n-3) and/or linoleic acid (LOA, 18:2 n-6), meet the requirements of fatty acids for the optimal growth of Nile tilapia when reared in the ideal temperature range (26–30 °C). However, dietary supplementation with n-3 long-chain PUFA (n-3 LC-PUFA), such as docosahexaenoic acid (DHA, 22:6 n-3), promotes further growth and utilization of nutrients when Nile tilapia are maintained for long periods at low temperatures such as 22 °C [14–16]. Several studies have reported that Nile tilapia are efficient in storing body n-3 LC-PUFA when available in the diet, especially in response to cold temperature, increasing fatty acid unsaturation in cell membranes to maintain their fluidity and permeability [15,18]. Such a mechanism is reflected in an increase in body PUFA (mainly n-3 PUFA) concentration [10,18].

The main sources of n-3 LC-PUFAs in aquafeeds are fish meal and oil, but with the stagnation of extractive fishing, the use of these feedstuffs has become unsustainable for aquaculture production [19]. Currently, there is a growing number of studies looking for possible substitutes for fish meal and oil, mostly using plant meals and/or plant oils. However, the fatty acid composition of plant oils can be a limiting factor when used as an exclusive lipid source, since they are commonly deficient in n-3 PUFA, presenting a high n-6:n-3 ratio [20]. In addition, the effects of plant oils on lipid metabolism and the health of fish may lead to imbalances in body fatty acids, affecting organ integrity [21].

Aurantiochytrium sp. is a heterotrophic microorganism found in marine habitats, belonging to the family Thraustochytridae, which can be considered as a possible novel substitute for fish oil because it is rich in DHA. Our team evaluated the inclusion of such a novel feed additive in diets of Nile tilapia at an optimal temperature (28 °C) [22,23] and at a suboptimal temperature (22 °C) [15]. Our findings show that for Nile tilapia kept at cold, suboptimal temperature, dietary supplementation with 10 g kg^{-1} *Aurantiochytrium* sp. meal (AM) improved growth by 16%, in addition to improving feed efficiency, specific growth rate, and apparent net protein retention when compared to fish fed a diet without supplementation [15]. When comparing two sources of DHA (AM and cod liver oil, CLO), at similar DHA concentrations, fish fed the diet supplemented with AM had significantly higher growth performance, feed efficiency, and protein utilization than fish fed the CLO-supplemented diet. However, the effect of supplementing this novel feed additive on the integrity and composition of important organs such as gut and hepatopancreas have not yet been evaluated in Nile tilapia. Nevertheless, histopathological changes were already reported in Nile tilapia reared at suboptimal growth temperatures [13]. Thus, in this study we aimed beyond the assessment of zootechnical gains, by evaluating body lipid profiles, somatic indices, and the histology of important organs. Such an approach will allow a more reliable assessment of 'healthy growth' when supplementing a novel feed additive, rich in DHA, to Nile tilapia, at a suboptimal low temperature.

2. Materials and Methods

2.1. Experimental Design and Diets

The experimental diets were formulated to meet the nutritional requirements of Nile tilapia [24,25] using practical ingredients (Table 1). Five experimental diets were obtained by supplementing with AM (ALL-G-RICH™, provided by Alltech®, Nicholasville, KY, USA), at concentrations of 0.0 (un-supplemented), 5, 10, 20, and 40 g kg^{-1} of the diet. Dietary treatments were named 0AM, 5AM, 10AM, 20AM, and 40AM, respectively. Additionally, a positive control diet was formulated, aiming at comparing two DHA-rich sources: a traditional source (CLO) versus a novel feed additive (AM). The CLO diet was only compared to the 10AM diet, since both contained the same amount of DHA in their final composition (~0.2 g kg^{-1} dry diet).

The extrusion parameters were the same as those described by Nobrega et al. [15]. After extrusion, the pellets (2 to 3 mm) were dried at 50 °C in a forced air circulation oven to 8% moisture, stored in tightly closed containers in the absence of light, and kept at 20 °C to prevent fatty acid oxidation.

The experiment was run in a completely randomized design comprised of six diets with five replicates for each dietary treatment.

2.2. Fish and Experimental Procedures

Nile tilapia juveniles, of GIFT (genetic improvement of farmed tilapia) lineage, sexually inverted to males, were acclimated to the laboratory conditions for five weeks in three 1000 L tanks, connected to a freshwater recirculation aquaculture system (RAS), and water temperature control (28 °C). The photoperiod was adjusted to 12 h.

Thereafter, groups of 25 fish were randomly stocked into 30 tanks with 100 L capacity and were acclimatized to experimental conditions at 28 °C for a week. In the second week, water temperature was lowered gradually from 28 °C to 22 °C (1 °C per day) and, in the third week of acclimation, water temperatures were maintained at 22 °C. During this acclimation period, fish were fed a negative control diet without supplementation of AM.

At the beginning of the trial, fish average body weight was 8.47 ± 0.19 g (average ± standard error). Fish were fed twice daily (10:00 and 18:00 h) to apparent satiation for 87 days.

Table 1. Formulation, proximate composition, and fatty acid profile of the experimental diets.

	Diets					
	0AM	**5AM**	**10AM**	**20AM**	**40AM**	**CLO**
Ingredient [a], g kg^{-1} dry diet						
Soybean meal	473.8	471.6	472.0	469.9	465.9	477.4
Corn	321.7	320.0	316.0	315.6	305.0	314.1
Poultry by-product meal	157.2	157.2	157.2	157.2	157.2	157.2
Vitamin and mineral premix [b]	28.3	28.3	28.3	28.3	28.3	28.3
Swine lard	19.0	17.9	16.5	9.0	-	-
Corn oil	-	-	-	-	3.60	3.00
ALL-G-RICH™	-	5.0	10.0	20.0	40.0	-
Cod liver oil	-	-	-	-	-	20.0
Composition, g 100 g^{-1} dry weight						
Gross energy, kcal kg^{-1}	4168	4251	4216	4257	4297	4168
Dry matter	89.47	90.32	89.22	89.74	90.66	90.32
Crude protein	36.40	36.30	36.08	35.93	35.85	36.20
Lipid	8.64	8.93	8.99	9.13	9.90	9.20
Ash	7.11	7.17	7.18	7.20	7.61	7.20
16:0 PAL [c]	1.55	1.63	1.77	2.03	2.43	1.16
18:1 n-9 OLA	2.56	2.31	1.93	1.89	1.49	1.89
18:2 n-6 LOA	1.92	1.81	1.75	1.76	1.71	1.81

Table 1. *Cont.*

	Diets					
	0AM	**5AM**	**10AM**	**20AM**	**40AM**	**CLO**
20:4 n-6 ARA	ND [e]	ND	ND	ND	ND	0.05
18:3 n-3 α-LNA	0.04	0.03	0.03	0.02	0.02	0.23
20:5 n-3 EPA	ND	ND	ND	0.01	0.02	0.17
22:5 n-3 DPA	ND	ND	ND	0.01	0.02	0.04
22:6 n-3 DHA	ND	0.09	0.20	0.38	0.75	0.23
Σ SFA [d]	2.14	2.17	2.24	2.57	2.98	1.29
Σ MUFA	3.02	2.72	2.28	2.26	1.80	2.65
Σ PUFA	1.98	2.00	2.04	2.37	2.70	2.79
Σ PUFA n-6	1.94	1.86	1.83	1.89	1.87	1.89
Σ LC-PUFA n-6	0.02	0.06	0.08	0.13	0.16	0.08
Σ PUFA n-3	0.04	0.13	0.21	0.47	0.84	0.83
Σ LC-PUFA n-3	ND	0.09	0.20	0.45	0.82	0.54
n-3:n-6	0.02	0.07	0.12	0.25	0.45	0.44

[a] Corn and soybean meal provided by Nicoluzzi Rações Ltd.a (Penha, Santa Catarina, Brazil). Poultry by-product meal was produced by Kabsa S.A. (Porto Alegre, Rio Grande do Sul, Brazil). Swine lard was produced by Seara Alimentos S.A. (Itajaí, Santa Catarina, Brazil). ALL-G-RICH™ was produced by Alltech Inc. (Nicholasville, Kentucky, USA) and imported by Alltech do Brasil Agroindustrial Ltd.a (Araucária, Paraná, Brazil). Corn oil "Suavit" was produced by Cocamar Ltd.a (Maringá, Paraná, Brazil). Cod liver oil "Möllers Tran" was produced by Orkla Health (Oslo, Østlandet, Norway). [b] Dicalcium phosphate (13.5 g kg^{-1}), choline bitartrate (3.0 g kg^{-1}), butylated hydroxytoluene (BHT) (1.0 g kg^{-1}), threonine (0.80 g kg^{-1}), and vitamin–micromineral premix (10.0 g kg^{-1}; produced by Cargill, Campinas, São Paulo), composition per kg: folic acid 420 mg, pantothenic acid 8333 mg, BHT 25,000 mg, biotin 134 mg, cobalt sulphate 27 mg, copper sulphate 1833 mg, iron sulphate 8000 mg, calcium iodate 92 mg, manganese sulphate 3500 mg, niacin 8.333 mg, selenite 100 mg, vitamin (vit.) A 1666,670 UI, vit. B$_1$ 2083 mg, vit. B$_{12}$ 5000 µg, vit. B$_2$ 4166 mg, vit. B$_6$ 3166 mg, ascorbic acid equivalent 66,670 mg, vit. D$_3$ 666,670 UI, vit. E 16,666 UI, vit. K$_3$ 833 mg, zinc sulfate 23,330 mg, inositol 50,000 mg, and calcium propionate 250,000 mg. [c] Fatty acids: PAL, palmitic acid; OLA, oleic acid; LOA, linoleic acid; ARA, arachidonic acid; α-LNA, alpha-linolenic acid; EPA, eicosapentaenoic acid; DPA, docosapentaenoic acid; DHA, docosahexaenoic acid. [d] Groups of fatty acids: SFA = saturated, MUFA = monounsaturated, PUFA = polyunsaturated (>2 double bonds), LC-PUFA = long-chain PUFA (>20 carbons). [e] Not detected (<0.05% total fatty acid).

2.3. Sample Collection

At the end of the experiment, fish were deprived of feed for 24 h and euthanized via overdose (200 mg L^{-1}) of the anesthetic Eugenol® (Biodinâmica Química and Farmacêutica Ltd., Ibiporã, PR, Brazil), followed by sectioning of the dorsal spine. At the beginning of the experiment, 90 fish (three groups of 30 fish) were euthanized with a lethal dose of Eugenol® and stored at −20 °C for analysis of the initial fatty acid body profile.

For fatty acid analysis and retention calculation, 15 fish per dietary treatment (three per experimental unit) were euthanized and stored at −20 °C. Additionally, the hepatopancreas and the anterior part of the intestine of 10 fish per treatment (two per experimental unit) were dissected and fixed in 10% buffered formalin for histological analyses. For the analyses of fatty acids and glycogen, the hepatopancreas of 15 fish (three fish from each experimental unit) were sampled, immediately frozen in liquid nitrogen, and stored at −80 °C until the analyses were performed.

Then, 10 fish per treatment (two fish per experimental unit) were dissected, and the viscera and hepatopancreas were weighed to calculate viscerosomatic index (VSI) and hepatosomatic index (HSI) indices, according to the following equations:

$$\text{VSI}(\%) = (\text{viscera weight} \div \text{body weight}) \times 100 \tag{1}$$

$$\text{HSI}(\%) = (\text{hepatopancreas weight} \div \text{body weight}) \times 100 \tag{2}$$

2.4. Proximate and Biochemical Composition Analyses

The proximate composition of the diets was conducted at the Fish Nutrition Lab (LABNUTRI, UFSC), following procedures standardized by the Association of Official Analytical Chemists [26]: moisture (drying at 105 °C to constant weight, method 950.01), crude protein (Kjeldahl, method 945.01), total lipid (Soxhlet, method 920.39C), and ash (incineration at 550 °C, method 942.05). Crude energy was determined in a calorimeter (PARR, model ASSY 6200), according to instructions from the manufacturer.

The fatty acid profiles of the experimental diets, fish, and hepatopancreas were analyzed using gas chromatography. The lipid extraction and chromatographic conditions were analyzed using the procedures described by Nobrega et al. [15]. Fatty acids detected and summed but not included in the tables: 10:0, 12:0, 14:0, 15:0, 18:0, 20:0, 22:0, 16:1 n-7, 18:1 n -9; 18:1 n-7, 20:1 n-9, 22:1 n-9, 24:1 n-9, 16:2 n-4, 18:3 n-6, 18:4 n-3, 20:2 n-6, 20:3 n-6, 20:4 n-3, and 22:3 n-3. The most important fatty acids for fish metabolism are included in the body and liver composition tables. The sum of the fatty acid group's SFA, monounsaturated (MUFA), PUFA, n-6 PUFA, n-3 PUFA, and LC-PUFA are also expressed. The n-3:n-6 ratio was calculated using the $\sum$n-3 PUFA:$\sum$n-6 PUFA.

Following the fatty acid profile analyses, we calculated the apparent body retention rates (ARRs) of LOA, α-LNA, and DHA, as well as the total n-3 PUFA and n-6 PUFA groups, following the methodology proposed by Glencross et al. (2003). The equation used was:

$$ARR = 100 \times \{(FAf - FAi) \times (FAc) - 1\} \tag{3}$$

where FAf is the absolute amount of a specific fatty acid in the fish at the end of the study, FAi is the absolute specific fatty acid in the fish at the initial time, and FAc is the total consumption of specific fatty acids over the study period.

Hepatopancreas glycogen analysis was performed at the Laboratory of Biomarkers and Aquatic and Immunochemical Contamination (LABCAI, UFSC), following the methodology proposed by Carrol et al. [27], with some modifications, including the addition of 1.0 mL of 10% trichloroacetic acid to 0.15 g of tissue, and centrifugation at 3000× *g* for 15 min at 5 °C. A 0.35 mL aliquot of the supernatant was added to previously refrigerated ethanol (1:5, *v:v*). The mixture was centrifuged at 5000× *g* for 30 min at 5 °C, the supernatant was discarded, and 0.2 mL of ethanol was added, and then the mixture was centrifuged again for 5 min. The pellet formed was dried at 50 °C and resuspended in 1.5 mL ultrapure water. After total dilution of the pellet, a 0.1 mL aliquot of the extract and 1.0 mL of 10% anthrone reagent were heated for 5 min at 90 °C, and the mixture was immediately immersed in ice. An aliquot of 0.2 mL of the final mixture was used for microplate reading at 620 nm in a spectrophotometer.

2.5. Histological Analyses

Histological analyses were performed at the Laboratory of Pathology and Health of Aquatic Organisms (AQUOS, UFSC). The organs (hepatopancreas and the anterior third of the intestines) were dehydrated in a series of increasing ethyl alcohol concentrations, clarified in xylol, and embedded in paraffin at 60 °C. Organs were cut to a thickness of 3 to 5 μm (PAT-MR10 microtome) and two cuts of each organ were evaluated. The slides were stained with hematoxylin and eosin (H&E). After staining, the slides were mounted in Entellan® medium and analyzed under a microscope, as described by Brum et al. [28].

Regarding intestinal morphology, we measured the height and width of the intestinal folds and quantified the number of folds and goblet cells using Zen Pro software (Zeiss, Jena, Germany). Histological alterations in the hepatopancreas were evaluated semiquantitatively by ranking the severity of tissue lesions, according to the modified method described by Schwaiger et al. [29]. The ranking was 0 (absence of alteration), 1 (mild alteration, corresponding to <25% of the tissue area), 2 (moderate alteration, 25% to 50% of the tissue area), and 3 (severe alteration, >50% of the tissue area).

The following alterations were considered in the hepatopancreas: cordonal appearance and hepatocyte size variation, pancreas with intact acini, hepatocyte ballooning, cholestasis, large vessel congestion in the pancreas and sinusoids, hepatocyte nucleus displacement, sinusoid dilatation, eosinophilic and mononuclear lymphocytic, macrosteatosis, microsteatosis, necrosis, nucleus with pyknosis, karyolysis and kariorrhexis, loss of pancreatic structure, and presence of bilirubin.

2.6. Statistical Analyses

To determine the optimal dietary concentration of AM, the dependent variables related to body composition and retention of fatty acids, hepatic glycogen, total fat and fatty acids in the hepatopancreas, somatic index, and intestinal measurements of fold height, fold width, fold count, and goblet cell count were subjected to polynomial regression analysis. To evaluate the same variables when comparing the CLO and 10AM diets, we used the Student's t-test. The histological changes evaluated in the intestine and hepatopancreas were analyzed using the non-parametric Kruskal–Wallis test, followed by Dunn's test. For all statistical analyses, Statistica 13.0 software (Statsoft Inc. Tulsa, OK, USA) was used and a significance level of 5% was adopted.

3. Results

3.1. Body Fatty Acid Composition and Apparent Retention

The increasing dietary levels of AM significantly ($p < 0.05$) affected fatty acid body composition, presenting a significant linear response (Table 2). The body content of DHA, total PUFA, n-3 PUFA, and n-3 LC-PUFA increased linearly. Docosahexaenoic acid concentration increased from 0.02 to 0.41 g 100 g^{-1} in fish fed 0AM and 40AM, respectively. The concentration of Σ LC-PUFA n-3 also ranged from 0.04 g 100 g^{-1} (0AM) to 0.48 g 100 g^{-1} (40AM). The body contents of eicosapentaenoic acid (EPA, 20:5 n-3) and docosapentaenoic acid (DPA, 22:4 n-6) were detected only in fish fed with the highest concentrations of AM, 20 and 40AM. In contrast, the body concentrations of arachidonic acid (ARA, 20:6 n-6), SFA, MUFA, n-6 PUFA, and n-6 LC-PUFA decreased linearly in fish fed increasing levels of AM.

Table 2. Whole body fatty acid composition of Nile tilapia juveniles fed increasing concentrations of *Aurantiochytrium* sp. meal (AM) for 87 days, at 22 °C [1].

Fatty Acids g 100 g^{-1} Dry Weight	Initial Fish	Diets					Pooled SEM [2]	*p* Value [3]
		0AM	5AM	10AM	20AM	40AM		
16:0 PAL	0.52	2.09	2.22	2.02	2.09	2.02	0.18	NS [4]
18:2 n-6 LOA	0.20	1.08	1.15	1.10	1.15	1.14	0.10	NS
20:4 n-6 ARA	0.03	0.11	0.09	0.08	0.08	0.07	0.01	<0.001
22:4 n-6 ADA	ND [5]	0.13	0.10	0.10	0.10	0.11	0.02	NS
18:3 n-3 α-LNA	0.01	0.05	0.06	0.06	0.06	0.06	0.01	NS
20:5 n-3 EPA	ND	ND	ND	ND	0.04	0.04	0.01	<0.001
22:5 n-3 DPA	ND	ND	ND	ND	0.03	0.03	0.00	<0.001
22:6 n-3 DHA	0.02	0.04	0.09	0.11	0.23	0.41	0.02	<0.001
Σ SFA [6]	0.78	3.05	3.21	2.89	2.78	2.50	0.26	0.007
Σ MUFA	0.88	4.31	4.39	4.08	4.02	3.67	0.36	<0.001
Σ PUFA	0.27	1.84	1.87	1.80	2.02	2.16	0.16	<0.001
Σ PUFA n-6	0.25	1.83	1.79	1.63	1.60	1.55	0.14	<0.001
Σ LC-PUFA n-6	0.03	0.35	0.30	0.29	0.28	0.27	0.03	<0.001
Σ PUFA n-3	0.04	0.15	0.20	0.22	0.40	0.60	0.03	<0.001
Σ LC-PUFA n-3	0.03	0.04	0.09	0.11	0.30	0.48	0.02	<0.001
n-3:n-6	0.18	0.10	0.12	0.14	0.24	0.32	0.01	<0.001

[1] Results are based on linear regression and expressed as the average of five replicates ($n = 3$ fish per replicate. [2] Standard error of means. [3] Where linear regression was significant, the following equations were obtained: ARA $y = -0.0091x + 0.0987$, $R^2 = 0.720$; EPA $y = 0.0119x - 0.002$, $R^2 = 0.723$; DPA $y = 0.0094x - 0.002$, $R^2 = 0.7709$; DHA $y = 0.0936x + 0.0361$, $R^2 = 0.9828$; SFA $y = -0.0566x + 3.0886$, $R^2 = 0.1346$; MUFA $y = -0.0566x + 3.0886$. $R^2 = 0.1346$; PUFA $y = 0.0866x + 1.8097$, $R^2 = 0.5196$; PUFA n-6 $y = -0.0866x + 1.8097$, $R^2 = 0.5196$; LC-PUFA n-6 $y = -0.0166x + 0.324$, $R^2 = 0.4313$; PUFA n-3 $y = 0.1157x + 0.1394$, $R^2 = 0.9656$; LC-PUFA n-3 $y = 0.1149x + 0.0321$, $R^2 = 0.9705$; n-3:n-6 $y = 0.0584x + 0.0967$, $R^2 = 0.9719$. [4] Not significant ($p > 0.05$). [5] Not detected (<0.05% of total fatty acids). [6] Groups of fatty acids: SFA = saturated, MUFA = monounsaturated, PUFA = polyunsaturated (>2 double bonds), LC-PUFA = long-chain PUFA (>20 carbons).

When body fatty acid composition was compared between DHA sources (CLO versus 10AM), fish fed the 10AM diet presented higher palmitic acid (PAL) content and n-6 LC-PUFA (Table 3). Conversely, fish fed the CLO diet showed higher concentrations of α-LNA,

DHA, n-3 PUFA, n-3 LC-PUFA, and n-3:n-6 ratio. EPA and DPA were detected only in fish fed the CLO diet.

Table 3. Whole body fatty acid composition of Nile tilapia juveniles fed two sources of docosahexaenoic acid (DHA) for 87 days, at 22 °C [1,2].

Fatty Acids g 100 g^{-1} Dry Weight	Diets		p Value
	10AM	CLO	
16:0 PAL	2.02 ± 0.14	1.72 ± 0.18	0.037
18:2 n-6 LOA	1.10 ± 0.08	1.01 ± 0.07	NS [3]
20:4 n-6 ARA	0.08 ± 0.00	0.08 ± 0.01	NS
18:3 n-3 α-LNA	0.06 ± 0.01	0.08 ± 0.01	<0.001
20:5 n-3 EPA	ND [4]	0.06 ± 0.01	-
22:5 n-3 DPA	ND	0.05 ± 0.00	-
22:6 n-3 DHA	0.11 ± 0.01	0.18 ± 0.01	<0.001
Σ SFA [5]	2.89 ± 0.21	2.54 ± 0.26	NS
Σ MUFA	4.08 ± 0.33	3.76 ± 0.34	NS
Σ PUFA	1.80 ± 0.13	1.83 ± 0.12	NS
Σ PUFA n-6	1.55 ± 0.11	1.46 ± 0.10	NS
Σ LC-PUFA n-6	0.29 ± 0.02	0.23 ± 0.02	<0.001
Σ PUFA n-3	0.22 ± 0.02	0.42 ± 0.03	<0.001
Σ LC-PUFA n-3	0.11 ± 0.01	0.29 ± 0.02	0.008
n-3:n-6	0.14 ± 0.00	0.29 ± 0.10	<0.001

[1] Results are based on a *t*-test and expressed as the average of five replicates (*n* = 3 fish per replicate), followed by the standard error. [2] Diets with similar contents of DHA (~0.2 g kg^{-1} DHA dry diet); 10AM = 10 g kg^{-1} *Aurantiochytrium* sp. meal and CLO = cod liver oil. [3] Not significant (*p* > 0.05). [4] Not detected (<0.05% of total fatty acids). [5] Groups of fatty acids: SFA = saturated, MUFA = monounsaturated, PUFA = polyunsaturated (≥2 double bonds), LC-PUFA = long-chain PUFA (≥20 carbons).

Adjustment of a polynomial quadratic regression was significant for the apparent retention rate (ARR) of body fatty acid (*p* < 0.05) (Table 4). According to the regression trend of DHA, there was a reduction in the body retention rate of this fatty acid with an increase in dietary levels of AM. The ARR of DHA was 57.96% and 29.59% in fish fed the lowest inclusion level (5AM) and the highest inclusion level (40AM), respectively. The percent retention of DHA decreased by approximately 28% between fish fed the highest and lowest inclusion. The diet without the inclusion of AM did not contain DHA (Table 1); therefore, the retention rate was not calculated for fish fed that dietary treatment. With increasing levels of dietary inclusion of AM, a positive tendency was observed in the ARR for α-LNA. However, ARR for n-6 and n-3 PUFA showed a negative trend, decreasing its retention with the increased dietary inclusion of AM. There was no significant difference in the ARR of LOA among fish fed the different inclusion levels of the additive.

Table 4. Apparent body retention rate (ARR) of fatty acids in Nile tilapia juveniles fed increasing concentrations of *Aurantiochytrium* sp. meal (AM) for 87 days, at 22°C [1].

ARR, %	Diets					Pooled SEM [2]	p Value [3]
	0AM	5AM	10AM	20AM	40AM		
18:2 n-6 LOA	38.52	38.29	37.72	36.02	36.21	1.83	NS [4]
18:3 n-3 α-LNA	102.60	107.21	116.81	138.67	174.98	7.20	<0.001
22:6 n-3 DHA	-	57.96	34.41	32.75	29.59	2.09	<0.001
Σ PUFA [5] n-6	54.19	49.01	47.30	42.97	41.25	2.27	0.004
Σ PUFA n-3	277.15	95.85	58.11	46.49	38.48	4.44	<0.001

[1] Results are based on quadratic regression and expressed as the average of five replicates (*n* = 3 fish per replicate. [2] Standard error of means. [3] When the polynomial regression was significant, the following equations were obtained: α-LNA y= 0.345x^2 + 17.306x + 100.631, R^2 = 0.761; DHA y = 4.909x^2 − 28.688x + 66.389, R^2 = 0.687; PUFA n-6 y = 1.181x^2 − 7.802x + 232.205, R^2 = 0.487; PUFA n-3 y = 33.967x^2 − 182.232x + 232.205, R^2 = 0.801. [4] Not significant (*p* > 0.05). [5] Total polyunsaturated ≥2 double bonds) fatty acids.

When comparing the two sources of fatty acids, 10AM versus CLO, the ARRs of α-LNA and n-6 PUFA were higher in fish fed 10AM (Table 5). In contrast, the ARRs of DHA and n-3 PUFA were higher in the CLO-fed fish.

Table 5. Apparent body retention rate (ARR) of fatty acids in Nile tilapia juveniles fed two sources of DHA for 87 days, at 22 °C [1,2].

ARR, %	Diets		*p* Value [3]
	10AM	**CLO**	
18:2 n-6 LOA	37.72 ± 5.74	32.49 ± 1.84	NS [4]
18:3 n-3 α-LNA	116.81 ± 19.64	17.47 ± 0.65	0.008
22:6 n-3 DHA	34.41 ± 5.59	44.52 ± 2.68	0.007
Σ PUFA [4] n-6	47.30 ± 6.90	39.32 ± 1.83	0.032
Σ PUFA n-3	58.11 ± 9.18	287.86 ± 41.73	<0.001

[1] Results are based on a *t*-test and expressed as the average of five replicates (*n* = 3 fish per replicate), followed by the standard error. [2] Diets with similar contents of DHA (~0.2 g kg^{-1} DHA dry diet); 10AM = 10 g kg^{-1} *Aurantiochytrium* sp. meal and CLO = cod liver oil. [3] Not significant (*p* > 0.05). [4] Total polyunsaturated fatty acids (≥2 double bonds).

3.2. Fatty Acid Composition in the Hepatopancreas

Total lipid and fatty acid composition in the hepatopancreas of Nile tilapia juveniles was influenced by dietary AM contents, presenting a significant linear response (*p* < 0.05) (Table 6). Linear regression best fitted the results. With increasing dietary AM levels, the concentration of total lipid in the liver decreased linearly, ranging from 30.69 to 23.65 g 100 g^{-1} between the treatments 0AM and 40AM. Some fatty acids, such as ARA, ADA (adrenic acid, 22:4 n-6), EPA, n-6 PUFA, and n-6 LC-PUFA, showed a similar response to total lipids. However, DHA, n-3 PUFA, n-3 LC-PUFA, and n-3:n-6 ratio hepatopancreas lipid contents increased linearly with the inclusion of AM. The DHA content in the hepatopancreas presented a significant variation of 0.15 to 1.05 g 100 g^{-1} between treatments 0AM and 40AM, equivalent to a 105% increase in the accumulation of DHA.

Table 6. Total lipid and fatty acid composition of hepatopancreas in Nile tilapia juveniles fed increasing concentrations of *Aurantiochytrium* sp. meal (AM) for 87 days, at 22 °C [1].

Fatty Acids g 100 g^{-1} Dry Weight	Diets					Pooled SEM [2]	*p* Value [3]
	0AM	**5AM**	**10AM**	**20AM**	**40AM**		
Total lipid	30.69	28.88	27.44	24.43	23.65	6.30	0.024
16:0 PAL	4.11	4.00	4.42	3.57	3.44	1.30	NS [4]
18:2 n-6 LOA	0.72	1.03	0.99	0.85	1.02	0.37	NS
20:4 n-6 ARA	0.50	0.51	0.50	0.46	0.38	0.09	0.002
22:4 n-6 ADA	0.83	0.72	0.64	0.51	0.43	0.15	<0.001
18:3 n-3 α-LNA	0.47	0.67	0.75	0.47	0.46	0.24	NS
20:5 n-3 EPA	0.09	0.08	0.08	0.07	0.06	0.02	0.008
22:6 n-3 DHA	0.15	0.40	0.56	0.80	1.05	0.08	<0.001
Σ SFA [5]	7.89	8.29	8.55	7.17	6.82	2.04	NS
Σ MUFA	8.21	8.39	9.36	7.28	7.06	2.87	NS
Σ PUFA	3.25	4.28	4.01	3.57	3.94	0.90	NS
Σ PUFA n-6	2.40	2.93	2.47	2.12	2.09	0.66	NS
Σ LC-PUFA n-6	1.61	1.91	1.48	1.27	1.07	0.33	<0.001
Σ PUFA n-3	0.85	1.35	1.54	1.45	1.70	0.32	<0.001
Σ LC-PUFA n-3	0.26	0.48	0.64	0.87	1.13	0.09	<0.001
n-3:n-6	0.36	0.46	0.62	0.69	0.82	0.11	<0.001

[1] Results are based on linear regression and expressed as the average of five replicates (*n* = 3 fish per replicate). [2] Standard error of means. [3] Where linear regression was significant, the following equations were obtained: Total lipids: y = −1.750x + 29.55, R^2 = 0.30; ARA: y = −0.0326x + 0.523, R^2 = 0.35; ADA: y = 0.774x − 0.0960, R^2 = 0.60; EPA: y = −0.00660x + 0.0858, R^2 = 0.30; DHA: y = 0.212x + 0.277, R^2 = 0.90; PUFA n-3: y = 0.162x + 1.137, R^2 = 0.41; LC-PUFA n-3: y = 0.209x + 0.360, R^2 = 0.91; LC-PUFA n-6: y = - 0.174x + 1.735, R^2 = 0.50; n-3:n-6: y = 0.110x + 0.427, R^2 = 0.77. [4] Not significant (*p* > 0.05). [5] Groups of fatty acids: SFA = saturated, MUFA = monounsaturated, PUFA = polyunsaturated (≥2 double bonds), LC-PUFA = long-chain PUFA (≥20 carbons).

The two different sources of DHA affected the total lipid and fatty acid profiles in the hepatopancreas (Table 7). Fish fed 10AM accumulated more lipids in the hepatopancreas. In addition, PAL, ADA, SFA, MUFA, and n-6 LC-PUFA also showed higher concentrations in fish fed 10AM. For fish fed the CLO diet, we observed higher concentrations of DHA, PUFA n-3, LC-PUFA n-3, and the n-3:n-6 ratio.

Table 7. Total lipid and fatty acid composition of hepatopancreas in Nile tilapia juveniles fed with two sources of DHA for 87 days, at 22 °C [1,2].

Fatty Acids g 100 g^{-1} Dry Weight	Diets		*p* Value
	10AM	**CLO**	
Total lipid	27.44 ± 3.5	22.70 ± 1.72	0.026
16:0 PAL	4.42 ± 1.09	3.08 ± 0.22	0.032
18:2 n-6 LOA	0.99 ± 0.25	0.73 ± 0.05	NS [3]
20:4 n-6 ARA	0.50 ± 0.02	0.48 ± 0.10	NS
22:4 n-6 ADA	0.64 ± 0.08	0.32 ± 0.03	<0.001
18:3 n-3 α-LNA	0.75 ± 0.33	0.48 ± 0.09	NS
20:5 n-3 EPA	0.08 ± 0.0	0.08 ± 0.01	NS
22:6 n-3 DHA	0.56 ± 0.05	0.84 ± 0.04	<0.001
Σ SFA [4]	8.55 ± 1.89	6.33 ± 0.38	0.033
Σ MUFA	9.36 ± 2.47	6.45 ± 0.66	0.034
Σ PUFA	4.01 ± 0.65	3.63 ± 0.29	NS
Σ PUFA n-6	2.47 ± 0.38	1.65 ± 0.11	NS
Σ LC-PUFA n-6	1.48 ± 0.07	1.03 ± 0.06	0.008
Σ PUFA n-3	1.54 ± 0.40	1.90 ± 0.16	0.017
Σ LC-PUFA n-3	0.64 ± 0.19	1.17 ± 0.14	0.008
n-3:n-6	0.62 ± 0.13	0.87 ± 0.03	0.008

[1] Results are based on t-tests and expressed as the average of five replicates (*n* = 3 fish per replicate), followed by the standard error. [2] Diets with similar contents of DHA (~0.2 g kg^{-1} DHA dry diet); 10AM = 10 g kg^{-1} *Aurantiochytrium* sp. meal and CLO = cod liver oil. [3] Not significant (*p* > 0.05). [4] Groups of fatty acids: SFA = saturated, MUFA = monounsaturated, PUFA = polyunsaturated (≥2 double bonds), LC-PUFA = long-chain PUFA (≥20 carbons).

3.3. Somatic Indexes and Hepatic Glycogen Concentration

Our findings reveal that VSI, HSI, and the hepatic glycogen were not significantly affected by increasing dietary AM (*p* > 0.05) (Table 8). Similarly, no differences were found in these same variables (*p* > 0.05) when comparing the two sources of DHA (Table 9).

Table 8. Somatic indexes, hepatic glycogen, and intestinal morphometry of Nile tilapia juveniles fed increasing concentrations of *Aurantiochytrium* sp. meal (AM) for 87 days, at 22 °C [1].

Variables	Diets					Pooled SEM [2]	*p* Value [3]
	0AM	**5AM**	**10AM**	**20AM**	**40AM**		
Viscerosomatic index	11.76	11.93	12.10	12.08	12.26	0.93	NS [4]
Hepatosomatic index	2.73	2.95	2.85	2.76	2.70	0.48	NS
Hepatic glycogen	5.43	4.88	6.17	5.63	5.57	1.31	NS
Intestinal morphometry							
Number of folds	39.56	39.50	46.67	45.33	46.22	5.26	0.019
Fold height, μm	420.5	479.0	449.5	404.1	392.0	81.50	NS
Fold width, μm	113.2	119.3	114.3	110.7	118.6	22.10	NS
Number of goblet cells	362.9	379.6	428.0	472.0	324.0	251.90	0.047

[1] Results are based on polynomial regression analysis and expressed as the average of five replicates (*n* = 2 fish per replicate). [2] Standard error of means. [3] When the polynomial regression was significant, the following equations were obtained: number of folds = 1.034x^2 + 5.806x + 39.046, R^2 = 0.310; number of goblet cells = −46.423x^2 + 181.967x + 341.074, R^2 = 0.138. [4] Not significant (*p* > 0.05).

Table 9. Somatic indexes, hepatic glycogen, and intestinal morphometry of Nile tilapia juveniles fed two sources of docosahexaenoic acid (DHA) for 87 days, at 22 °C [1,2].

Variables	Diets		*p* Value
	10AM	**CLO**	
Viscerosomatic index	12.10 ± 0.49	11.41 ± 0.71	NS [3]
Hepatosomatic index	2.85 ± 0.35	2.93 ± 0.20	NS
Hepatic glycogen	6.17 ± 0.66	5.58 ± 0.82	NS
Intestinal morphometry			
Number of folds	45.60 ± 1.60	41.60 ± 1.60	NS
Fold height, μm	449.56 ± 20.29	440.81 ± 17.58	NS
Fold width, μm	114.38 ± 4.30	120.04 ± 5.86	NS
Number of goblet cells	428.00 ± 97.43	400.36 ± 273.50	NS

[1] Results based on *t*-tests and are expressed as the average of five replicates (n = 2 fish per replicate), followed by the standard error. [2] Diets with similar contents of DHA (~0.2 g kg^{-1} DHA dry diet); 10AM = 10 g kg^{-1} *Aurantiochytrium* sp. meal and CLO = cod liver oil. [3] Not significant ($p > 0.05$).

3.4. Morphology and Histological Changes in the Intestine and Hepatopancreas

The intestinal morphology data of fish fed increasing levels of AM were evaluated using polynomial regression. A quadratic trend was observed regarding the number of intestinal folds and the supplementation with AM, which increased from 39.5 intestinal folds in fish fed 0AM and 5AM to 46.6 in fish fed 10AM (Table 8, Figure 1B,F). The inclusion of AM did not influence the height and width of the intestinal folds. However, the number of goblet cells also increased with the addition of levels up to 20AM; however, in fish fed 40AM there was a decrease in goblet cells (Table 8, Figure 1G,I). No differences were found in the morphometric variables when comparing the two sources of DHA (Table 9).

In the hepatopancreas, the following variables were not affected by increasing supplementation of AM or by feeding the CLO diet, presenting only mild to moderate changes in the tissue: cordonal aspect, intact acini, balloon aspect of hepatocytes, cholestasis, congestion in large vessels, congestion in the pancreas and sinusoids, displacement of the nucleus of hepatocytes, dilatation of the venous sinus, mononuclear and eosinophilic infiltrates, hypertrophy of hepatocytes, and nucleus with pyknosis. However, fish fed the 0AM diet presented the highest hepatocyte size variation, whereas those fed 40AM presented the lowest (Table 10, and Figure 2A,C,D). The intensity of macrosteatosis was highest in fish fed 0AM and did not differ significantly between fish fed 5AM and 10AM. Fish fed 20AM, 40AM, and CLO diets showed a low intensity of macrosteatosis, and they differed significantly from 0AM-fed fish. Fish fed the 0AM diet were the only ones showing microsteatosis in the hepatopancreas (Figure 2A). Hepatopancreas necrosis was significantly lower in fish fed 40AM than in those fed the diet 0AM (Figure 2B).

Fish fed the 5AM diet showed the lowest number of hepatocyte nuclei with karyolysis and differed from fish fed both the 10AM and 20AM diets (Figure 3A,C, respectively), which presented the greatest intensity of this alteration. Meanwhile, fish fed the 5AM diet also had less intensity of karyorrhex than fish fed both the 20AM and 40AM diets (Figure 3B). The loss of the nuclei by hepatocytes was more intense in fish fed the 0AM diet. Loss of the nuclei of the pancreatic acinos was observed in fish fed the diets 5AM and 10AM (Figure 3D,E). Only fish fed the 5AM and 10AM diets showed loss of the pancreatic acini nuclei (Figure 3F). The presence of macrophages with bilirubin was identified only in fish fed the CLO diet (Figure 3F).

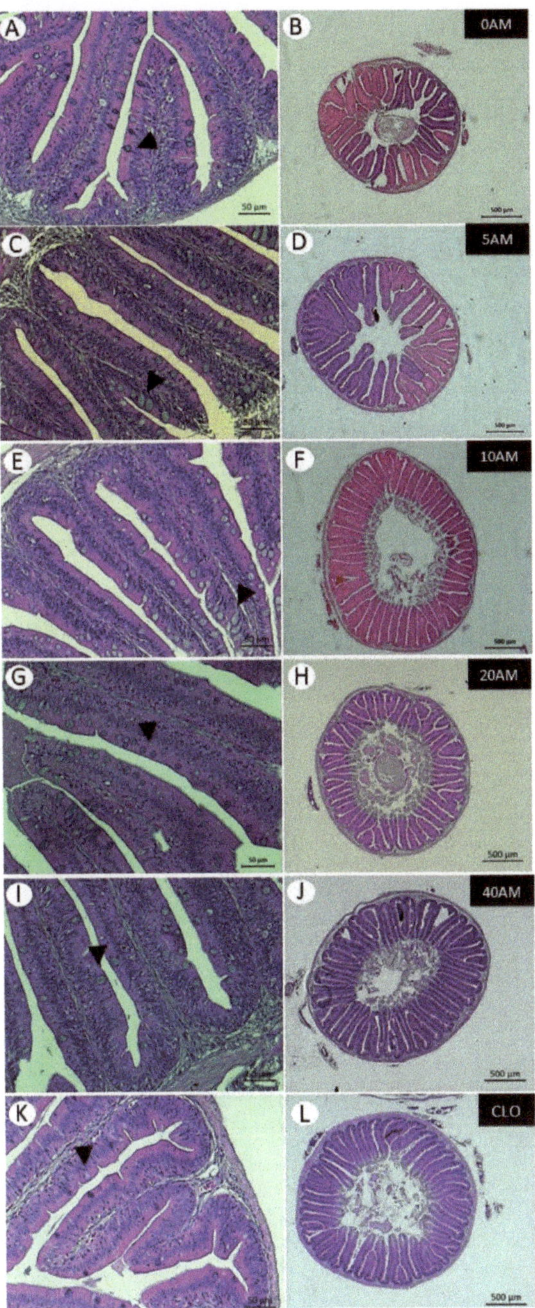

Figure 1. Histological changes in the anterior third of the intestines of juvenile Nile tilapia when fed different levels of AM or CLO, for 87 days, at 22 °C. Figures (**A**,**C**,**E**,**G**,**I**,**K**) represent a partial cross-section of the intestine of fish fed 0AM, 5AM, 10AM, 20AM, 40AM, and CLO diets, respectively. Goblet cells are identified with the arrowhead. Bar: 50 μm. Figures (**B**,**D**,**F**,**H**,**J**,**L**) represent a cross-section of the whole intestine. Bar: 500 μm. Staining: H&E.

Table 10. Morphological alterations in hepatopancreas of Nile tilapia juveniles, when fed with increasing concentrations of *Aurantiochytrium* sp. meal (AM) or cod liver oil (CLO) for 87 days, at 22°C [1].

Variable	Diets						*p* Value [2]
	0AM	5AM	10AM	20AM	40AM	CLO	
Cell size variation	1.83 ± 0.69 [a]	1.00 ± 0.49 [ab]	0.92 ± 0.41 [ab]	1.08 ± 0.76 [ab]	0.45 ± 0.66 [b]	1.10 ± 0.30 [ab]	0.0001
Hypotrophy of hepatocyte nucleus	1.00 ± 0.0 [a]	0.08 ± 0.28 [b]	0.00 ± 0.0 [b]	0.00 ± 0.0 [b]	0.00 ± 0.0 [b]	0.00 ± 0.0 [b]	<0.0001
Macrosteatosis	1.50 ± 1.16 [a]	0.92 ± 0.9 [ab]	1.08 ± 0.7 [ab]	0.42 ± 0.67 [b]	0.36 ± 0.82 [b]	0.54 ± 0.82 [b]	0.0288
Microsteatosis	0.83 ± 0.85 [a]	0.00 ± 0.0 [b]	0.00 ± 0.0 [b]	0.00 ± 0.0 [b]	0.00 ± 0.0 [b]	0.00 ± 0.0 [b]	<0.0001
Necrosis	1.92 ± 0.28 [a]	1.75 ± 0.74 [ab]	1.58 ± 0.67 [ab]	1.33 ± 0.65 [ab]	1.18 ± 0.60 [b]	1.45 ± 0.68 [ab]	0.0376
Nuclei with karyolysis	1.58 ± 0.51 [ab]	1.17 ± 0.39 [b]	1.92 ± 0.67 [a]	1.92 ± 0.29 [a]	1.81 ± 0.60 [ab]	1.81 ± 0.60 [ab]	0.0066
Nuclei with karyorrhexis	1.58 ± 0.51 [bc]	1.41 ± 0.51 [c]	1.83 ± 0.58 [abc]	1.92 ± 0.29 [ab]	2.09 ± 0.54 [a]	1.82 ± 0.60 [abc]	0.0419
Loss of hepatocyte nucleus	0.92 ± 0.29 [a]	0.25 ± 0.45 [b]	0.17 ± 0.58 [b]	0.25 ± 0.62 [b]	0.00 ± 0.00 [b]	0.27 ± 0.65 [b]	<0.0001
Loss of nucleus in pancreatic acini	0.00 ± 0.00 [b]	0.58 ± 0.67 [a]	0.25 ± 0.45 [a]	0.00 ± 0.0 [b]	0.00 ± 0.00 [b]	0.00 ± 0.00 [b]	0.015
Macrophage with bilirubin	0.00 ± 0.00 [b]	0.00 ± 0.0 [b]	0.00 ± 0.00 [b]	0.00 ± 0.00 [b]	0.00 ± 0.00 [b]	0.55 ± 0.52 [a]	<0.0001

[1] Histological alterations in the hepatopancreas were evaluated semi-quantitatively by ranking the severity of the tissue lesions. The ranking was 0 (absence of alteration), 1 (mild alteration, corresponding to <25% tissue area), 2 (moderate alteration, 25% to 50% tissue area), and 3 (severe alteration, >50% tissue area). [2] Results are based on the non-parametric Kruskal–Wallis test and expressed as the average of five replicates ($n = 2$ fish per replicate), followed by the standard error. [a,b,c] Values followed by different letters are significantly different ($p < 0.05$).

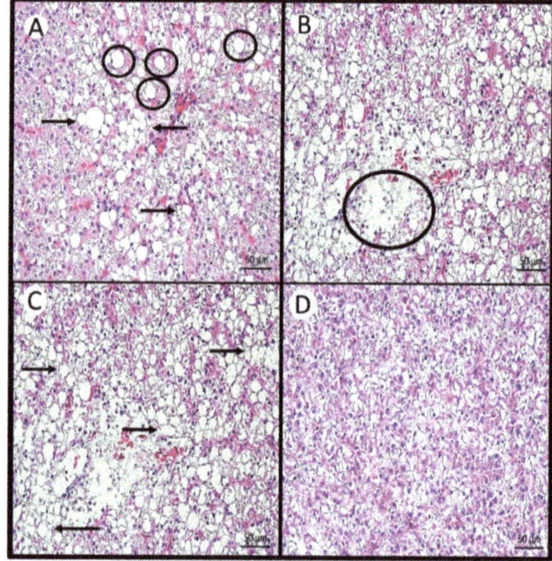

Figure 2. Histological changes in the hepatopancreas of Nile tilapia juveniles fed different levels of *Aurantiochytrium* sp. meal (AM) or a cod liver oil (CLO)-supplemented diet for 87 days, at 22 °C. (**A,B**) High intensity of macrosteatosis (circle) and microsteatosis (arrow) in fish fed the 0AM diet. (**B**) High intensity of necrosis (circle) in fish fed the 0AM diet. (**C**) High loss of hepatocyte nucleus (arrow) in fish fed the 0AM diet. (**D**) Smaller variation in hepatocyte size in the hepatopancreas and lower intensity of macrosteatosis, microsteatosis, and necrosis in fish fed the 40AM diet.

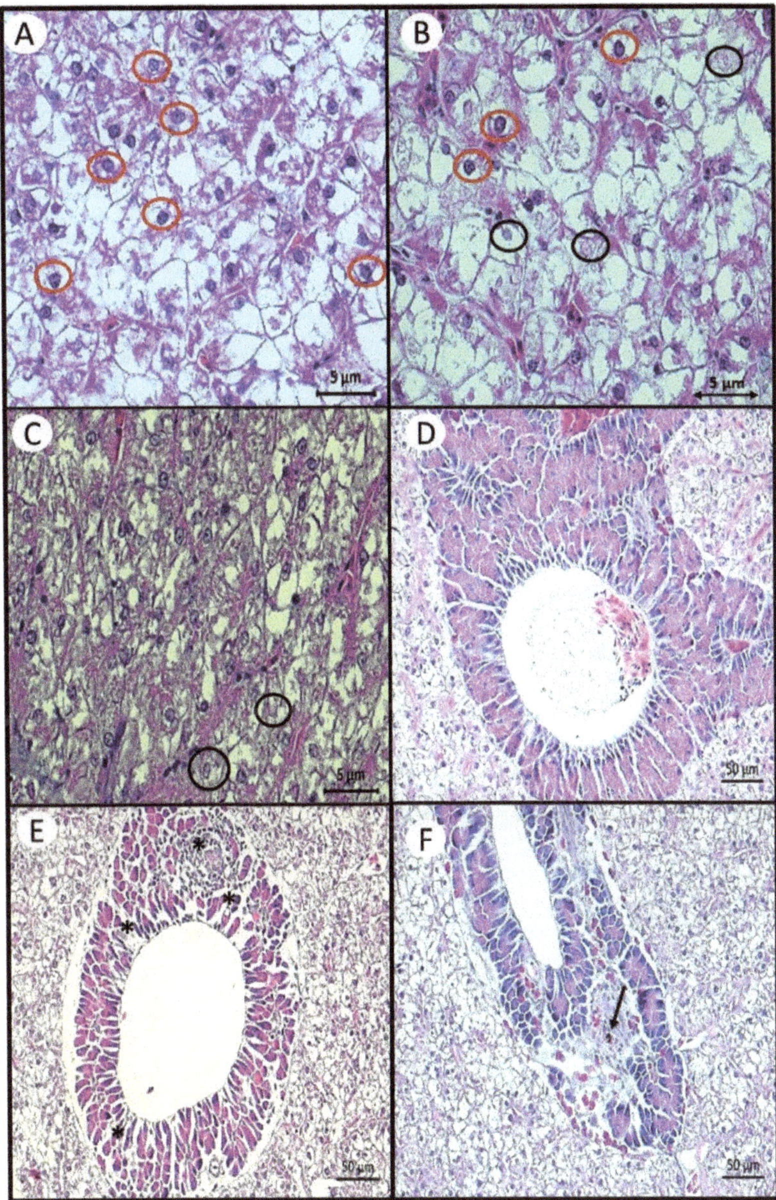

Figure 3. Histological changes in the hepatopancreas of Nile tilapia juveniles fed different levels of *Aurantiochytrium* sp. meal (AM) or a cod liver oil (CLO)-supplemented diet for 87 days, at 22 °C. (**A**) High intensity of nuclei with karyolysis (red circle) in fish fed the 20AM diet. (**B**) High intensity of nuclei with karyorrhexis (black circle) in fish fed the 40AM diet. Nuclei with karyolysis are indicated by red circles. (**C**) Lower intensity of nuclei with karyolysis and karyorrex in fish fed the 5AM diet. (**D**) Nuclei in pancreatic acini without alterations in fish fed the 0AM diet. (**E**) Loss of nuclei in pancreatic acini in fish fed the 5AM diet (alteration indicated by asterisks). (**F**) Macrophage with bilirubin in fish fed the CLO diet (alteration indicated by arrow).

4. Discussion

4.1. Body Fatty Acid Composition and Apparent Retention

It has been well established in fish that the tissue fatty acid composition generally reflects dietary fatty acid composition [30–32]. Based on the regression analysis, the higher the inclusion of AM in the diets, the higher the accumulation of DHA in the whole body of the fish. Our findings also show an increase in the whole-body n-3 fatty acid content, followed by an increase in the dietary inclusion of AM, a source of DHA. Several authors cited an accumulation of DHA in the body and muscle composition of fish, especially at a low temperature [15,18,33,34]. The DHA is considered a fatty acid of high biological value, with a close link between DHA and the composition of structural phospholipids in cells [17]. Thus, DHA has a different catabolism from other fatty acids, being more preserved in cell phospholipids and less catabolized to produce energy [35]. Due to its high flexibility provided by its chemical arrangement, DHA maintains a cell membrane bilayer with the balance between fluidity and rigidity needed to accommodate rapid conformational changes in cell membrane proteins. This structure brings fluidity to the membrane at low temperatures [17]. In addition to preserving DHA, which can be modulated directly by the diet, fish also have desaturase and elongate enzymes that are responsible for desaturating LOA to ARA, LNA to EPA, and EPA to DHA. Indeed, Nile tilapia have been reported to express the genes for Δ-4 and Δ-6 desaturases, thus enabling the conversion of EPA into DHA by two different pathways, which are mediated by both desaturases [17,36].

On the other hand, during the β-oxidation of fatty acids, fish preferentially use SFA and MUFA, with shorter chain lengths and numbers of unsaturation, as energy sources [37,38]. This observation may explain the linear reduction in PAL, SFA, and MUFA in fish fed 10AM, 20AM, and 40AM, which in turn, could be preferentially degraded for energy production in order to preserve the fatty acids of greatest biological value, such as PUFAs. The decrease in ARA, n-6 PUFA, and n-6 LC-PUFA groups can be explained by the competition of desaturase and elongase enzymes, which compete for the substrates of n-3 and n-6 series fatty acids and have a greater affinity for n-3 PUFA substrates [39]. Thus, the reduction in bioconversion from n-6 PUFA to n-6 LC-PUFA may have occurred because of the inhibition of the enzymatic activity of desaturases and elongases in the presence of higher levels of DHA in the diets [32,39].

In the present study, the higher concentration of DHA in the bodies of fish fed the CLO diet than in fish fed the diet containing 10AM probably occurred due to the bioconversion of its precursors (LNA, EPA, and DPA) which were detected at higher levels only in fish fed the CLO diet. Through the study of the fatty acid balance, it was possible to understand that for Nile tilapia, the bioconversion of LNA to DHA is efficient, and the biosynthesis of DHA from EPA can be more direct or faster than the production of EPA from LNA [32]. Our study corroborates such findings, where the CLO diet contained more LNA and EPA in its composition, favoring their bioconversion to DHA.

The ARR of DHA and n-3 PUFA was influenced by the levels of dietary inclusion; the higher the dietary concentration of these fatty acids, the lower the retention. Other authors have also reported a similar pattern, where the higher the concentration of a fatty acid in the diet, the lower its relative deposition [18,39]. In our study, as also reported by Brignol et al. [22] at 28 °C, the ARR of α-LNA increased with the inclusion of AM. Increasing the dietary concentration of DHA following the inclusion of AM possibly prevented the elongation and desaturation of α-LNA for the production of LC-PUFA, thus preserving and retaining α-LNA in the membranes. In addition, diets containing higher inclusions of AM contained less α-LNA content. Consequently, the lower the presence of an essential fatty acid in the diet, the higher its body retention. However, contrary to the report by Brignol et al. [22], at 28 °C, where there was an increase in n-6 PUFA body retention with the inclusion of AM, we did not observe such a response in our study, at 22 °C. Here, PUFA n-6 retention decreased significantly with the inclusion of AM, following the body composition trend that also decreased, probably due to the competition of desaturases and elongases with the n-3 substrate and/or due to β-oxidation of the PUFA n-6, in addition to the greater conservation of the n-3 fatty acids to maintain membrane functionality at low temperatures.

When comparing the two sources of DHA, fish fed CLO showed higher retention of the following fatty acids: LOA, α-LNA, DHA, n-3 PUFA, and n-6 PUFA at 28 °C [22]. However, in our study, at 22 °C, the retention of α-LNA and n-6 PUFA was higher in fish fed the AM-supplemented diet. The higher retention of α-LNA can be explained by the low content of this fatty acid in the diet supplemented with AM, which may also have led to an increased retention of n-6 PUFA, as there was little n-3 substrate for inhibition of the desaturases and elongases in the n-6 pathway. Despite having similar amounts of DHA, the higher DHA, total PUFA, and n-3 PUFA retention in fish fed the CLO diet was probably due to the higher synthesis of DHA from precursor fatty acids, detected only in the CLO diet [17,22].

4.2. Fatty Acid Composition in the Hepatopancreas

The greater the dietary inclusion of AM, the lower the amount of total lipid in the liver, in addition to EPA, ARA, ADA, PUFA n-6, and LC-PUFA n-6. However, the increasing inclusions of AM increased the composition of DHA, PUFA, LC-PUFA n-3, and the n-3:n-6 ratio. The fatty acid composition of hepatopancreas largely reflected that of the diet, which is consistent with studies in Atlantic salmon [40].

The hepatopancreas is the main organ that regulates lipid metabolism, including both the synthesis and degradation of fatty acids, where several regulating enzymes show varied affinities for the different fatty acids available in that organ [37,41]. In addition, the hepatopancreas functions as an important energy reservoir, often in the form of triacylglycerols [21]. The high intake of PUFAs (mainly EPA and DHA) prevents the accumulation of lipids by inducing lipid oxidation, [41,42], inhibiting lipogenic metabolism, and stimulating the synthesis of lipoproteins [43–46].

The significant increase in DHA in the hepatopancreas, about 105% between 0AM- and 40AM-fed fish, suggests the important role of this fatty acid in cell membrane function at a low temperature. Additionally, the decrease in EPA content in the hepatopancreas may be linked to the bioconversion of EPA to DHA [17,32]. In addition, the content of n-6 PUFA decreased, suggesting the preferential route of activity of the desaturase and elongase enzymes by the n-3 series fatty acids. Indeed, Chen et al. [32] reported that increasing dietary inclusion of LNA in Nile tilapia raised at the optimal temperature could block or at least slow down n-6 LC-PUFA biosynthesis from LOA. Therefore, there is competition for accessing the Δ-6 desaturase and elongase between substrates of the n-3 and n-6 series. Although we have not evaluated such enzymes in our study, this substrate competition is well known for tilapia [32].

4.3. Morphology and Histological Changes in the Intestine and Hepatopancreas

Different levels of dietary supplementation with AM affected the histology of the hepatopancreas and intestine. Fish fed the highest levels of AM showed a significant increase in the number of intestinal folds. The increase in the absorption area, caused by the increase in the number of folds, could improve the digestive and absorptive processes, suggesting a more efficient use of nutrients [47]. Our results regarding the increase in the number of villi in fish fed the 10AM diet corroborate the findings of Nobrega et al. [15], where fish fed the 10AM diet reached the highest weight gain. Such high growth may be associated with an increase in the nutrient absorption area. The influence of diet-derived substances on intestinal epithelial function, including barrier integrity, is likely to be important [48].

A recent review on mammals [48] reported that n-3 LC-PUFA, especially DHA, contributes to maintaining the integrity of the intestinal epithelial barrier by exerting anti-inflammatory effects and accelerating recovery from intestinal inflammation. However, few studies have addressed the relationship between dietary fatty acids, especially LC-PUFA n-3, and intestinal health in fish. Dietary supplementation with *Schizochytrium* sp. (a

DHA-producing marine heterotrophic microorganism from the same family as *Aurantiochytrium* sp.) in juvenile mirror carp (*Cyprinus carpio* var. specularis) resulted in a higher intestinal fold height in fish fed 30 and 60 g kg^{-1} supplementation if compared to fish fed the non-supplemented diet [49]. The dietary replacement of fish oil and fish meal by heterotrophic microorganisms results in a good response to intestinal integrity, since such microorganisms also contain additional bioactive cell wall compounds such as β-glucans, β-carotenes, flavonoids, nucleotides, and water-soluble peptides [50–52] which can affect nutrient availability and growth performance, but also enhance the well-being of fish by improving gut health and thus nutrient assimilation and immune competence [52].

The rising levels of AM also resulted in a significant increase in the number of goblet cells in fish fed diets 0AM to 20AM. An increase in the number of goblet cells was also reported in the intestine of Atlantic salmon, reared at an optimal temperature for the species (10.2 °C), with an increase in dietary inclusion of 6 to 15 g kg^{-1} *Schizochytrium* sp. meal [52]. Goblet cells produce mucus, which plays an important role in immunity. Besides serving as a mechanical barrier, making it difficult for pathogenic bacteria to adhere, mucus contains several components of the innate immune response, such as lysozymes, immunoglobulins, complement system proteins, lectins, and several other antimicrobial components [53]. However, in fish fed the highest AM level (40AM), there was a reduction in the number of goblet cells. A previous study evaluated different dietary sources of n-3 LC-PUFA (fish oil, EPA-enriched oil, and DHA-enriched oil) for the carnivorous marine fish pompano, *Trachinotus ovatus*, raised at an optimal temperature and reported that high dietary EPA or DHA levels caused a depressed expression of Muc13 mRNA in the intestines [54]. Mucins are the major constituent of the mucous layer and are produced by goblet cells [55]. Therefore, more research needs to be carried out to assess how dietary DHA interacts with mucus production in the intestine. Furthermore, the decrease in goblet cells does not seem to be related to inflammation in the intestine, since treatments with higher AM inclusions showed a greater number of villi, showing its positive effect on intestinal morphometry.

Histological analysis of the hepatopancreas showed a similar pattern to that observed in the total lipid content of this organ. Likewise, the intensity of macrosteatosis, microsteatosis, necrosis, and loss of the nucleus of hepatocytes were lower in the hepatopancreas of fish fed the highest inclusions of AM. Therefore, the highest intensity of histological and lipid composition changes in fish fed the control diet, without the inclusion of AM but with a lipid base of corn oil and swine lard, corroborates the findings for other fish species when fish oil is replaced with vegetable oils [21,56]. Lima de Andrade et al. [13] also found histopathological changes in tilapia juveniles fed diets with different LOA:LNA (n6/n3 = 12.02 and n6/n3 = 3.85), at temperatures of 30 and 20 °C. The frequency of cell displacement injury was lower in fish fed the diet containing the low n6/n3. This may be associated with lower lipid deposition. In addition, the liver of fish fed n6/n3 = 3.85 showed low cytoplasmic vacuolization at both temperatures [13]. Therefore, an imbalance in dietary fatty acids can modify the function and morphology of this organ. It is possible to establish a relationship between the type of dietary fatty acids and the appearance of steatosis, that is, LOA > LNA > oleic acid [21].

In a three-month trial, the marine fish, *Sparus aurata*, fed diets based on fish oil, canola, flaxseed, or a mix of these oils, showed uniformity in cell size and little lipid accumulation. However, fish fed soybean oil presented steatosis foci, with hepatocytes containing numerous lipid vacuoles [21]. In another study testing rearing temperatures (optimum 12 °C and suboptimal 5 °C) and lipid sources (soy oil replacing fish oil, at 50% and 100%) in Atlantic salmon, the suboptimal temperature positively influenced the deposition of fat in liver cells and the intestines [56]. In addition, the diet with 100% soy oil resulted in the highest accumulation of fat in the liver at 5 °C [56]. According to the authors, the accumulation of fat at low temperatures could be explained by the reduction in the activity of the enzymes involved in the esterification of fatty acids into triacylglycerol and phospholipids for very low-density lipoprotein (VLDL) production.

On the other hand, despite appearing at a low intensity, the presence of nuclei with karyolysis and karyorrhexis was lower in fish fed the lowest level of AM (5AM). It is possible that a minimal inclusion of n-3 PUFA would be sufficient to prevent these changes. Karyolysis is the complete dissolution of chromatin in a cell that is dying due to enzymatic degradation, and is usually associated with karyorrhexis, which occurs mainly as a result of necrosis [57]. This is the first time that such a change has been reported with an increase in dietary PUFAs. These changes have already been described, in greater intensity than found in our study, for several species of fish, when exposed to contaminating agents [58–60].

Several studies have shown that a drop in temperature causes fasting induced by cold, heat stress, and metabolic depression in fish [61]. To mitigate these effects, adequate nutrition is suggested in this study. We verified that AM, as a source of DHA in Nile tilapia, provides the body with the accumulation of DHA, an important fatty acid for adequate metabolic functioning of fish, when subjected to low suboptimal temperatures. In addition, increasing the inclusion of AM decreased the hepatopancreas lipid content, increased the hepatopancreas concentration of n-3 fatty acid series, and promoted significant improvements in the morphophysiology of the hepatopancreas, preventing signs of macrosteatosis, microsteatosis, and necrosis, seen in fish fed a practical diet, without any DHA supplementation. These data provide evidence for the physiological need for DHA supplementation in Nile tilapia diets at suboptimal temperatures and the potential for the development of specific winter aquafeeds for the species, with the aim of improving growth performance and physiological well-being.

In addition to its important role in fish performance and metabolism, DHA appears to play an important role in the cardiac, cardiovascular, brain, and visual functions of humans [62]. Currently, a greater intake of n-3 fatty acids is desirable for reducing the risk of many of the highly prevalent chronic diseases in Western societies, as well as in developing countries. *Aurantiochytrium* sp. meal supplementation in Nile tilapia diets can also provide a source of n-3 fatty acids for human consumption, adding value to this freshwater species.

More research should be carried out to further evaluate the effects of dietary AM supplementation in the intestinal functions of fish. We have registered a positive effect on intestinal health; however, a wide range of analyses including intestinal microbiota composition and gut-associated lymphoid tissues (GALT) immune responses would provide further understanding about the effect of such a novel additive. Furthermore, future studies should evaluate the dietary supplementation of AM in growing-out conditions to validate our findings in a controlled lab situation.

5. Conclusions

Nile tilapia maintained at 22 °C responded linearly to increasing dietary inclusions of AM regarding body composition. The inclusion level of 40 g kg^{-1} of AM promoted the best results in improving the body and hepatopancreas n-3 fatty acid profile, decreasing hepatopancreas lipid content, and significantly improving the morphophysiology of the hepatopancreas. When comparing the two DHA sources, CLO allowed the highest n-3 fatty acid body content and retention, possibly due to the increased synthesis of DHA from fatty acid precursors. By representing a novel and renewable DHA source, the positive biological responses of Nile tilapia to dietary supplementation with AM make this additive an excellent candidate to replace fish oil in winter diets.

Author Contributions: Conceptualization: R.O.B., R.O.N., D.D.S. and D.M.F.; methodology: R.O.B., R.O.N., D.D.S. and D.M.F.; software: R.O.B.; validation, R.O.B. and R.O.N.; formal analysis: R.O.B.; investigation, R.O.B. and R.O.N.; resources: D.M.F. and J.E.P.; data curation: R.O.B., R.O.N., D.D.S., J.E.P. and D.M.F.; writing—original draft preparation: R.O.B.; writing—review and editing: R.O.N., D.D.S., J.E.P. and D.M.F.; supervision: D.D.S. and D.M.F.; project administration, D.M.F.; funding acquisition, D.M.F. All authors have read and agreed to the published version of the manuscript.

Funding: Alltech Inc. (USA), Coordenação de Aperfeicoamento de Pessoal de Nível Superior (CAPES-Brazil, Finance Code 001), and National Council for Scientific and Technological Development (CNPq-Brazil, grant # 313185/2020-4).

Institutional Review Board Statement: The study was conducted according to the guidelines of the National Council for Control of Animal Experimentation (CONCEA) and was approved by the Ethics Committee on Animal Use of the Federal University of Santa Catarina (CEUA/UFSC), protocol code 5665210917 in the meeting of 03/14/2018.

Informed Consent Statement: Not applicable.

Data Availability Statement: The data that support the findings of this study are available from the corresponding author upon reasonable request.

Acknowledgments: We thank Alltech Inc. for funding this study (USA). The additive evaluated was produced and sold by Alltech for use in animal diets under the tradename ALL-G-RICH^{TM}. We also thank CAPES-Brazil (Finance Code 001) for granting fellowships to the first and second authors and CNPq-Brazil, which granted a fellowship to the last author. Thanks are also due to Nicoluzzi Rações Ltd.a (Penha, Santa Catarina, Brazil) and Kabsa Exportadora S.A. (Porto Alegre, RS, Brazil) for donating ingredients for the manufacturing of the experimental diets, as well as to Lucas Cardoso NEPAQ-UFSC-Brazil for the assistance with the histological analysis and Jacó Joaquim Mattos LABCAI-UFSC-BRAZIL for the assistance with the glycogen analysis.

Conflicts of Interest: This study was funded by Alltech, Inc. (USA) through a joint research alliance between Alltech (USA) and the Universidade Federal de Santa Catarina (UFSC-Brazil). One of the authors is a consultor for Alltech through a committee containing both Alltech staff and UFSC staff, where the research project was developed and approved.

References

1. Martinez-Palacios, C.A.; Charez-Sanchez, M.C.; Ross, L.G. The effects of water temperature on food intake, growth and body composition of *Cichlasoma urophthalmus* (Gunther) juveniles. *Aquac. Res.* **1996**, *27*, 455–461. [CrossRef]
2. Fatma, S.; Ahmed, I. Effect of water temperature on protein requirement of *Heteropneustes fossilis* (Bloch) fry as determined by nutrient deposition, hemato-biochemical parameters and stress resistance response. *Fish. Aquat. Sci.* **2020**, *23*, 1–14. [CrossRef]
3. Barton, B.A. Stress in fishes: A diversity of responses with particular reference to changes in circulating corticosteroids. *Integr. Comp. Biol.* **2002**, *42*, 517–525. [CrossRef]
4. Panase, P.; Saenphet, S.; Saenphet, K. Biochemical and physiological responses of Nile tilapia *Oreochromis niloticus* Lin subjected to cold shock of water temperature. *Aquac. Rep.* **2018**, *11*, 17–23. [CrossRef]
5. Donaldson, M.R.; Cooke, S.J.; Patterson, D.A.; Macdonald, J.S. Cold shock and fish. *J. Fish Biol.* **2008**, *73*, 1491–1530. [CrossRef]
6. FAO. *The State of World Fisheries and Aquaculture 2020. Sustainability in Action*; FAO: Rome, Italy, 2020. [CrossRef]
7. Peixe-BR Anuário da Piscicultura 2020. *Assoc. Bras. Piscic.* **2020**, *1*, 1–136.
8. Sebastião, F.A.; Pilarski, F.; Kearney, M.T.; Soto, E. Molecular detection of *Francisella noatunensis* subsp. orientalis in cultured Nile tilapia (*Oreochromis niloticus* L.) in three Brazilian states. *J. Fish Dis.* **2017**, *40*, 1731–1735. [CrossRef] [PubMed]
9. Da Silva, B.C.; Pereira, A.; Marchiori, N.D.C.; Mariguele, K.H.; Massago, H.; Klabunde, G.H.F. Cold tolerance and performance of selected Nile tilapia for suboptimal temperatures. *Aquac. Res.* **2021**, *52*, 1071–1077. [CrossRef]
10. Ma, X.Y.; Qiang, J.; He, J.; Gabriel, N.N.; Xu, P. Changes in the physiological parameters, fatty acid metabolism, and SCD activity and expression in juvenile GIFT tilapia (*Oreochromis niloticus*) reared at three different temperatures. *Fish Physiol. Biochem.* **2015**, *41*, 937–950. [CrossRef]
11. Azaza, M.S.; Dhraïef, M.N.; Kraïem, M.M. Effects of water temperature on growth and sex ratio of juvenile Nile tilapia *Oreochromis niloticus* (Linnaeus) reared in geothermal waters in southern Tunisia. *J. Therm. Biol.* **2008**, *33*, 98–105. [CrossRef]
12. Nobrega, R.O.; Banze, J.F.; Batista, R.O.; Fracalossi, D.M. Improving winter production of Nile tilapia: What can be done? *Aquac. Rep.* **2020**, *18*, 100453. [CrossRef]
13. Lima de Almeida, C.A.; Lima de Almeida, C.K.; de Fátima Ferreira Martins, E.; Gomes, Â.M.; da Anunciação Pimentel, L.; Pereira, R.T.; Fortes-Silva, R. Effect of the dietary linoleic/α-linolenic ratio (n6/n3) on histopathological alterations caused by suboptimal temperature in tilapia (*Oreochromis niloticus*). *J. Therm. Biol.* **2019**, *85*, 102386. [CrossRef]
14. Corrêa, C.F.; Nobrega, R.O.; Mattioni, B.; Block, J.M.; Fracalossi, D.M. Dietary lipid sources affect the performance of Nile tilapia at optimal and cold, suboptimal temperatures. *Aquac. Nutr.* **2017**, *23*, 1016–1026. [CrossRef]
15. Nobrega, R.O.; Batista, R.O.; Corrêa, C.F.; Mattioni, B.; Filer, K.; Pettigrew, J.E.; Fracalossi, D.M. Dietary supplementation of *Aurantiochytrium* sp. meal, a docosahexaenoic-acid source, promotes growth of Nile tilapia at a suboptimal low temperature. *Aquaculture* **2019**, *507*, 500–509. [CrossRef]
16. Abdel-Ghany, H.M.; El-Sayed, A.F.M.; Ezzat, A.A.; Essa, M.A.; Helal, A.M. Dietary lipid sources affect cold tolerance of Nile tilapia (*Oreochromis niloticus*). *J. Therm. Biol.* **2019**, *79*, 50–55. [CrossRef] [PubMed]

17. Glencross, B.D. Exploring the nutritional demand for essential fatty acids by aquaculture species. *Rev. Aquac.* **2009**, *1*, 71–124. [CrossRef]
18. Corrêa, C.F.; Nobrega, R.O.; Block, J.M.; Fracalossi, D.M. Mixes of plant oils as fish oil substitutes for Nile tilapia at optimal and cold suboptimal temperature. *Aquaculture* **2018**, *497*, 82–90. [CrossRef]
19. FAO. *The State of Fisheries and Aquaculture in the World 2018*; FAO: Rome, Italy, 2018.
20. Turchini, G.M.; Torstensen, B.E.; Ng, W.K. Fish oil replacement in finfish nutrition. *Rev. Aquac.* **2009**, *1*, 10–57. [CrossRef]
21. Caballero, M.J.; Izquierdo, M.S.; Kjørsvik, E.; Fernández, A.J.; Rosenlund, G. Histological alterations in the liver of sea bream, *Sparus aurata* L., caused by short- or long-term feeding with vegetable oils. Recovery of normal morphology after feeding fish oil as the sole lipid source. *J. Fish Dis.* **2004**, *27*, 531–541. [CrossRef] [PubMed]
22. Brignol, F.D.; Fernandes, V.A.G.; Nobrega, R.O.; Corrêa, C.F.; Filler, K.; Pettigrew, J.; Fracalossi, D.M. *Aurantiochytrium* sp. meal as DHA source in Nile tilapia diet, part II: Body fatty acid retention and muscle fatty acid profile. *Aquac. Res.* **2019**, *50*, 707–716. [CrossRef]
23. Fernandes, V.A.G.; Brignol, F.D.; Filler, K.; Pettigrew, J.; Fracalossi, D.M. *Aurantiochytrium* sp. meal as DHA source in Nile tilapia diet, part I: Growth performance and body composition. *Aquac. Res.* **2019**, *50*, 390–399. [CrossRef]
24. Furuya, W.M.; Furuya, V.R.B.; Nagae, M.Y.; Graciano, T.S.; Michelato, M.; Xavier, T.O.; Vidal, L.V. Nutrição de Tilápias no Brasil. *Sci. Agrar. Parana.* **2012**, *11*, 19–34. [CrossRef]
25. NRC. *Nutrient Requirements of Fish and Shrimp*; National Academic Press (National Research Council): Washington, DC, USA, 2011.
26. AOAC. *Official Methods of Analysis*, 16th ed.; Association of Official Analytical Chemists: Washington, DC, USA, 1999. Available online: http://www.sciepub.com/reference/180240 (accessed on 16 July 2021).
27. Carroll, N.V.; Longley, W.; And Roe, J.H. The determination of glycogen in liver and muscle by use of anthrone reagent. *J. Biol. Chem.* **1956**, *220*, 583–593. [CrossRef]
28. Brum, A.; Cardoso, L.; Chagas, E.C.; Chaves, F.C.M.; Mouriño, J.L.P.; Martins, M.L. Histological changes in Nile tilapia fed essential oils of clove basil and ginger after challenge with *Streptococcus agalactiae*. *Aquaculture* **2018**, *490*, 98–107. [CrossRef]
29. Schwaiger, J.; Wanke, R.; Adam, S.; Pawert, M.; Hönnen, W.; Triebskorn, R. The use of histopathological indicators to evaluate contaminant-related stress in fish. *J. Aquat. Ecosyst. Stress Recover.* **1997**, *6*, 75–86. [CrossRef]
30. Tonial, I.B.; Stevanato, F.B.; Matsushita, M.; De Souza, N.E.; Furuya, W.M.; Visentainer, J.V. Optimization of flaxseed oil feeding time length in adult Nile tilapia (*Oreochromis niloticus*) as a function of muscle omega-3 fatty acids composition. *Aquac. Nutr.* **2009**, *15*, 564–568. [CrossRef]
31. Qiu, H.; Jin, M.; Li, Y.; Lu, Y.; Hou, Y.; Zhou, Q. Dietary lipid sources influence fatty acid composition in tissue of large yellow croaker (*Larmichthys crocea*) by regulating triacylglycerol synthesis and catabolism at the transcriptional level. *PLoS ONE* **2017**, *12*, e0169985. [CrossRef] [PubMed]
32. Chen, C.; Guan, W.; Xie, Q.; Chen, G.; He, X.; Zhang, H.; Guo, W.; Chen, F.; Tan, Y.; Pan, Q. n-3 essential fatty acids in Nile tilapia, *Oreochromis niloticus*: Bioconverting LNA to DHA is relatively efficient and the LC-PUFA biosynthetic pathway is substrate limited in juvenile fish. *Aquaculture* **2018**, *495*, 513–522. [CrossRef]
33. Liu, C.; Ge, J.; Zhou, Y.; Thirumurugan, R.; Gao, Q.; Dong, S. Effects of decreasing temperature on phospholipid fatty acid composition of different tissues and hematology in Atlantic salmon (*Salmo salar*). *Aquaculture* **2020**, *515*, 734587. [CrossRef]
34. Mufatto, L.M.; Nobrega, R.O.; Menoyo, D.; Fracalossi, D.M. Dietary ratios of n-3/n-6 fatty acids do not affect growth of Nile tilapia at optimal temperatures (28°C) nor at temperatures that simulate the onset of winter (22°C). *Aquac. Nutr.* **2019**, *25*, 646–661. [CrossRef]
35. Khripach, V.A.; Zhabinskii, V.N.; de Groot, A.E. *Biosynthesis and Metabolism*; Elsevier Inc.: Amsterdam, The Netherlands, 1999; ISBN 9780128112304.
36. Oboh, A.; Kabeya, N.; Carmona-Antoñanzas, G.; Castro, L.F.C.; Dick, J.R.; Tocher, D.R.; Monroig, O. Two alternative pathways for docosahexaenoic acid (DHA, 22:6n-3) biosynthesis are widespread among teleost fish. *Sci. Rep.* **2017**, *7*, 1–10. [CrossRef]
37. Kiessling, K.H.; Kiessling, A. Selective utilization of fatty acids in rainbow trout (*Oncorhynchus mykiss* Walbaum) red muscle mitochondria. *Can. J. Zoo.* **1993**, *71*, 248–251. [CrossRef]
38. Salini, M.J.; Turchini, G.M.; Glencross, B.D. Effect of dietary saturated and monounsaturated fatty acids in juvenile barramundi *Lates calcarifer*. *Aquac. Nutr.* **2017**, *23*, 264–275. [CrossRef]
39. Glencross, B.D.; Hawkins, W.E.; Curnow, J.G. Restoration of the fatty acid composition of red seabream (*Pagrus auratus*) using a fish oil finishing diet after grow-out on plant oil based diets. *Aquac. Nutr.* **2003**, *9*, 409–418. [CrossRef]
40. Betancor, M.B.; Howarth, F.J.E.; Glencross, B.D.; Tocher, D.R. Influence of dietary docosahexaenoic acid in combination with other long-chain polyunsaturated fatty acids on expression of biosynthesis genes and phospholipid fatty acid compositions in tissues of post-smolt Atlantic salmon (*Salmo salar*). *Comp. Biochem. Physiol. Part B Biochem. Mol. Biol.* **2014**, *172–173*, 74–89. [CrossRef]
41. Henderson, R.J. Fatty acid metabolism in freshwater fish with particular reference to polyunsaturated fatty acids. *Arch. Anim. Nutr.* **1996**, *49*, 5–22. [CrossRef] [PubMed]
42. Duplus, E.; Glorian, M.; Forest, C. Fatty acid regulation of gene transcription. *J. Biol. Chem.* **2000**, *275*, 30749–30752. [CrossRef] [PubMed]
43. Shikata, T.; Iwanaga, S.; Shimeno, S. Metabolic Response of Acclimation Temperature in Carp. *Fish. Sci.* **1995**, *61*, 512–516. [CrossRef]

44. Menoyo, D.; Izquierdo, M.S.; Robaina, L.; Ginés, R.; Lopez-Bote, C.J.; Bautista, J.M. Adaptation of lipid metabolism, tissue composition and flesh quality in gilthead sea bream (*Sparus aurata*) to the replacement of dietary fish oil by linseed and soyabean oils. *Br. J. Nutr.* **2004**, *92*, 41–52. [CrossRef]

45. Alvarez, M.J.; Díez, A.; López-Bote, C.; Gallego, M.; Bautista, J.M. Short-term modulation of lipogenesis by macronutrients in rainbow trout (*Oncorhynchus mykiss*) hepatocytes. *Br. J. Nutr.* **2000**, *84*, 619–628. [CrossRef] [PubMed]

46. Regost, C.; Arzel, J.; Robin, J.; Rosenlund, G.; Kaushik, S.J. Total replacement of fish oil by soybean or linseed oil with a return to fish oil in turbot (*Psetta maxima*) 1. Growth performance, flesh fatty acid profile, and lipid metabolism. *Aquaculture* **2003**, *217*, 465–482. [CrossRef]

47. Ferreira, P.M.F.; Caldas, D.W.; Salaro, A.L.; Sartori, S.S.R.; Oliveira, J.M.; Cardoso, A.J.S.; Zuanon, J.A.S. Intestinal and liver morphometry of the Yellow Tail Tetra (*Astyanax altiparanae*) fed with oregano oil. *An. Acad. Bras. Cienc.* **2016**, *88*, 911–922. [CrossRef] [PubMed]

48. Durkin, L.A.; Childs, C.E.; Calder, P.C. Omega-3 polyunsaturated fatty acids and the intestinal epithelium—A review. *Foods* **2021**, *10*, 99. [CrossRef] [PubMed]

49. Xiao, F.; Xing, J.; Li, H.; Xu, X.; Hu, Z.; Ji, H. Effects of the defatted *Schizochytrium* sp. on growth performance, fatty acid composition, histomorphology and antioxidant status of juvenile mirror carp (*Cyprinus carpio* var. *specularis*). *Aquac. Res.* **2021**, *52*, 3062–3076. [CrossRef]

50. Shah, M.R.; Lutzu, G.A.; Alam, A.; Sarker, P.; Kabir Chowdhury, M.A.; Parsaeimehr, A.; Liang, Y.; Daroch, M. Microalgae in aquafeeds for a sustainable aquaculture industry. *J. Appl. Phycol.* **2018**, *30*, 197–213. [CrossRef]

51. Gupta, A.; Barrow, C.J.; Puri, M. Omega-3 biotechnology: Thraustochytrids as a novel source of omega-3 oils. *Biotechnol. Adv.* **2012**, *30*, 1733–1745. [CrossRef]

52. Kousoulaki, K.; Østbye, T.K.K.; Krasnov, A.; Torgersen, J.S.; Mørkøre, T.; Sweetman, J. Metabolism, health and fillet nutritional quality in Atlantic salmon (*Salmo salar*) fed diets containing n-3-rich microalgae. *J. Nutr. Sci.* **2015**, *4*, 1–13. [CrossRef]

53. Subramanian, S.; MacKinnon, S.L.; Ross, N.W. A comparative study on innate immune parameters in the epidermal mucus of various fish species. *Comp. Biochem. Physiol. B Biochem. Mol. Biol.* **2007**, *148*, 256–263. [CrossRef]

54. You, C.; Chen, B.; Zhang, M.; Shao, Y.; Wang, S.; Chen, C.; Lin, L.; Huang, Y.; Zhou, M.; Dong, Y.; et al. Evaluation of different dietary n-3 lc-pufa on the growth, intestinal health and microbiota profile of golden pompano (*Trachinotus ovatus*). *Aquac. Nutr.* **2021**, *27*, 953–965. [CrossRef]

55. Maranduba, C.M.D.C.; De Castro, S.B.R.; De Souza, G.T.; Rossato, C.; Da Guia, F.C.; Valente, M.A.S.; Rettore, J.V.P.; Maranduba, C.P.; De Souza, C.M.; Do Carmo, A.M.R.; et al. Intestinal microbiota as modulators of the immune system and neuroimmune system: Impact on the host health and homeostasis. *J. Immunol. Res.* **2015**, *2015*, 931574. [CrossRef]

56. Ruyter, B.; Moya-Falcón, C.; Rosenlund, G.; Vegusdal, A. Fat content and morphology of liver and intestine of Atlantic salmon (*Salmo salar*): Effects of temperature and dietary soybean oil. *Aquaculture* **2006**, *252*, 441–452. [CrossRef]

57. Bibbo, M.; Wilbur, D. *Comprehensive Cytopathology*; Elsevier: Amsterdam, The Netherlands, 2008.

58. Brraich, O.S.; Kaur, M. Histopathological alterations in the kidneys of *Labeo rohita* due to lead toxicity. *J. Environ. Biol.* **2017**, *38*, 257–262. [CrossRef]

59. Li, L.; Xie, P.; Guo, L.; Ke, Z.; Zhou, Q.; Liu, Y.; Qiu, T. Field and laboratory studies on pathological and biochemical characterization of microcystin-induced liver and kidney damage in the phytoplanktivorous bighead carp. *Sci. World J.* **2008**, *8*, 121–137. [CrossRef]

60. Råbergh, C.M.I.; Bylund, G.; Eriksson, J.E. Histopathological effects of microcystin-LR, a cyclic peptide toxin from the cyanobacterium (blue-green alga) Microcystis aeruginosa on common carp (*Cyprinus carpio* L.). *Aquat. Toxicol.* **1991**, *20*, 131–145. [CrossRef]

61. Ibarz, A.; Padrós, F.; Gallardo, M.Á.; Fernández-Borràs, J.; Blasco, J.; Tort, L. Low-temperature challenges to gilthead sea bream culture: Review of cold-induced alterations and "Winter Syndrome". *Rev. Fish Biol. Fish.* **2010**, *20*, 539–556. [CrossRef]

62. Ghasemi Fard, S.; Wang, F.; Sinclair, A.J.; Elliott, G.; Turchini, G.M. How does high DHA fish oil affect health? A systematic review of evidence. *Crit. Rev. Food Sci. Nutr.* **2019**, *59*, 1684–1727. [CrossRef] [PubMed]

Article

Effects of Single or Combined Administration of Dietary Synbiotic and Sodium Propionate on Humoral Immunity and Oxidative Defense, Digestive Enzymes and Growth Performances of African Cichlid (*Labidochromis lividus*) Challenged with *Aeromonas hydrophila*

Omid Safari [1,*], Mehrdad Sarkheil [1], Davar Shahsavani [2] and Marina Paolucci [3]

[1] Department of Fishery, Faculty of Natural Resources and Environment, Ferdowsi University of Mashhad, Mashhad 91773-1363, Iran; sarkheil@um.ac.ir

[2] Department of Food Hygiene and Aquaculture, School of Veterinary Medicine, Ferdowsi University of Mashhad, Mashhad 91773-1363, Iran; davar@um.ac.ir

[3] Department of Sciences and Technologies, University of Sannio, 82100 Benevento, Italy; paolucci@unisannio.it

* Correspondence: omidsafari@um.ac.ir; Tel.: +98-513-880-5466

Citation: Safari, O.; Sarkheil, M.; Shahsavani, D.; Paolucci, M. Effects of Single or Combined Administration of Dietary Synbiotic and Sodium Propionate on Humoral Immunity and Oxidative Defense, Digestive Enzymes and Growth Performances of African Cichlid (*Labidochromis lividus*) Challenged with *Aeromonas hydrophila*. Fishes **2021**, 6, 63. https://doi.org/10.3390/fishes6040063

Academic Editors: Maria Angeles Esteban, Bernardo Baldisserotto and Eric Hallerman

Received: 23 October 2021
Accepted: 10 November 2021
Published: 15 November 2021

Publisher's Note: MDPI stays neutral with regard to jurisdictional claims in published maps and institutional affiliations.

Abstract: The aim of the present study was to investigate the potential effects of dietary synbiotic (SYN) (*Pediococcus acidilactici* + Galactooligosaccharides; 10 g kg^{-1}), sodium propionate (SP; 5, 10 and 20 g kg^{-1}) and a combination of SYN + SP on the growth performance, humoral immunity, antioxidant responses and disease resistance against *Aeromonas hydrophila* of African cichlid (*Labidochromis lividus*) fingerlings (0.52 ± 0.05 g) in a feeding trial lasting 63 days. A completely randomized design was run with eight treatments, including 0 (control) and supplemented diets containing SYN + SP (e.g., 10 + 5, 10 + 10, 10 + 20, 0 + 5, 0 + 10, 0 + 20 and 10 + 10). The lowest feed conversion ratio value was observed in fish fed the 5 g kg^{-1}-SP and 10 g kg^{-1}-SYN ($p < 0.05$). The highest values of protein efficiency ratio and protein productive value were recorded in fish fed the 10 g kg^{-1}-SYN ($p < 0.05$). Fish fed the 10 g kg^{-1}-SYN diet had the highest activities of immunity (lysozyme, immunoglobulin) and antioxidant responses (glutathione peroxidase and superoxide dismutase) ($p < 0.05$). After 28 days post-challenge, the highest survival rate (57%) was recorded in the diet containing 10 g kg^{-1} SYN and 5 g kg^{-1} SP. The results indicated that the single administration of SYN or combined with SP, especially at the level of 5 g kg^{-1} of diet, enhanced the survival and growth performances, humoral immune response, antioxidant and digestive enzymes of African cichlid.

Keywords: synbiotic; acidifier; organic salt; humoral immune response; antioxidant enzymes; digestive enzymes; disease resistance

1. Introduction

Nowadays, the rearing techniques of ornamental fish are highly developed in the aquaculture industry. The ornamental fish industry fulfills approximately 90% of the freshwater traded organisms worldwide [1]. Because of the economic importance of aquarium fish, improving the health conditions, welfare and disease resistance of cultured fish are key factors in achieving sustainable production. During the last decades, the administration of environment-friendly immunostimulants (probiotics, prebiotics, parabiotics, synbiotics, organic salts and phyto-products), known as feed additives, has been considered to promote the growth indices and reduce microbial infections in aquatic animals [2–6].

Synbiotics, the combined form of probiotics and prebiotics, provides beneficial effects to the host by improving growth performance, digestive enzyme activities, disease resistance and increasing immune responses of aquatic animals [7–9]. The mucosal immunity of angelfish, *Pterophyllum scalare*, was improved via feeding with Artemia and synbiotics (*P. acidilactici* and fructooligosaccharide) [10]. Safari et al. [11] also showed that mucus immune responses and bactericidal activity in crayfish (*Astacus leptodactylus leptodactylus*) fed synbiotics (*P. acidilactici* + mannanoligosaccharide) were higher than those fed single probiotic and prebiotic. The function of synbiotics is highly related to the probiotic species matched with a special prebiotic. In this regard, the degree of polymerization of a prebiotic substrate with a special probiotic species and the production of major by-products of the fermentation process can affect the efficiency of synbiotic-supplemented diets [4].

Probiotics are defined as live microorganisms, which improve the health and immune response through balancing intestinal flora of host animals [12,13]. Recently, Gram-positive bacteria (lactic acid bacteria (LAB) and *Bacillus* sp.), Gram-negative bacteria (Aeromonas, Pseudomonas and *Vibrio* sp.) and yeast were investigated extensively in aquafeeds [9,14]. *Pediococcus acidilactici* is a Gram-positive bacterium, and its beneficial effects have been reported on growth indices of aquaculture species [15–17]. Dietary probiotic *P. acidilactici* was able to modulate the gut microbiota and up-regulate mucosal antibody immunoglobulin T in rainbow trout (*Oncorhynchus mykiss*) [18]. Prebiotics, non-digestible feed ingredients, beneficially affect the host causing microbial changes in the gastrointestinal tract, with subsequent physiological cascading processes [19]. Galactooligosaccharide (GOS) is an oligosaccharide that is mainly composed of galactose and glucose molecules [20]. The profitable effects of GOS as the promising prebiotic on growth indices have been reported in different fish species [20–22]. It was shown that dietary GOS at a level of 2% increased lactic acid bacteria (LAB) levels in the intestine of Caspian white (*Rutilus frisii kutum*) and Caspian roach (*Rutilus caspicus*) fingerlings after 6 weeks [23].

Short-chain organic acids, known as acidifiers, are regarded as one of the by-products of the fermentation process in the digestive tract. Recently, several studies have reported the effects of dietary organic acids (e.g., acetate, butyrate, lactate, propionate) and their salts on growth indices of aquatic animals [24,25]. Dietary sodium propionate enhanced mucosal immune responses and glutathione peroxidase (GPX) gene expression as an antioxidant enzyme in the liver of common carp, *Cyprinus carpio* [11]. Silva et al. [26] reported that the supplementation of propionate salt to a commercial shrimp diet at 2 g kg^{-1} significantly enhanced feed intake. The efficiency of dietary organic salts supplementation on aquatic species depends on the type of organic acid, dose, diet production method, nutrition history and ontogeny stage [24]. To the best of our knowledge, there is no literature on the simultaneous administration of synbiotics and acidifiers in aquafeeds. Therefore, the aim of the present study was to test the potential effects of different levels of dietary sodium propionate as an acidifier, synbiotic (*P. acidilactici* + galactooligosaccharides) and their combinations on the humoral immune status, serum antioxidant enzymes and digestive enzymes activities and growth performance, as well as intestinal microbiota of African cichlid (*Labidochromis lividus*) fingerlings.

2. Materials and Methods

2.1. Experimental Diets

The synbiotic (SYN) applied in this study was prepared by using *Pediococcus acidilactici* (Bactocell®,Lallemand Inc., Montreal, QC, Canada; 7.59 log CFU g^{-1}) as a probiotic and galactooligosaccharides (GOS) as a prebiotic. Sodium propionate (SP) ($C_3H_5NaO_2$) as an acidifier was purchased from Sigma-Aldrich Chemical Co. (USA).

Eight experimental diets were prepared by adding different concentrations of SYN and SP to the basal diet (Table 1). The control (basal) diet was produced with a twin-screw extruder (AquaSadra Co. Mashhad, Iran) with preconditioning (30 °C) and three temperature zones (60, 90 and 130 °C). The experimental diets were as follows: (1) control, basal diet without SYN and SP; (2) (10 + 5) SYN (10 g kg^{-1}) + SP (5 g kg^{-1}); (3) (10 + 10) SYN (10 g kg^{-1}) + SP (10 g kg^{-1}); (4) (10 + 20) SYN (10 g kg^{-1}) + SP (20 g kg^{-1}); (5) (0 + 5) SP (5 g kg^{-1}); (6) (0 + 10) SP (10 g kg^{-1}); (7) (0 + 20) SP (20 g kg^{-1}); (8) (10 + 0) SYN

(10 g kg^{-1}). To produce the above-mentioned test diets, all feed additives (SYN, GOS and SP) were replaced with filler existing in the control diet.

Table 1. The feed ingredients and chemical composition of the control diet-fed juvenile African cichlid (*Labidochromis lividus*).

Feed Ingredient	Content (g kg^{-1})
Fishmeal [1]	350
Spirulina meal [2]	20
Soybean meal [1]	73
Corn gluten [1]	100
Wheat flour [1]	240
Fish oil [1]	70
Canola oil [1]	70
Soy lecithin [1]	5
Choline chloride (70%) [3]	4
Vitamin C (stay) [3]	5
Vitamin premix [3,*]	15
Mineral premix [3,*]	15
Antifungus [1]	3
Filler (Carboxymethyl cellulose) [4]	30
Chemical composition (g kg^{-1})	
Dry matter	899.39
Crude protein	421
Crude fat	282
Crude fiber	65
Nitrogen free extract	131.39
Gross energy (Mj kg^{-1})	15.58

[1] Saramad Fish Aquafeed Co, Iran; [2] ACECR, Iran; [3] Kimia Roshd Co. Iran; [4] Sigma, Germany. * Mineral premix contains (mg Kg^{-1}) Mg, 100; Zn, 60; Fe, 40; Cu, 5; Co, 0.1; I, 0.1; Antioxidant, 100. * Vitamin premix contains (mg Kg^{-1}) E, 30; K, 3; Thiamine, 2; Riboflavin, 7; Pyridoxine, 3; Pantothenic acid, 18; Niacin, 40; Folacin, 1.5; Choline, 600; Biotin, 0.7 and Cyanocobalamin, 0.02.

2.2. Feeding Trial

Five hundred healthy African cichlid (*Labidochromis lividus*) fingerlings were purchased from a local ornamental fish farm. The fish were acclimatized in four glass aquariums (200 L) for two weeks. Afterward, fish (0.52 ± 0.05 g) were randomly distributed into 24 glass aquariums (20 fish per 150-L aquarium). The fish were fed ad libitum on the experimental diets three times a day (8:00, 12:00 and 16:00) for 63 days. During the feeding trial, the water in each aquarium was renewed (20% per day). Water temperature, dissolved oxygen and pH were monitored daily and maintained at 27.5 ± 1 °C, 6.4 ± 0.3 mg L^{-1} and 8.2 ± 0.4, respectively [27,28]. The feeding experiment was carried out in triplicate. All experiments were done according to FUM animal ethics.

2.3. Growth Indices

After the feeding trial period, the weight and length of each fish were individually measured. Growth indices were calculated based on the standard formulas as follows:

Weight gain: $W_f - W_i$

Specific growth rate (SGR; % Body weight day^{-1}) = $[(LnW_f - LnW_i)/T] \times 100$

Condition factor (CF) = $W_f/L_f^3 \times 100$

Feed conversion ratio (FCR) = (Feed consumed/W_{gain})

Protein efficiency ratio (PER) = $(W_f - W_i)/$Crude protein intake

Protein production value % (PPV) = (Whole-body protein gain)/Protein consumption) $\times 100$

Survival rate (%) = $(N_f/N_i) \times 100$

Where:

W$_i$: Initial weight; W$_f$: Final weight; L$_f$: Final length; N$_i$: Initial number of fish; N$_f$: Final number of fish; T: Time period (day).

2.4. Immunological and Antioxidant Parameter Analyses

To analyze the serum immune parameters, at the end of the feeding trial, the fish, starved for 24 h (*n*= nine per treatment), were randomly sampled from each aquarium at the end of the feeding trial. The serum samples were prepared based on the method described in Safari and Sarkheil [6]. The total immunoglobulins were measured according to Siwicki and Anderson [29]. Briefly, immunoglobulins were precipitated with polyethylene glycol (12%), incubated (25 °C for 2 h) under constant shaking, and finally centrifuged (3000× *g* for 15 min). Then, the supernatant was removed, and the precipitated protein content was measured. Immunoglobulins were calculated by subtracting the precipitated proteins from the total proteins in the serum.

Lysozyme (LYZ) activity in the serum was estimated by determining the level of lysis of the lysozyme-sensitive Gram-positive bacterium, *Micrococcus lysodeikticus* (Sigma), according to the procedure described by Kumari et al. [30]. Briefly, 15 μL of the serum sample was poured into a plate (96 well) in triplicate. Thereafter, *M. lysodeikticus* suspension (150 μL) was prepared using 0.02 M sodium acetate buffer with pH 5.8 (0.02 mg L^{-1}) and transferred to each well. The absorbance was measured at a wavelength of 450 nm using the spectrophotometer (HACH DR/5000, Hach Co., Colorado, USA) at a one-hour interval, and the difference was calculated. The LYZ activity was expressed as U mL^{-1}.

Alternative hemolytic complement activity (ACH50) was evaluated as reported in Yano [31]. Briefly, the serum samples were diluted in ethylene glycol tetra acetic acid–magnesium–gelatin veronal buffer (EGTA–Mg–GVB) to a volume of 250 μL in test tubes. Following, 100 mL of rabbit red blood cells (RaRBC) were dispersed into each tube and incubated at 20 °C for 90 min. Then, NaCl (3.15 mL) was added to each test tube and centrifuged (1600× *g* for 5 min). The optical density (OD) of the obtained supernatant was read at a wavelength of 414 nm. The ACH50 activity (U mL^{-1}) was calculated based on the volume of serum, which causes the 50% lysis of the RaRBC. Superoxide dismutase (SOD) and glutathione peroxidase (GPX) enzymes were measured using commercial kits (Cusabio Biotech Co., Ltd.; Wuhan, China) following the manufacturer's instructions.

2.5. Digestive Enzyme Analyses

At the end of the feeding trial period, three fish starved for 24 h were randomly sampled from each aquarium. The fish were anesthetized (cloves extract, 500 mg L^{-1}), and the intestine was isolated and rinsed with cold distilled water (4 °C) [32]. The intestine was mixed with 0.2 M NaCl (1:5; *w*/*v*) [33] and homogenized on ice using a DI 18 Disperser homogenizer. Following, the samples were centrifuged (15,000× *g* for 15 min at 4 °C), and the supernatants were frozen (−80 °C). The enzyme activity was measured using a microplate scanning spectrophotometer (HACH DR/4000, USA) and expressed as U mg^{-1} protein min^{-1}.

The α-amylase activity was measured according to the 3, 5-dinitrosalicylic acid method, as reported in Worthington [34], and read at a wavelength of 540 nm. Protease activity was estimated using the casein-hydrolysis method described by Hidalgo et al. [34]. Briefly, 0.05 mL of supernatant was mixed with 0.125 mL of casein and 0.125 mL of buffer (0.1 M Tris-HCl, pH 9.0) and incubated (37 °C for 1 h). Afterward, the reaction was stopped by adding 0.3 mL of trichloroacetic acid (TCA) (8% *w*/*v*) solution. Then, the samples were centrifuged (1800× *g* for 10 min) after incubation at 4 °C for 1 h. In the end, the supernatant was measured at a wavelength of 280 nm. Lipase activity was determined according to the method reported by Gawlicka et al. [33], with 4-nitrophenylmyristate (0.4 mM) as a substrate at 25 °C. The absorbance was measured at a 405 nm wavelength.

2.6. Microbiota Assays

The microbial counts of total bacterial aerobic, lactobacillus, fungi and Escherichia coli were determined in the digestive tract of the fish. For this purpose, the fish from each aquarium (n = nine per treatment) were randomly sampled, anesthetized with ice and washed (with benzalkolium chloride 0.1% for 60 min). Thereafter, the digestive tract was removed and homogenized in the presence of NaCl (0.9 w/v) using a homogenizer (DI 18 Disperser) [34,35]. The homogenate was centrifuged at $5000 \times g$, 4 °C for five min. Then, a 100 µL aliquot of each prepared sample was plated onto a plate count agar (Merck, Darmstadt, Germany), plate de Man, Rogosa and Sharpe media (Merck, Darmstadt, Germany), Potato Dextrose Agar (Merck, Darmstadt, Germany) and Mac Conkey Agar (Merck, Darmstadt, Germany) to determine total aerobic bacterial count, lactobacillus count, fungi count and *E. coli* count, respectively. Finally, the plates were incubated (25 °C for 5 days), and those containing 30–300 colonies were used for bacterial counting as colony-forming units per gram (CFU g^{-1}) [4,5].

2.7. Proximate Analysis

The analysis of dry matter (oven drying, 105 °C), crude protein ($n \times 6.25$, Kjeldahl system: Buchi Labortechnik AG, Flawil, Switzerland), crude fat (Soxtec System HT 1043: Foss Tecator, AB, USA), ash (muffle furnace, 550 °C), gross energy (Parr bomb calorimetry model 1266, Parr Instrument Co., Moline, IL, USA) and crude fiber (after digestion with H_2SO_4 and NaOH) analysis of feedstuffs, diets and carcasses were performed according to standard methods [36]. Nitrogen-free extract (NFE) was calculated by subtracting dry matter minus crude protein, crude fat, crude fiber and ash contents.

2.8. Bacterial Exposure Challenge

After the feeding trial, 12 fish from each tank (36 fish per group) were selected and intraperitoneally injected with 100 µL of phosphate-buffered saline solution (PBS) containing 1×10^7 live *Aeromonas hydrophila*, ATCC 49040. The injected fishes were fed with a control diet. The number of dead fish was recorded daily for 28 days.

2.9. Statistical Analysis

Data were presented as mean $\pm$ standard deviation (SD). The data were analyzed using SPSS software (Version, 19). The normality assumption of the data was tested using Kolmogorov–Smirnov test. Significant differences between the means were determined by one-way analysis of variance (ANOVA) followed by Tukey's test. The significant difference was accepted at a level of $p < 0.05$.

3. Results

3.1. Growth Performance and Survival Rate

The growth performance, feed utilization parameters and survival rate of African cichlid (*L. lividus*) after the feeding trial period of 63 days are shown in Table 2. The final weight and weight gain of fish fed the supplemented diets increased significantly compared to the control group, except for the groups fed 10 and 20 g kg^{-1}-SP ($p < 0.05$). The specific growth rate % (SGR) was higher in the supplemented dietary groups with respect to the control group, except for the 10 and 20 g kg^{-1}-SP dietary groups ($p < 0.05$). Fish fed the supplemented diets had a lower food conversion ratio (FCR) than the control, except for the 10 and 20 g kg^{-1}-SP dietary groups ($p < 0.05$). The lowest FCR value was observed in the 5 g kg^{-1}-SP and 10 g kg^{-1}-SYN dietary groups ($p < 0.05$). There was no significant difference between the condition factor % (CF) of fish fed the experimental diets ($p > 0.05$). The protein efficiency ratio (PER) value was significantly higher in the fish fed supplemented diets than the control, except for the 20 g kg^{-1}-SP dietary group ($p < 0.05$). The highest PER value was recorded in the 10 g kg^{-1}-SYN dietary group ($p < 0.05$). Fish fed the supplemented diets showed a higher protein production value % (PPV) than the control ($p < 0.05$). Fish fed the 10 g kg^{-1}-SYN diet had the highest PPV % ($p < 0.05$). The survival rate (%) of fish fed the supplemented diets was higher than the control, except for the 10 and 20 g kg^{-1}-SP dietary groups ($p < 0.05$). As shown in Table 3, the highest crude protein content of carcasses was measured in the fish fed with the diet containing 10 g kg^{-1} SYN and 0 g kg^{-1} SP ($p < 0.05$). The highest crude fat content of carcasses was observed in the fish fed the control and 0 g kg^{-1} SYN and 20 g kg^{-1} SP diets.

3.2. Immunological Assay

The variations of hemato-immunological indices in the serum of fish fed the experimental diets are shown in Figure 1. The lysozyme (LYZ) activity increased significantly in the fish fed supplemented diets compared to the control group ($p < 0.05$). The total immunoglobulin (Ig) level in all supplemented dietary groups was significantly higher than the control, except for the 20 g kg^{-1}-SP dietary group ($p < 0.05$). Fish fed the 10 g kg^{-1}-SYN diet had the highest LYZ and Ig activity ($p < 0.05$). The alternative hemolytic complement activity (ACH50) decreased significantly in fish fed the supplemented diets compared to the control ($p < 0.05$). The lowest ACH50 activity was observed in the 10 g kg^{-1}-SYN dietary group ($p < 0.05$).

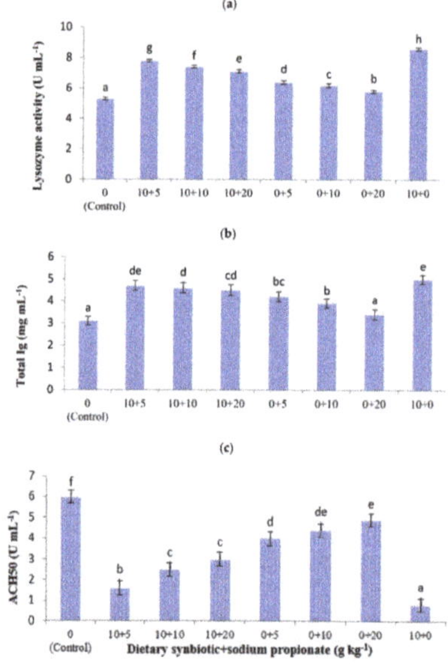

Figure 1. Lysozyme (U mL^{-1}) (**a**), total immunoglobulin (mg mL^{-1}) (**b**) and alternative hemolytic complement activity (ACH50) (U mL^{-1}) (**c**) activities in the serum of African cichlid (*L. lividus*) fed diets supplemented with different levels of synbiotic (0 and 10 g kg^{-1}) and sodium propionate (0, 5, 10 and 20 g kg^{-1}) for 63 days. Bars with different letters are significantly different (mean $\pm$ SD; ANOVA, $p < 0.05$; $n = 3$). 10 + 5 = SYN (10 g kg^{-1}) + SP (5 g kg^{-1}); 10 + 10 = SYN (10 g kg^{-1}) + SP (10 g kg^{-1}); 10 + 20 = SYN (10 g kg^{-1}) + SP (20 g kg^{-1}); 0 + 5 = SP (5 g kg^{-1}); 0 + 10 = SP (10 g kg^{-1}); 0 + 20 = SP (20 g kg^{-1}); 10 + 0 = SYN (10 g kg^{-1}).

3.3. Antioxidant Enzyme Assay

Figure 2 shows the activity of the antioxidant enzymes in the serum of African cichlid fed diets supplemented with SYN and SP. The glutathione peroxidase (GPx) and superoxide dismutase (SOD) activity levels increased significantly in the serum of fish fed the supplemented diets compared to the control ($p < 0.05$). The highest GPx and SOD levels were observed in the 10 mg kg^{-1}-SYN dietary group ($p < 0.05$).

Table 2. Growth performance, feed utilization parameters and survival rate of African cichlid (*L. lividus*) fed diets supplemented with different levels of synbiotic and sodium propionate for 63 days (mean $\pm$ SD, $n = 3$).

	Dietary Synbiotic + Sodium Propionate Level (g kg^{-1})							
	0 (Control)	10 + 5	10 + 10	10 + 20	0 + 5	0 + 10	0 + 20	10 + 0
Initial weight (g)	0.65 $\pm$ 0.05	0.58 $\pm$ 0.04	0.67 $\pm$ 0.11	0.69 $\pm$ 0.05	0.65 $\pm$ 0.03	0.7 $\pm$ 0.02	0.68 $\pm$ 0.03	0.65 $\pm$ 0.09
Final weight (g)	3.49 $\pm$ 0.09 [a]	3.83 $\pm$ 0.09 [b]	3.81 $\pm$ 0.13 [b]	3.82 $\pm$ 0.07 [b]	3.60 $\pm$ 0.52 [b]	3.59 $\pm$ 0.06 [a]	3.63 $\pm$ 0.07 [a]	3.86 $\pm$ 0.04 [b]
Weight gain (g)	2.83 $\pm$ 0.058 [a]	3.25 $\pm$ 0.13 [b]	3.14 $\pm$ 0.026 [b]	3.13 $\pm$ 0.025 [b]	2.95 $\pm$ 0.50 [b]	2.89 $\pm$ 0.055 [a]	2.94 $\pm$ 0.10 [a]	3.20 $\pm$ 0.11 [b]
(% BW day^{-1})	2.65 $\pm$ 0.10 [a]	2.99 $\pm$ 0.16 [d]	2.77 $\pm$ 0.22 [bcd]	2.71 $\pm$ 1.02 [bc]	2.69 $\pm$ 0.19 [cd]	2.59 $\pm$ 0.03 [a]	2.65 $\pm$ 0.09 [ab]	2.82 $\pm$ 0.22 [cd]
Feed conversion ratio	1.76 $\pm$ 0.03 [c]	1.46 $\pm$ 0.06 [b]	1.43 $\pm$ 0.01 [b]	1.43 $\pm$ 0.02 [b]	1.16 $\pm$ 0.21 [a]	1.73 $\pm$ 0.03 [c]	1.59 $\pm$ 0.05 [c]	1.24 $\pm$ 0.04 [a]
Condition factor	1.26 $\pm$ 0.01	1.43 $\pm$ 0.07	1.62 $\pm$ 0.09	1.49 $\pm$ 0.07	1.26 $\pm$ 0.1	1.26 $\pm$ 0.04	1.47 $\pm$ 0.04	1.49 $\pm$ 0.3
PER	1.3 $\pm$ 0.14 [a]	2.4 $\pm$ 0.15 [e]	2.2 $\pm$ 0.16 [de]	2 $\pm$ 0.14 [cd]	2 $\pm$ 0.15 [bc]	1.7 $\pm$ 0.13 [b]	1.4 $\pm$ 0.15 [a]	2.8 $\pm$ 0.14 [f]
PPV	47 $\pm$ 0.14 [a]	59 $\pm$ 0.16 [f]	55 $\pm$ 0.17 [e]	53 $\pm$ 0.2 [d]	53 $\pm$ 0.18 [d]	52 $\pm$ 0.19 [c]	49 $\pm$ 0.19 [b]	69 $\pm$ 0.14 [g]
Survival rate (%)	83.33 $\pm$ 3.60 [a]	97.91 $\pm$ 3.60 [b]	95.83 $\pm$ 7.21 [b]	93.75 $\pm$ 6.25 [b]	97.91 $\pm$ 3.60 [b]	91.66 $\pm$ 9.54 [ab]	81.25 $\pm$ 12.5 [a]	100 $\pm$ 0.00 [b]

10 + 5 = SYN (10 g kg^{-1}) + SP (5 g kg^{-1}); 10 + 10 = SYN (10 g kg^{-1}) + SP (10 g kg^{-1}); 10 + 20 = SYN (10 g kg^{-1}) + SP (20 g kg^{-1}); 0 + 5 = SP (5 g kg^{-1}); 0 + 10 = SP (10 g kg^{-1}); 0 + 20 = SP (20 g kg^{-1}); 10 + 0 = SYN (10 g kg^{-1}). The values with different letters in the same row are significantly different (ANOVA, $p < 0.05$).

Table 3. Carcass proximate compositions of African cichlid (*L. lividus*) fed diets supplemented with different levels of synbiotic (0 and 10 g kg^{-1}) and sodium propionate (0, 5, 10 and 20 g kg^{-1}) for 63 days (mean $\pm$ SD, $n = 3$).

	Dietary Synbiotic + Sodium Propionate Level (g kg^{-1})							
	0 (Control)	10 + 5	10 + 10	10 + 20	0 + 5	0 + 10	0 + 20	10 + 0
Dry matter (%)	26.88 $\pm$ 0.76	26.63 $\pm$ 0.86	26.98 $\pm$ 0.47	27.08 $\pm$ 0.56	27.00 $\pm$ 0.36	26.78 $\pm$ 0.56	27.05 $\pm$ 0.66	27.03 $\pm$ 0.56
Crude protein (%)	16.07 $\pm$ 0.37 [a]	16.95 $\pm$ 0.32 [f]	16.84 $\pm$ 0.42 [e]	16.67 $\pm$ 0.23 [d]	16.69 $\pm$ 0.23 [d]	16.53 $\pm$ 0.23 [c]	16.27 $\pm$ 0.27 [b]	17.04 $\pm$ 0.22 [g]
Crude lipid (%)	5.44 $\pm$ 0.05 [e]	4.12 $\pm$ 0.08 [a]	4.60 $\pm$ 0.07 [c]	4.93 $\pm$ 0.13 [d]	4.92 $\pm$ 0.10 [d]	4.89 $\pm$ 0.09 [d]	5.43 $\pm$ 0.09 [e]	4.42 $\pm$ 0.11 [b]
Ash (%)	3.16 $\pm$ 0.15 [a]	3.38 $\pm$ 0.32 [c]	3.41 $\pm$ 0.29 [c]	3.39 $\pm$ 0.20 [c]	3.18 $\pm$ 0.21 [a]	3.19 $\pm$ 0.23 [a]	3.25 $\pm$ 0.35 [b]	3.43 $\pm$ 0.22 [c]

10 + 5 = SYN (10 g kg^{-1}) + SP (5 g kg^{-1}); 10 + 10 = SYN (10 g kg^{-1}) + SP (10 g kg^{-1}); 10 + 20 = SYN (10 g kg^{-1}) + SP (20 g kg^{-1}); 0 + 5 = SP (5 g kg^{-1}); 0 + 10 = SP (10 g kg^{-1}); 0 + 20 = SP (20 g kg^{-1}); 10 + 0 = SYN (10 g kg^{-1}). The values with different letters in the same row are significantly different (ANOVA, $p < 0.05$).

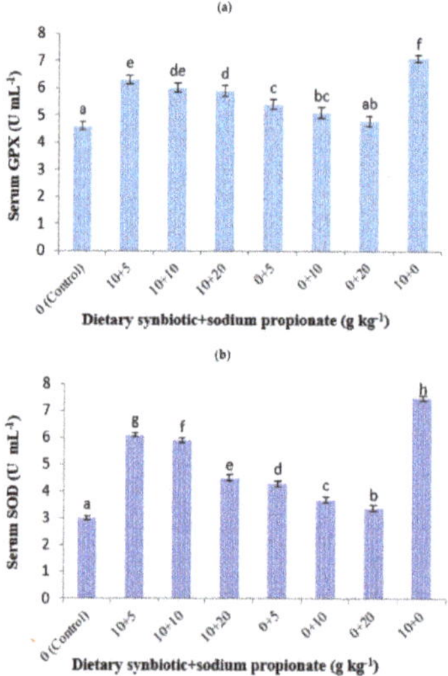

Figure 2. Glutathione peroxidase (GPx) (U mL^{-1}) (**a**) and superoxide dismutase (SOD) (U mL^{-1}) (**b**) activities in serum of African cichlid (*L. lividus*) fed diets supplemented with different levels of synbiotic (0 and 10 g kg^{-1}) and sodium propionate (0, 5, 10 and 20 g kg^{-1}) for 63 days. Bars with different letters are significantly different (mean $\pm$ SD; ANOVA, $p < 0.05$; $n = 3$). 10 + 5 = SYN (10 g kg^{-1}) + SP (5 g kg^{-1}); 10 + 10 = SYN (10 g kg^{-1}) + SP (10 g kg^{-1}); 10 + 20 = SYN (10 g kg^{-1}) + SP (20 g kg^{-1}); 0 + 5 = SP (5 g kg^{-1}); 0 + 10 = SP (10 g kg^{-1});0 + 20 = SP (20 g kg^{-1}); 10 + 0 = SYN (10 g kg^{-1}).

3.4. Digestive Enzymes Activity

The activity of the digestive enzymes in African cichlid fed experimental diets is shown in Table 4. The protease activity increased in all supplemented dietary groups compared to the control, except for the 20 g kg^{-1}-SP dietary group ($p < 0.05$). The α-amylase activity was significantly higher in fish fed the supplemented diets than in the control group ($p < 0.05$). Fish fed the supplemented diets showed a higher lipase activity than the control, except for the 20 g kg^{-1}-SP diet ($p < 0.05$). The highest digestive enzymes activity was observed in the 10 g kg^{-1}-SYN dietary group ($p < 0.05$).

3.5. Microbiota Assay

Total aerobic bacteria count (Log CFU g^{-1}) of experimental fish did not show significant differences and ranged from 6.20 to 6.48 (Table 5). The highest Lactobacillus count and the lowest *E. coli* count ($p < 0.05$) were observed in the fish fed the diet supplemented with 10 g kg^{-1} SYN and 5 g kg^{-1} SP (Table 3).

3.6. Pathogen Resistance

Single and combined administration of dietary synbiotic and sodium propionate levels enhanced the resistance of African cichlid to pathogen infection compared to those fed the control diet (Figure 3). After 28 days post-challenge, the highest survival rate (57%) was recorded in the diet containing 10 g kg^{-1} SYN and 5 g kg^{-1} SP (Figure 3).

Table 4. Digestive enzymes activity (U mg protein^{-1} min^{-1}) of African cichlid (*L. lividus*) fed diets supplemented with different levels of synbiotic (0 and 10 g kg^{-1}) and sodium propionate (0, 5, 10 and 20 g kg^{-1}) for 63 days (mean ± SD, $n = 3$).

	Dietary Synbiotic + Sodium Propionate Level (g kg^{-1})							
	0 (Control)	10 + 5	10 + 10	10 + 20	0 + 5	0 + 10	0 + 20	10 + 0
Protease	8.1 ± 0.14 [a]	10.2 ± 0.16 [d]	9.5 ± 0.17 [c]	9.4 ± 0.2 [c]	9.3 ± 0.18 [bc]	9 ± 0.19 [b]	8.4 ± 0.19 [a]	10.5 ± 0.16 [d]
α-amylase	3.9 ± 0.14 [a]	7 ± 0.15 [g]	6.3 ± 0.16 [f]	5.5 ± 0.14 [e]	5.2 ± 0.15 [d]	4.9 ± 0.13 [c]	4.5 ± 0.15 [b]	8.2 ± 0.15 [h]
Lipase	4.3 ± 0.19 [a]	6.4 ± 0.2 [e]	6.1 ± 0.22 [d]	6.4 ± 0.23 [cd]	6.1 ± 0.23 [bc]	5.9 ± 0.2 [b]	4.6 ± 0.23 [a]	8.4 ± 0.15 [f]

10 + 5 = SYN (10 g kg^{-1}) + SP (5 g kg^{-1}); 10 + 10 = SYN (10 g kg^{-1}) + SP (10 g kg^{-1}); 10 + 20 = SYN (10 g kg^{-1}) + SP (20 g kg^{-1}); 0 + 5 = SP (5 g kg^{-1}); 0 + 10 = SP (10 g kg^{-1}); 0 + 20 = SP (20 g kg^{-1}); 10 + 0 = SYN (10 g kg^{-1}). The values with different letters in the same row are significantly different (ANOVA, $p < 0.05$).

Table 5. Total aerobic bacteria count (TAB; Log CFU g^{-1}), lactobacillus count (LAB; Log CFU g^{-1}), *E. coli* count (Log CFU g^{-1}) and fungi count (Log CFU g^{-1}) of intestines extracted from African cichlid (*L. lividus*) fed diets supplemented with different levels of synbiotic (0 and 10 g kg^{-1}) and sodium propionate (0, 5, 10 and 20 g kg^{-1}) for 63 days (mean ± SD, $n = 3$).

Bacteria Count (Log CFU g^{-1})	Dietary Synbiotic + Sodium Propionate Level (g kg^{-1})							
	0 (Control)	10 + 5	10 + 10	10 + 20	0 + 5	0 + 10	0 + 20	10 + 0
Total aerobic	6.28 ± 1.09	6.20 ± 1.21	6.48 ± 1.30	6.38 ± 1.47	6.41 ± 1.31	6.47 ± 1.53	6.39 ± 1.52	6.48 ± 1.42
Lactobacillus	0.42 ± 1.13 [a]	3.31 ± 1.09 [e]	3.12 ± 1.27 [d]	2.93 ± 1.35 [c]	2.89 ± 1.07 [c]	1.67 ± 1.43 [b]	1.71 ± 1.29 [b]	3.09 ± 1.18 [d]
E. coli count	0.51 ± 0.32 [e]	0.25 ± 0.28 [a]	0.43 ± 0.31 [d]	0.37 ± 0.21 [c]	0.38 ± 0.18 [c]	0.31 ± 0.25 [b]	0.32 ± 0.26 [b]	0.42 ± 0.28 [d]
Fungi count	0.42 ± 0.43	0.39 ± 0.24	0.37 ± 0.21	0.38 ± 0.24	0.41 ± 0.29	0.43 ± 0.13	0.36 ± 0.19	0.36 ± 0.19

10 + 5 = SYN (10 g kg^{-1}) + SP (5 g kg^{-1}); 10 + 10 = SYN (10 g kg^{-1}) + SP (10 g kg^{-1}); 10 + 20 = SYN (10 g kg^{-1}) + SP (20 g kg^{-1}); 0 + 5 = SP (5 g kg^{-1}); 0 + 10 = SP (10 g kg^{-1});0 + 20 = SP (20 g kg^{-1}); 10 + 0 = SYN (10 g kg^{-1}). The values with different letters in the same row are significantly different (ANOVA, $p < 0.05$).

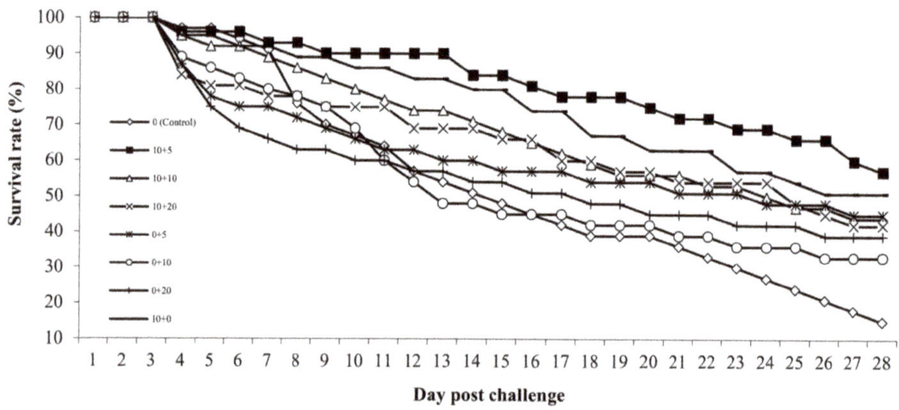

Figure 3. Survival rate (%) of African cichlid (*L. lividus*) injected with *Aeromonas hydrophila* during the 28 days post challenge. Fish were fed experimental diets with different levels of synbiotic (0 and 10 g kg^{-1}) and sodium propionate (0, 5, 10 and 20 g kg^{-1}) with three replicates. 10 + 5 = SYN (10 g kg^{-1}) + SP (5 g kg^{-1}); 10 + 10 = SYN (10 g kg^{-1}) + SP (10 g kg^{-1}); 10 + 20 = SYN (10 g kg^{-1}) + SP (20 g kg^{-1}); 0 + 5 = SP (5 g kg^{-1}); 0 + 10 = SP (10 g kg^{-1});0 + 20 = SP (20 g kg^{-1}); 10 + 0 = SYN (10 g kg^{-1}).

4. Discussion

The effects of a single administration of dietary probiotic *P. acidilactici*, GOS as prebiotic and SP on the growth performance, immune status, digestive enzyme activity and the intestinal microbiota of different fish and shellfish species have been reported [4,11,18,20,22]. Several studies have been conducted to improve the survival and growth performance of aquatic animals through the administration of safe and eco-friendly feed-supplements, such as synbiotics and acidifiers [5,10,37,38]. To the best of our knowledge, there is not enough data on the efficiency of the combination of synbiotics and acidifiers on the growth performance of fish. The results of the current study reveal that the single administration of SYN (*P. acidilactici* + GOS) or combined with different levels of SP promoted the survival rate (%), growth performance and feed utilization indices, including final weight, weight gain, SGR%, FCR and PER and PPV values of fish after 63 days. The enhancement of growth performance could be attributed to the stimulation of digestive enzyme activities in the gastrointestinal tract [39], resulting in the better absorption of different nutrients, such as proteins and lipids [40]. The single administration of SP at the level of 5 g kg^{-1} of diet improved the survival rate and growth performance of fish, which may be due to the higher activity of digestive enzymes in this SP-dietary group. The effects of the different levels (0%, 0.25%, 0.5%, 1% and 2%) of SP on the growth performance of Caspian white (*R. frisii kutum*) fry showed that the final weight, WG and SGR% were higher in fish fed 0.25% dietary SP [37]. Kuhlmann et al. [41] also found an increase of 19% in the yield of shrimp; *L. vannamei* fed a 0.5% KDF-supplemented diet. However, it is not clear how organic acids could enhance digestive enzyme activities, nutrient retention efficiency, and finally, the growth performance of aquatic species. Some researchers related these biological responses to reduce the pH value of the diets supplemented with organic acids, improve gut microbiota of digestive tract via increment in lactic acid bacteria count and the reduction of Gram-negative bacteria (e.g., *E. coli*) [24,25]. Nonetheless, the positive effects of the combination SYN and SP in the diet of African cichlid were confirmed in the present study. Further studies need to evaluate the associated physiological pathways.

The results indicated that the lysozyme (LYZ) activity and total immunoglobulin (Ig) level increased significantly in fish fed diets supplemented with SYN and different levels of SP, except the 20 g kg^{-1}-SP group. The highest increase in the Ig level and LYZ activity was observed in fish fed a diet supplemented with SYN (*P. acidilactici* + GOS). It was found that the LYZ activity increased in rainbow trout (*O. mykiss*) fed a *P. acidilactici*-supplemented diet (at 2.4×10^6 CFU g^{-1}) for 4 weeks [18]. It was reported that the total Ig was higher in zebrafish (*Danio rerio*) fed 1% or 2% GOS than the control group, while the LYZ activity showed no change compared to control after eight weeks [22]. Guerreiro et al. [20] found that dietary GOS (1%) incorporation had no significant effect on Ig level and LYZ activity in white sea bream (*Diplodus sargus*) juveniles. In the present study, the increase of total Ig and LYZ activity in African cichlid is probably related to the simultaneous administration of *P. acidilactici* and GOS. Rahimnejad et al. [42] also reported that the serum LYZ activity and ACH50 level increased in juvenile rockfish (*Sebastes schlegeli*) fed synbiotics (1% GOS and 6.3 log CFU g^{-1} *P. acidilactici*) for 8 weeks. In contrast, the finding of the present study showed that fish fed supplemented diets, especially SYN-supplemented diets, had lower ACH50 activity than the control group. This finding is in accordance with the finding of another study, which found that ACH50 decreased in White Sea bream (*D. sargus*) fed GOS compared to fish fed a control diet [20]. The serum Ig level and LYZ activity decrease paralleled the increase in SP level from 5 to 20 g kg^{-1} of diet, while the ACH50 increased with increasing SP of diets. Evaluation of the effects of different levels (0%, 0.25%, 0.5%, 1% and 2%) of dietary SP on Caspian white fish (*R. frisii kutum*) fry (2 g) humoral immune responses showed that lysozyme and ACH50 activities were higher in 0.25% and 0.5% treatments than other treatments [37]. In contrast, Safari et al. [11] found that the serum total Ig level and lysozyme activity increased with the elevation of the SP level from 0% to 2% in diets of common carp (*C. carpio*) (~25 g). These contradictory findings may be due to the difference in fish species and life stages [24]. However, further studies need to elucidate the cause of such different results.

The results of the present study revealed that GPx and SOD activities increased remarkably in 10 g kg^{-1}-SYN and 10 g kg^{-1}-SYN + 5 g kg^{-1}-SP groups. Similarly, dietary synbiotics (GOS+ *P. acidilactici*) increased liver antioxidant enzymes (CAT, GST and GR) activity in rainbow trout (*O. mykiss*) [43]. The plasma SOD activity was higher in juvenile rockfish (*S. schlegeli*) fed a diet supplemented with synbiotic (1% GOS and 6.3 log CFU g^{-1} *P. acidilactici*) than those fed *P. acidilactici* and GOS-supplemented diets [42]. On the contrary, the single administration of dietary GOS had no significant effect on antioxidant enzymatic activity, including SOD and GPx, in white sea bream (*D. sargus*) [20]. The results of another study also showed that SOD and GPx activities in erythrocyte hemolysate of common carp (*C. carpio*) were not affected by dietary probiotic *P. acidilactici* [44]. It has been reported that dietary SP increased the expression of the GPx gene in the liver of common carp (*C. carpio*) [11]. The findings of the current study also indicated the elevation of SOD and GPx activities in fish fed SP-supplemented diets compared to the control group. The inclusion of SP at the level of 5 g kg^{-1} diet had a more positive effect on the activity of these enzymes. To interpret the results, it is important to consider fish species, ontogeny stages (larvae, fry, adult and broodstock), diet regimes (herbivorous, omnivorous and carnivorous), nutritional history and basal diet formulation.

The elevated digestive enzyme activities have been reported in fish and shellfish fed probiotics, prebiotics and their combination as synbiotics [9,45]. In the present study, the maximum digestive enzymes activity, including protease, α-amylase and lipase, was observed in fish fed diets supplemented with SYN (*P. acidilactici* + GOS). It was shown that the digestive enzyme activities enhanced in zebra fish (*D. rerio*) fed diets supplemented with different levels of *P. acidilactici* [46]. In contrast, dietary prebiotic GOS had no positive effect on digestive enzyme activity, such as protease, trypsin and lipase, in white seabream (*D. sargus*) after 12 weeks [20]. To date, few studies have investigated the effects of dietary organic acids on the digestive enzyme activity of aquatic animals [24]. Increased digestive enzymes activity has been reported in hybrid tilapia and white shrimp (*Litopenaeus vannamei*) fed citric acid-supplemented diets [47,48] and in green terror (*Andinoacara rivulatus*) fed apple cider vinegar-supplemented diets [25]. In the current study, dietary SP, especially at the levels of 5 and 10 g kg^{-1} diet, enhanced the activity of digestive enzymes compared to the control. However, the combination of SP with SYN and elevation of its level from 5 to 20 g kg^{-1} diet led to the decrease in digestive enzyme activities compared to the single administration of SYN. Although the exact cause of this effect is not clear, it needs more studies to identify the physiological pathways.

Lactic acid bacteria (LAB) are considered beneficial microorganisms within the gastrointestinal (GI) tract of fish. It is well known that they have the ability to stimulate the digestive function, host GI development, immune responses and disease resistance in fish species [49,50]. In the present study, the African cichlid fed an SYN-supplemented diet showed the highest lactobacillus count. Dietary supplementation with 2% GOS increased the LAB level and the ratio of LAB to TVC in the gut microbiota of Caspian roach (*R. caspicus*) and Caspian white (*R. frisii kutum*) fish fingerlings after 6 weeks [23]. The population of LAB also increased significantly in the intestinal tract of angelfish (*Pterophyllum scalare*) fed adult *Artemia franciscana* enriched with synbiotic (*P. acidilactici* + GOS) for 7 weeks [10]. Feeding beluga larvae (*Huso huso*) with *Artemia urmiana* nauplii enriched with *P. acidilactici* for 9 h also had a significant effect on the LAB level in the digestive tract of fish [51]. It was found that the addition of 3 g kg^{-1} K-diformate (KDF) to plant protein-based diets stimulated the growth of LAB in the gastrointestinal tract of tilapia [52]. Wassef et al. [53] reported that the inclusion of SP at the levels of 0.2% and 0.3% beneficially modified the distal intestine microbiota of European seabass (*Dicentrarchus labrax*) fry. In contrast, the supplementation of the diet with 2% Na-butyrate had no significant effects on the gut bacterial community of African catfish (*Clarias gariepinus*) [54]. In this trial, dietary supplementation with singular and combination use of synbiotics and sodium propionate improved the survival rate of *A. hydrophila*-injected African cichlid. Moreover, lactobacillus count in all dietary groups was higher than in the control group. In the present study, we observed the stimulating effects of using synbiotics and sodium propionate on growth performance and intestinal lactobacillus count of African cichlid. However, it needs further investigations to be confirmed.

5. Conclusions

The single administration of dietary SYN (*P. acidilactici* + GOS) and different levels of SP and their combination improved the survival rate, growth performance and lactic acid bacteria count in the gastrointestinal tract and enhanced selected humoral immune responses, antioxidant enzymes and digestive enzyme activities of African cichlid. Twenty-eight days after *Aeromonas hydrophila* injection, the survival rate of fish fed the diets containing SYN and/or SP were higher than those of fish fed the control diet. According to the results, the best inclusion level of SP as an acidifier was 5 g kg^{-1} of diet. These findings suggested that the supplementation of African cichlid-diets with SYN (*P. acidilactici* + GOS) and SP could be an effective way to improve fish health and reduce production costs.

Author Contributions: O.S. ran the experiment and analyzed the samples. M.S. collected the samples and wrote the manuscript. D.S. analyzed the samples and edited the manuscript. M.P. prepared the materials and kits and edited the manuscript. All authors have read and agreed to the published version of the manuscript.

Funding: This research was funded by the Research and Technology Deputy of Ferdowsi University of Mashhad under the 5%-Special Election Support Agreement of FUM (no: 73209; 16 March 2019).

Institutional Review Board Statement: The authors declare that the experiments were performed according to FUM animal ethics and the Directive 2010/63/EU of the European Parliament and of the Council of 22 September 2010 on the protection of animals used for scientific purposes.

Data Availability Statement: The data that support the findings of this study are available from the corresponding author upon reasonable request.

Acknowledgments: The authors thanks from Masoomeh Meraban Sang Atash for technical support.

Conflicts of Interest: The authors declare no conflict of interest. The funders had no role in the design of the study; in the collection, analyses, or interpretation of data; in the writing of the manuscript, or in the decision to publish the results.

References

1. Ladisa, C.; Bruni, M.; Lovatelli, A. Overview of Ornamental Species *Aquaculture*. *FAO Aquac. Newsl.* **2017**, *56*, 39–40.
2. Ngo, V.H. The use of medicinal plants as immunostimulants in aquaculture: A review. *Aquaculture* **2015**, *446*, 88–96.
3. Vallejos-Vidal, E.; Reyes-López, F.; Teles, M.; MacKenzie, S. The response of fish to immunostimulant diets. *Fish. Shellfish. Immunol.* **2016**, *56*, 34–69. [CrossRef]
4. Safari, O.; Paolucci, M.; Ahmadniaye Motlagh, H.R. Effects of synbiotics on immunity and disease resistance of narrow-clawed crayfish, *Astacus leptodactylus leptodactylus* (Eschscholtz, 1823). *Fish. Shellfish Immunol.* **2017**, *64*, 392–400. [CrossRef] [PubMed]
5. Safari, O.; Paolucci, M. Effect of in vitro selected synbiotics (galactooligosaccharide and mannanoligosaccharide with or without *Enterococcus faecalis*) on growth performance, immune responses and intestinal microbiota of juvenile narrow clawed crayfish, *Astacus leptodactylus leptodactylus* Eschscholtz, 1823. *Aquac. Nutr.* **2018**, *24*, 247–259.
6. Safari, O.; Sarkheil, M. Dietary administration of eryngii mushroom (*Pleurotus eryngii*) powder on haemato-immunological responses, bactericidal activity of skin mucus and growth performance of koi carp fingerlings (*Cyprinus carpio koi*). *Fish. Shellfish Immunol.* **2018**, *80*, 505–513. [CrossRef]
7. Wang, Y.B.; Tian, Z.; Yao, J.; Li, W. Effect of probiotics, *Enterococcus faecium*, on tilapia (*Oreochromis niloticus*) growth performance and immune response. *Aquaculture* **2008**, *277*, 203–207. [CrossRef]
8. Merrifield, D.L.; Dimitroglou, A.; Foey, A.; Davies, S.J.; Baker, R.T.M.; Bogwald, J. The current status and future focus of probiotic and prebiotic applications for salmo¬nids. *Aquaculture* **2010**, *302*, 1–18. [CrossRef]
9. Ghasempour Dehaghani, P.; Javaheri Baboli, M.; Taghavi Moghadam, A.; Ziaei-Nejad, S.; Pourfarhadi5, M. Effect of synbiotic dietary supplementation on survival, growth performance, and digestive enzyme activities of common carp (*Cyprinus carpio*) fingerlings. *Czech. J. Anim. Sci.* **2015**, *60*, 224–232. [CrossRef]
10. Azimirad, M.; Meshkini, S.; Ahmadifard, N.; Hoseinifar, S.H. The effects of feeding with synbiotic (*Pediococcus acidilactici* and fructooligosaccharide) enriched adult Artemia on skin mucus immune responses, stress resistance, intestinal microbiota and performance of angelfish (*Pterophyllum scalare*). *Fish. Shellfish Immunol.* **2016**, *54*, 516–522. [CrossRef]
11. Safari, R.; Hoseinifar, S.H.; Nejadmoghadam, S.; Khalili, M. Non-specific immune parameters, immune, antioxidant and growth-related genes expression of common carp (*Cyprinus carpio* L.) fed sodium propionate. *Aquac. Res.* **2017**, *48*, 1–9. [CrossRef]
12. Fuller, R. Probiotic in man and animals. *J. Appl. Microbiol.* **1989**, *66*, 365–378.
13. Harikrishnan, R.; Balasundaram, C.; Heo, M.S. Lactobacillus sakei BK19 enriched diet enhances the immunity status and disease resistance to streptococcosis infection in kelp grouper, *Epinephelus bruneus*. *Fish. Shellfish Immunol.* **2010**, *29*, 1037–1043. [CrossRef]
14. Nayak, S.K. Probiotics and immunity: A fish perspec¬tive. *Fish. Shellfish Immunol.* **2010**, *29*, 2–14. [CrossRef] [PubMed]
15. Merrifield, D.L.; Bradley, G.; Harper, G.M.; Baker, R.T.M.; Munn, C.B.; Davies, S.J. Assessment of the effects of vegetative and lyophilized *Pediococcus acidilactici* on growth, feed utilization, intestinal colonization and health parameters of rainbow trout (*Oncorhynchus mykiss* Walbaum). *Aquac. Nutr.* **2011**, *17*, 73–79. [CrossRef]
16. Neissi, A.; Rafiee, G.; Nematollahi, M.; Safari, O. The effect of *Pediococcus acidilactici* bacteria used as probiotic supplement on the growth and non-specific immune responses of green terror, *Aequidens rivulatus*. *Fish. Shellfish Immunol.* **2013**, *35*, 1976–1980. [CrossRef]
17. Safari, R.; Hoseinifar, S.H.; Kavandi, M. Modulation of antioxidant defence and immune response in zebra fish (*Danio rerio*) using dietary sodium propionate. *Fish. Physiol. Biochem.* **2016**, *42*, 1733–1739. [CrossRef] [PubMed]
18. Al-Hisnawi, A.; Rodiles, A.; Rawling, M.D.; Castex, M.; Waines, P.; Gioacchini, G.; Carnevali, O. Dietary *robiotic Pediococcus acidilactici* MA18/5M modulates the intestinal microbiota and stimulates intestinal immunity in rainbow trout (*Oncorhynchus mykiss*). *J. World Aquacult Soc.* **2019**, *50*, 1–19. [CrossRef]
19. Gibson, G.R.; Roberfroid, M.B. Dietary modulation of the colonic microbiota: Introducing the concept of prebiotics. *Nutr. J.* **1995**, *125*, 1401–1412. [CrossRef]
20. Guerreiro, I.; Couto, A.; Machado, M.; Castro, C.; Pousão-Ferreira, P.; Oliva-Teles, A. Prebiotics effect on immune and hepatic oxidative status and gut morphology of white sea bream (*Diplodus sargus*). *Fish. Shellfish Immunol.* **2016**, *50*, 168–174. [CrossRef]
21. Ringo, E.; Song, S.K. Application of dietary supplements (synbiotics and probiotics in combination with plant products and β-glucans) in aquaculture. *Aquac. Nutr.* **2016**, *22*, 4–24. [CrossRef]
22. Yousefi, S.; Hoseinifar, S.H.; Paknejad, H.; Hajimoradloo, A. The effects of dietary supplement of galactooligosaccharide on innate immunity, immune related genes expression and growth performance in zebrafish (*Danio rerio*). *Fish. Shellfish Immunol.* **2018**, *73*, 192–196. [CrossRef]
23. Hoseinifar, S.H.; Doan, H.V.; Ashouri, G. Galactooligosaccharide effects as prebiotic on intestinal microbiota of different fish species. *J. Agron. Anim. Ind.* **2019**, *14*, 266–278. [CrossRef]
24. Ng, W.K.; Koh, C.B. The utilization and mode of action of organic acids in the feeds of cultured aquatic animals. *Rev. Aquacult.* **2017**, *9*, 342–368. [CrossRef]

25. Ahmadniaye Motlagh, H.; Sarkhei, M.; Safari, O.; Paolucci, M. Supplementation of dietary apple cider vinegar as an organic acidifier on the growth performance, digestive enzymes and mucosal immunity of green terror (*Andinoacara rivulatus*). *Aquac. Res.* **2019**, *51*, 1–9. [CrossRef]

26. Silva, B.C.; Vieira, F.N.; Mourino, J.L.P.; Ferreira, G.S.; Seiffert, W.Q. Salts of organic acids selection by multiple characteristics for marine shrimp nutrition. *Aquaculture* **2013**, *384–387*, 104–110. [CrossRef]

27. Erdogan, F.; Erdogan, M.; Gümüş, E. Effects of dietary protein and lipid levels on growth performances of two African Cichlids (*Pseudotropheus socolofi* and *Haplochromis ahli*). *Turk. J. Fish. Aquat. Sci.* **2012**, *12*, 635–640.

28. Sarkheil, M.; Ameri, M.; Safari, O. Application of alginate-immobilized microalgae beads as biosorbent for removal of total ammonia and phosphorus from water of African cichlid (*Labidochromis lividus*) recirculating aquaculture system. *Environ. Sci. Pollut. Res.* **2021**, 1–13. [CrossRef] [PubMed]

29. Siwicki, A.K.; Anderson, D.P.; Rumsey, G.L. Dietary intake of immunostimulants by rainbow trout affects non-specific immunity and protection against furunculosis. *Vet. Immunol. Immunopathol.* **1994**, *41*, 125–139. [CrossRef]

30. Kumari, J.; Sahoo, P.K.; Swain, T.; Sahoo, S.K.; Sahu, A.K.; Mohanty, B.R. Seasonal variation in the innate immune parameters of the Asian catfish *Clarias batrachus*. *Aquaculture* **2006**, *252*, 121–127. [CrossRef]

31. Yano, T. Assay of Hemolytic Complement Activity. In *Techniques in Fish Immunology*; Stolen, J.S., Fletcher, T.C., Anderson, D.P., Hattari, S.C., Rowley, A.F., Eds.; SOS Publications: Trenton, NJ, USA, 1992; pp. 131–141.

32. Yanbo, W.; Zirong, X. Effect of probiotics for common carp (*Cyprinus carpio*) based on growth performance and digestive enzyme activities. *Anim. Feed Sci. Technol.* **2006**, *127*, 283–292. [CrossRef]

33. Gawlicka, A.; Parent, B.; Horn, M.H.; Ross, N.; Opstad, I.; Torrissen, O.J. Activity of digestive enzymes in yolk-sac larvae of Atlantic halibut (*Hippoglossus hippoglossus*): Indication of eadiness for first feeding. *Aquaculture* **2000**, *184*, 303–314. [CrossRef]

34. Worthington, C. Worthington Enzyme Manual Related Biochemical. *Freehold. N. J.* **1991**, *34*, 212–215.

35. Hidalgo, M.C.; Urea, E.; Sanz, A. Comparative study of digestive enzymes in fish with different nutritional habits: Proteolytic and amylase activities. *Aquaculture* **1999**, *170*, 267–283. [CrossRef]

36. Association of Official Analytical Chemists (AOAC). *Official Methods of Analysis*, 16th ed.; Association of Official Analytical Chemists Arlington: Washington, DC, USA, 2005.

37. Hoseinifar, S.H.; Zoheiri, F.; Caipang, C.M. Dietary sodium propionate improved performance, mucosal and humoral immune responses in Caspian white fish (*Rutilus frisii kutum*) fry. *Fish. Shellfish Immunol.* **2016**, *55*, 523–528. [CrossRef]

38. Sewaka, M.; Trullas, C.; Chotiko, A.; Rodkhum, C.; Chansue, N.; Boonanuntanasarn, S.; Pirarat, N. Efficacy of synbiotic *Jerusalem artichoke* and *Lactobacillus rhamnosus* GG-supplemented diets on growth performance, serum biochemical parameters, intestinal morphology, immune parameters and protection against *Aeromonas veronii* in juvenile red tilapia (*Oreochromis* spp.). *Fish. Shellfish Immunol.* **2018**, *86*, 260–268. [CrossRef] [PubMed]

39. Eslamloo, K.; Falahatkar, B.; Yokoyama, S. Effects of dietary bovine lactoferrin on growth, physiological performance, iron metabolism and non-specific immune responses of Siberian sturgeon *Acipenser baeri*. *Fish. Shellfish Immunol.* **2012**, *32*, 976–985. [CrossRef]

40. Haroun, E.; Goda, A.; Kabir, M. Effect of dietary probiotic biogen supplementation as a growth promoter on growth performance and feed utilization of Nile tilapia *Oreochromis niloticus* (L.). *Aquac. Res.* **2006**, *37*, 1473–1480. [CrossRef]

41. Kuuhlmann, K.J.; Jintasataporn, O.; Leucksteadt, C. Dietary potassium-diformate (KDF) improves growth performance of white-leg shrimp *Litopenaeus vannamei* under controlled conditions. *Int. Aquafeed* **2011**, *3*, 19–22.

42. Rahimnejad, S.; Guardiola, F.A.; Leclercq, E.; Esteban, M.A.; Castex, M.; Sotoudeh, E.; Lee, S.M. Effects of dietary supplementation with *Pediococcus acidilactici* MA18/5M, galactooligosaccharide and their synbiotic on growth, innate immunity and disease resistance of rockfish (*Sebastes schlegeli*). *Aquaculture* **2018**, *482*, 36–44. [CrossRef]

43. Hoseinifar, S.H.; Hoseini, S.M.; Bagheri, D. Effects of galactooligosaccharide and *Pediococcus acidilactici* on antioxidant defence and disease resistance of rainbow trout, *Oncorhynchus mykiss*. *Ann. Anim. Sci.* **2017**, *17*, 217–227. [CrossRef]

44. Mirshahpanah, M.; Salari, A.; Shahsavani, D.; Baghshani, H.; Nematy, M. Effect of *Pediococcus acidilatici* on intestinal microbiota and the oxidative parameters of blood and muscles in Common Carp (*Cyprinus carpio*). *J. Nutr. Fasting Health* **2019**, *7*, 37–43. [CrossRef]

45. Daniels, C.; Merrifield, D.; Boothroyd, D.; Davies, S.; Factor, J.; Arnold, K. Effect of dietary Bacillus spp. and man-nan oligosaccharides (MOS) on European lobster (*Homarus gammarus* L.) larvae growth performance, gut morphology and gut microbiota. *Aquaculture* **2010**, *304*, 49–57. [CrossRef]

46. Arani, M.M.; Salati, A.P.; Safari, O.; Keyvanshokooh, S. Dietary supplementation effects of Pediococcus acidilactici as probiotic on growth performance, digestive enzyme activities and immunity response in zebrafish (*Danio rerio*). *Aquac. Nutr.* **2019**, *25*, 854–861. [CrossRef]

47. Li, J.S.; Li, J.L.; Wu, T.T. Effects of non-starch polysaccharides enzyme, phytase and citric acid on activities of endogenous digestive enzymes of tilapia (*Oreochromis niloticus* × *Oreochromis aureus*). *Aquac. Nutr.* **2009**, *15*, 415–420. [CrossRef]

48. Su, X.; Li, X.; Leng, X.; Tan, C.; Liu, B.; Chai, X. The improvement of growth, digestive enzyme activity and disease resistance of white shrimp by the dietary citric acid. *Aquac. Int.* **2014**, *22*, 1823–1835. [CrossRef]

49. Li, X.; Ringo, E.; Hoseinifar, S.H.; Lauzon, H.; Birkbeck, H.; Yang, D. Adherence and colonisation of microorganisms in the fish gastrointestinal tract. *Rev. Aquacult.* **2018**, *11*, 603–618. [CrossRef]

50. Ringo, E.; Hoseinifar, S.H.; Ghosh, K.; Doan, H.V.; Beck, B.R.; Song, S.K. Lactic Acid Bacteria in Finfish—An Update. *Front. Microbiol.* **2018**, *9*, 1818. [CrossRef] [PubMed]

51. Zare, A.; Azari-Takami, G.; Taridashti, F.; Khara, H. The effects of *Pediococcus acidilactici* as a probiotic on growth performance and survival rate of great sturgeon, *Huso huso* (Linnaeus, 1758). *Iran. J. Fish. Sci.* **2017**, *16*, 150–161.

52. Wang, A.R.; Ran, C.; Ringo, E.; Zhou, Z. Progress in fish gastrointestinal microbiota research. *Rev. Aquacult.* **2018**, *10*, 626–640. [CrossRef]

53. Wassef, E.A.; Saleh, N.E.; Abdel-Meguid, N.E.; Barakat, K.M.; Abdel-Mohsen, H.H.; El-bermawy, N.M. Sodium propionate as a dietary acidifier for European seabass (*Dicentrarchus labrax*) fry: Immune competence, gut microbiome, and intestinal histology benefits. *Aquac. Int.* **2019**, *28*, 1–17. [CrossRef]

54. Owen, M.A.G.; Waines, P.; Bradley, G.; Davies, S. The Effect of Dietary Supplementation of Sodium Butyrate on the Growth and Microflora of *Clarias gariepinus* (Burchell 1822). In Proceedings of the 12th International Symposium on Fish Nutrition and Feeding, Biarritz, France, 28 May–1 June 2006; p. 149.

MDPI
St. Alban-Anlage 66
4052 Basel
Switzerland
Tel. +41 61 683 77 34
Fax +41 61 302 89 18
www.mdpi.com

Fishes Editorial Office
E-mail: fishes@mdpi.com
www.mdpi.com/journal/fishes